Biota
The Biodiversity Database Manager

Biota

The Biodiversity Database Manager

Robert K. Colwell
University of Connecticut

Sinauer Associates, Inc., Publishers
Sunderland, Massachusetts

THE COVER
The cover illustration, by Shahid Naeem, is an original digital painting, created using Painter (Fractal Designs, Inc.) on a Macintosh computer. Copyright Shahid Naeem, 1996. Used with the artist's permission. Cover design by Craig Malone.

Biota: The Biodiversity Database Manager
© 1996 by Robert K. Colwell
All rights reserved.

For information or to order, address:
  Sinauer Associates, Inc.
  P.O. Box 407
  Sunderland, Massachusetts 01375-0407 U.S.A.
  FAX: 413-549-1118
  Internet: publish@sinauer.com

Biota: The Biodiversity Database Manager is not in any way affiliated with the U. S. Department of Agriculture Systematic Entomology Laboratory's Biosystematic Information on Terrestrial Arthropods (BIOTA) nomenclatorial database.

4th Dimension, 4D Server, 4D Client, 4D Tools, 4D Backup, and 4D Engine are trademarks or registered trademarks of ACI/ACI US, Inc. Apple and Macintosh and System 7 are trademarks or registered trademarks of Apple Computer, Inc. Paradox, Quattro, and dBASE and are registered trademarks of Borland International, Inc. Windows, Microsoft Word, Microsoft Excel, Microsoft Access, and FoxPro are trademarks or registered trademarks of Microsoft Corp. CompuServe is a registered trademark of CompuServe, Inc. All other product names mentioned herein are used for identification purposes only, and may be the trademarks or registered trademarks of their respective companies.

ISBN 0-87893-128-7

Printed in Canada

Printed on recycled paper with 20% post-consumer waste

*To my parents,
Robert Pulliam Colwell
and Eleanor Knight Colwell,
who taught me to build my own barn.*

Brief Contents

Before You Begin xxi

Part 1: Introducing Biota 1

Chapter 1: Biodiversity Data and Relational Databases 1

Chapter 2: Key Concepts 5

Chapter 3: A Brief Tutorial with the Demo Database 21

Part 2: Express Route 59

Chapter 4: Quick Start 59

Chapter 5: Overview: Biota Tools and Features 63

Part 3: Data Entry and Information Retrieval 89

Chapter 6: Entering Data 89

Chapter 7: Record Codes 97

Chapter 8: Special Data Types 109

Chapter 9: Working with Records in Output and Input Screens 119

Chapter 10: Input—Table by Table 143

Chapter 11: Finding and Updating Records 185

Chapter 12: Creating, Finding, and Updating Specimen Series 227

Chapter 13: Printing Labels 243

Chapter 14: Printing Reports 265

Part 4: Customizing Biota 281

Chapter 15: User-Defined Auxiliary Fields 281

Chapter 16: Setting Default Entries, Using Entry Choice Lists, and Renaming Fields 313

Part 5: Special Tools and Features 327

Chapter 17: The Specimen Loan System 327

Chapter 18: Images 347

Chapter 19: Determination Histories 367

Chapter 20: The Synonymy System 379

Chapter 21: Host–Guest Relations 391

Chapter 22: Temporary Taxa for Approximate Determinations 405

Chapter 23: Security: Passwords and Access Privileges 409

Part 6: Exporting and Importing Data 433

Chapter 24: Exporting Data 433

Chapter 25: Biota and the Internet 473

Chapter 26: Importing Data 485

Appendices 521

Appendix A: Biota Structure 521

Appendix B: The Lists Table 524

Appendix C: Setting Preferences 526

Appendix D: Troubleshooting and Support 527

Appendix E: Using Biota4D with 4D Server or 4th Dimension 538

Appendix F: Data File Backup, Recovery, Compacting, and Segmenting 541

Appendix G: Biota Data File Conversion: Macintosh and Windows 550

Appendix H: Barcodes 552

Appendix I: Biota Menus: Quick Reference 554

Index 564

Contents

Before You Begin xxi

Biota Documentation xxi
   Read the ReadMe File xxi
   Visit the Biota Website xxi
How to Use This Book xxi
   Typographic and Terminological Conventions xxii
Hardware and Software Requirements xxiii
   Biota Versions and Flavors xxiii
   Operating Systems xxiv
   Hardware Requirements xxiv
   Installing Biota on Your Computer xxv
Should You Cite Biota? xxv

PART 1 Introducing Biota 1

Chapter 1: Biodiversity Data and Relational Databases 1

Biodiversity Data 1
Why Use a Relational Database? 1
   Flatfiles 1
   The Relational Model 2
   Linking Fields 2
   Efficiency of Data Input and Updating 3
   Efficiency of Relational Searches 4

Chapter 2: Key Concepts 5

Tables 5
Fields and Records 6
Key Fields 6
   Duplicate Key Errors 7
Linking Fields: Parent and Child Records 7
Data Files 8
   Opening Biota Data Files 8
   Creating an Empty Biota Data File 9
   Saving Changes in a Biota Data File 10
   To Find Out Which Data File Is Currently Open 11
Selections 11
Record Sets 12
   The Record Set Options Screen 13
   How Biota Uses Record Sets 15
   To Display a Record Set 15
   To Empty All Record Sets 16
Record Set Pointer Files 16
   Saving a Record Set Pointer File: First Method 16
   Saving a Record Set Pointer File: Second Method 17
   Loading a Record Set Pointer File 18
   Loaded Record Set Pointer Files and the Current Record Set 19

Chapter 3: A Brief Tutorial with the Demo Database 21

Installing Biota and Opening the Demo Data File 21
What's in the Demo Data File? 23
Lesson 1. Displaying and Editing Existing Records 23
   Step by Step 23
   Overview 34
Lesson 2. Creating New Records One Table at a Time 34
   Step by Step 34
   Overview 43
Lesson 3. Creating New Records "On the Fly" 44
   Step by Step 44
   Overview 47
Lesson 4. Finding Records with the Tree Hierarchy 48
   Step by Step 48
   Overview 51
Lesson 5. Finding Records in One Table Based on a Set of Records in Another Table 52

Step by Step 52
Overview 56
On Your Own—What to Try Next 57

PART 2
Express Route 59

Chapter 4: Quick Start 59

Quick-Start Strategy 59
Idiosyncrasies 60
  Opening Biota Data Files 60
  Saving Changes in a Biota Data File 60
  Hidden Window Cursor: 61
  Moving Dialog Windows (Macintosh) 61
  Ordering of Biota Table and Field Names in Certain Editors 61

Chapter 5: Overview: Biota Tools and Features 63

Operational Features 63
  Multitasking 63
  Macintosh Biota Flavors 63
  Windows Biota Flavors 64
  Server Option with Mixed-Platform Clients (Macintosh and Windows) 64
Biodiversity Information Types 64
  Specimen-Based Data 64
  Living Organisms 64
  "Lot"-Based Data 64
  Site-Based Species Lists with No Specimens 65
Tables and Fields 65
  Core Tables and Core Fields 65
  Peripheral Tables 65
  Field Aliases 65
  Alias Help 66
  Auxiliary Fields 66
  Creating and Displaying Auxiliary Fields 66
  Recursive Specimen Relations (Host Records) 67
Information Input 67
  Record Sets 67
  Images 67
  Automatic Prompting for Parent Record Entry 67
  Wildcard Entry for Linking Fields 68
  Lookup Tool for Entering Record Codes in Linking Fields 69
  Entry Choice Lists 69
  Field Value Defaults 69
  Using an Existing Record as a Template for a New One 69
  Automated Entry of Specimen Record Series 70
  Automatic Record Code Assignment 70
  Barcode Entry of Record Codes 70
Record Retrieval and Manipulation: General Tools 71
  Multifunction Output Screens 71
  Simultaneous Display of Records from Any Number of Tables 71
  Displaying Record Sets 71
  Finding Records Using the Taxonomic Hierarchy 71
  Ad Hoc Queries for All Tables 72
  Comparing and Retrieving Images 72
Special Tools for Finding Records in One Table Based on a Set of Records in Another Table 73
  Finding Records for Lower Taxa Based on Higher Taxa 73
  Finding Records for Higher Taxa Based on Lower Taxa 74
  Finding Place Records (Collection or Locality) for Specimens or Species 74
  Finding Specimen or Species Records for Places (Collections or Localities) 75
  Sequential, Cross-Hierarchy Searches 75
Finding Records by Record Code 75
  Single Specimen Records 75
  Unordered Specimen Series 75
  Ordered Specimen Series 75
  Finding Individual Records by Species Code, Collection Code, or Locality Code 76
Special Tools for Updating Records 76
  Updating Specimen Determinations and Other Specimen Information: Unordered Specimen Series 76
  Updating Specimen Determinations and Other Specimen Information: Ordered Specimen Series 76
  Updating Records (Any Table) by Importing Information from Text Files 77
  The Find and Replace Tool 77
  Keeping Track of Synonymies 77
Tools for Maintaining Database Security and Integrity 78
  User Password System 78
  User Access (Privilege) Levels 79
  Data File Password Link 79
  Spanish Dialog Screens 79
  Automatic Updating of Linking Fields in Child Records 79
  Finding Orphan and Childless Records 80
  Record Deletion Control 80
  Record Creation Control 80

Automatic Recording of Specimen Determination Histories 80
Duplicate Species Checking 81
Reports and Labels 81
Printed Reports 81
Specimen Label Printing for Fluid-Preserved, Pinned, Slide-Mounted, or Herbarium Specimens 82
Exporting Text Data for Creating Custom Locality/Collection Labels 82
Custom Label Printing 83
Species Labels 83
Specimen Loan System 83
Lending Specimens 83
Recording Returns 83
Loan Forms 83
Printed Report of Specimens Loaned 83
Text Flatfile Export of Specimens Loaned 83
Importing Data 84
The Import Editor 84
Exporting Data 84
The Export Editor 84
Exporting Specimen Flatfiles 85
Exporting Taxonomic Flatfiles 85
Exporting Notes 85
Exporting Auxiliary Field Values 85
Exporting Character Matrices in the NEXUS Format 86
Exporting Collections × Species Matrices 86
Exporting Specimens Examined Lists 86
Exporting Images 86
Custom Export 86
Exporting Web Pages 86

PART 3 Data Entry and Information Retrieval 89

Chapter 6: Entering Data 89
Screen Colors and Textures 89
Entry Areas and Default Entry Order 90
Keyboard Equivalents for On-Screen Buttons 91
Option Window (Radio) Buttons 91
Today Buttons 91
Double-Bordered (Default) Buttons 92
Using Entry Choice Lists 92
Entering Data in Linking Fields 92
"On-the-Fly" Creation of Linked Records 93
Table-by-Table Creation of Linked Records 95
A Powerful Shortcut: Wildcard Data Entry for Linking Fields 96

Chapter 7: Record Codes 97
Why Does Biota Require and Display Record Codes? 97
Guidelines and Suggestions for Designing Record Code Systems 98
Specimen Codes 98
Sequential Specimen Codes 99
Unified Record Code Systems: Specimen, Collection, and Locality Codes 99
Locality Codes 100
Species Codes 100
Sorting Records by Record Codes 101
Using Entry Choice Lists to Enter Record Code Prefixes 102
Assigning New Record Codes Automatically during Data Entry 102
The Format of Automatically Generated Record Codes 102
Changing the Defaults for Record Code Prefixes and Lengths 102
Assigning New Record Codes Automatically: Step by Step 104
Setting Default Prefixes for Recognizing Specimen and Species Record Codes 105
Using Alphanumeric Specimen Codes with the Series Tools 105
Setting Up for Automatic Record Code Prefix Recognition 106
Substituting Abbreviated Barcode Prefixes for Long Barcode Prefixes in the Data File 107

Chapter 8: Special Data Types 109
Intermediate Taxonomic Levels (Subtaxa and Supertaxa) 109
The Dilemma of Intermediate Levels 109
How Biota Handles Intermediate Taxonomic Levels 110
Default Names for Intermediate Taxon Fields 110
Renaming an Intermediate Taxon Field 111
Defining Additional Intermediate Levels 111
Updating Intermediate Taxa 111
Dates 111
Date Formats in Biota: U.S. Format for Input and Display, International Format for Labels 111
Entering Dates 112
Date Displays on Output Screens 112
Partial Dates 112
How Biota Handles Partial Dates 113
Importing and Exporting Partial Dates 113

Collection Date Ranges 113
Collection Date Ranges on Labels and in Exported Text Files 114
Geographical Coordinates: Latitude and Longitude 115
Options for Recording Geographical Coordinates in Biota 115
Entry and Display of Coordinate Data in the Locality Input Screen 116
Setting the Display Format for the Latitude and Longitude in the Locality Output Screen 116
Longitude and Latitude on Printed Labels and in Exported Text Files 117

Chapter 9: Working with Records in Output and Input Screens 119

Displaying Records in a Standard Output Screen 119
Viewing, Editing, Printing, or Deleting Individual Records from an Input Screen 120
To Return to the Output Screen 121
Editing and Saving Changes to the Record 122
Printing a Single Record Displayed in an Input Screen 122
Deleting a Record Displayed in the Input Screen 122
Moving Up the Table Hierarchies from an Input Screen: Full Record Buttons 124
Displaying Records with the Full Record Button 124
Editing Records Displayed with the Full Record Button 125
Moving Down the Table Hierarchies From an Input Screen: Child Records Buttons 125
Displaying Records with a Child Records Button 126
Using the Add Record Button 127
Using an Existing Record as a Template for a New Record 127
Creating a Sub-Selection of Records 128
Deleting a Group of Records from the Output Screen 129
Sorting Records in Output Screens 131
Using Fields from Related Tables for Sorting Records 134
Guidelines for Using Fields from Related Tables as Sort Criteria 135
Using Formulas to Sort Dates by Day, Month, or Year 136
Using a Formula to Sort Numbers in an Alphanumeric Field 139
Printing Reports or Creating Text Files Based on Records in an Output Screen 139
Printing a Report Based on Records in an Output Screen 139
Exporting a Text Flatfile Based on Records in an Output Screen 141

Chapter 10: Input—Table by Table 143

Specimen Input 143
Entering Specimen Data: Step by Step 144
Collection Input 152
Entering Collection Data: Step by Step 153
Locality Input 160
Entering Locality Data: Step by Step 160
Species Input 164
Entering Species Data: Step by Step 164
Genus, Family, Order, Class, Phylum, and Kingdom Input 168
Differences among the Higher Taxon Input Screens 168
A Special Problem for Genera: Legitimate Duplicate Generic Names 169
Entering Data in a Higher Taxon Input Screen: Step by Step 170
Personnel and Project Name Input 172
Entering Individual Personnel Records 173
Entering a New Group Personnel Record 174
Displaying, Changing, or Reordering Group Membership for an Existing Group Personnel Record 177
Entering the Project Name Record 177
Notes Input 177
Entering a New Note 179
Viewing, Editing, or Deleting a Note 181
Carrying Notes 183

Chapter 11: Finding and Updating Records 185

Finding and Displaying All Records for a Table 185
Finding and Updating Records Using Tools from the Tree Menu 186
Finding Lower Taxa for Higher Taxa or Higher Taxa for Lower Taxa 187
Finding and Updating Records by Navigating the Taxonomic Hierarchy 187
The Search Editor, a General-Purpose Tool for Finding Records Based on Content 192
Using the Search Editor: Step by Step 192
Using Fields from Related Tables in the Search Editor 197
Special Tools for Finding Records in One Table Based on a Set of Records in Another Table 200
Finding Records for Lower Taxa Based on Higher Taxa 201
Finding Records for Higher Taxa Based on Lower Taxa 205
Finding Place Records (Collection or Locality) for Specimens or Species 208

Finding Specimen or Species Records for Places (Collections or Localities) 210

Sequential, Cross-Hierarchy Searches 213

Finding Records by Record Codes 213

Finding Host and Guest Specimens and Collections 215

Finding Childless and Orphan Records 215

Finding Childless Records 215

Finding Orphan Records 216

Automatically Updating Child Records by Changing a Parent Record 218

Updating Records Using the Find and Replace Tool 220

Updating Records Using the Import Editor 225

Chapter 12: Creating, Finding, and Updating Specimen Series 227

Using the Input Specimen Series and Input and Identify Specimen Series Tools 228

Specimen Series Input: Step by Step 229

Using the Find Specimen Series and Find and Identify Specimen Series Tools 234

Finding (or Finding and Identifying) Specimen Series: First Steps 236

If You Choose "In Any Order" 236

If You Choose "In Consecutive Order" 238

Finding (or Finding and Identifying) Specimen Series: Final Steps 242

Chapter 13: Printing Labels 243

Printing Procedures 243

Label Option Windows 245

Sort Options 245

Data Options 245

Collection Labels 246

Information Included on Collection Labels 246

Exporting or Printing Collection Labels: Step by Step 247

Pin and Vial Specimen Labels 248

Information Included on Pin and Vial Labels 248

Printing Pin or Vial Specimen Labels: Step by Step 250

Slide Specimen Labels 251

Information Included on Slide Labels 251

Printing Slide Specimen Labels: Step by Step 253

Herbarium Specimen Labels 255

Information Included on Herbarium Specimen Labels 255

Printing Herbarium Specimen Labels: Step by Step 256

Species Labels 259

Information Included on Species Labels 259

Printing Species Labels: Step by Step 259

Designing and Printing Custom Labels 261

Using the Label Editor: Step by Step 261

Chapter 14: Printing Reports 265

Printing Procedures 265

Printing an Individual Record 266

Printing a Report Based on a Selection of Records 266

Printing a Report Based on the Active Record Set for a Table 266

Printing a Report Based on a Record Set Pointer File 266

Printing a Standard or Custom Report Based on the Records in an Output Screen 266

Printing a Specimen Count by Taxon Report 269

Designing and Printing Reports with the Quick Report Editor 269

Setting Up Columns in the Quick Report Layout 270

Specifying a Formula in a Column Instead of a Field Name 272

Resizing a Column 272

Adding or Removing Sort Criteria 273

Displaying All Values in a Sorted Column 273

Computing and Displaying Summary Statistics for All Records in the Selection 274

Computing and Displaying Summary Statistics for Sorted Groups of Records in the Selection 275

Framing, Fonts, and Text Styles 276

Hiding a Row or Column in the Printed Report 276

Adding Page Headers and Footers to the Report 276

Saving and Loading a Quick Report Layout 277

Printing a Report or Dismissing the Quick Report Editor Without Printing 277

An Example of a Quick Report 278

Using the Quick Report Graph Editor 279

PART

4 Customizing Biota 281

Chapter 15: User-Defined Auxiliary Fields 281

Core Fields and Auxiliary Fields 281

How Auxiliary Fields Work 282

Creating, Editing, and Ordering Auxiliary Field Names 283

Opening the Field Name Editor 283

Creating New Auxiliary Fields (First Method) 284

Creating New Auxiliary Fields (Second Method) 286

Editing the List of Auxiliary Field Names 288

Reordering Existing Auxiliary Field Names Alphabetically 291
Reordering Existing Auxiliary Field Names in a Specific Order 293
Entering and Displaying Data in Auxiliary Fields 295
  Entering New Data in Auxiliary Fields 298
  Carrying Auxiliary Fields 298
  Displaying Auxiliary Fields and Their Values for a Selection of Records 298
  Displaying Auxiliary Fields in the Standard Format: Records as Rows, Auxiliary Fields as Columns 300
  Displaying Auxiliary Fields in the Transposed Format: Auxiliary Fields as Rows, Records as Columns 302
Printing Auxiliary Fields 304
  Printing Auxiliary Fields in Matrix (Row-by-Column) Format 305
  Printing Auxiliary Fields in Standard Triplet Format (Record Code, Auxiliary Field Name, Auxiliary Field Value) 307
  Printing Auxiliary Fields in Transposed Triplet Format (Auxiliary Field Name, Record Code, Auxiliary Field Value) 307
Exporting Auxiliary Field Values 308
  Exporting Auxiliary Fields Values in the NEXUS Format 310

Chapter 16: Setting Default Entries, Using Entry Choice Lists, and Renaming Fields 313

Setting Default Entries: Field Value Defaults 313
  To Set Field Value Defaults 313
Using Entry Choice Lists 315
  Activating or Deactivating Choice Lists 316
  Adding, Deleting, or Modifying Items in a Choice List 317
  Saving New Lists or Recording Changes in Existing Lists 318
  Undoing All Changes Made in Choice Lists during the Current Biota Session 318
  Using an Entry Choice List to Enter Data in a Record 319
  Transferring Choice Lists to a Different Biota Data File 319
Changing the Names of Fields: The Core Field Alias System 323
  Advantages and Disadvantages of Using Aliases 323
  Renaming Core fields: Setting Aliases 324
  Checking Field Aliases 325
  Clearing All Aliases and Resetting to Defaults 326

PART 5 Special Tools and Features 327

Chapter 17: The Specimen Loan System 327

How Biota Keeps Track of Specimen Loans 327
Recording a New Loan 328
Previewing, Printing, and Exporting Loan Records 333
Displaying an Existing Loan 337
Recording Specimen Returns 339
  Using the Specimen Record Set to Record Returns 339
  Two Ways to Use the Loan Records Screens to Record Returns 341

Chapter 18: Images 347

Image Characteristics in Biota 347
  Image Sources 347
  Image Size and Shape 348
  Image Color 348
  Image File Formats 348
  Image Compression 349
Creating a New Image Record by Pasting or Importing an Image 350
Displaying, Printing, Exporting, or Copying an Image Record 355
Deleting an Image Record 358
Changing the Order of Image Records for a Species 359
Displaying Thumbnail Images in the Species Output Screen 363
Comparing Images in the Image Input Screen 365
Exporting Groups of Images to Disk Files 365

Chapter 19: Determination Histories 367

Determination History Records 367
Enabling or Disabling the Determination History System 369
Displaying the Determination History for a Specimen Record 369
How and When Changes in Determination are Recorded 372
  Changing a Determination in an Individual Specimen Record 372
  Changing a Determination in a Specimen Record Series 373

Changing a Determination in a Species Record 374
Changing a Determination in a Genus Record 374
Changing a Determination Using the Synonymy Tool 375
Displaying, Editing, or Deleting Determination History Records 375
Exporting and Importing Determination History Records 377

Chapter 20: The Synonymy System 379

How Biota Keeps Track of Species Synonymies 380
Displaying the Synonymy Status of a Species Record 381
The Synonymy Status Display 385
   To Enable or Disable Full Synonymy Display 385
   If Full Synonymy Display Is Not Enabled 385
   If Full Synonymy Display Is Enabled 386
   The Synonymy Display after Using the Synonymy Screen 387
Declaring a Species a Junior Synonym and Transferring Its Specimens 388
Clearing All Synonymies 391

Chapter 21: Host–Guest Relations 393

How Biota Handles Links Between Specimens 393
   Host–Guest Relations: An Example 394
   Recording Information in the Guest Collection Record 395
   Host Information in Guest Specimen Labels and Printed Reports 396
Creating a Host Specimen Link 396
Creating Guest Collection Records Automatically 398
   Creating Guest Collection Records Automatically: Step by Step 398
   Creating Guest Collection Records Automatically: Error Messages 400
Finding Host and Guest Records 401
   Finding Host Specimens and Host Collections 401
   Finding Guest Specimens and Guest Collections 402
   Finding Host Specimens for Guest Specimens 404
   Finding Guest Specimens for Host Specimens 404

Chapter 22: Temporary Taxa for Approximate Determinations 405

Biota's Convention for Temporary Taxon Records 406
Automatic Creation of Temporary Taxon Records 406
   Enabling or Disabling Automatic Creation of Temporary Taxon Records 406
How Biota Creates Temporary Taxon Records 407
Eliminating Unused Temporary Taxon Records 407
Temporary Taxa on Determination Labels 408

Chapter 23: Security: Passwords and Access Privileges 409

Activating and Deactivating the User Password System 409
   Activating the User Password System 410
   Deactivating the User Password System 411
Launching a Password-Protected Copy of Biota 413
Changing Your User Password 413
Using the Password Editor 415
   Users, User Names, Passwords, and Access Privilege Levels 415
   Opening and Closing the Password Editor 417
   Editing a User's Password and Profile 418
   Adding a New User Record 420
   Assigning Users to and Removing Users from Access Groups 420
   Pitfalls to Beware and Features to Ignore in the Password Editor 424
Moving User Names, User Passwords and Access Group Assignments to a New Copy of Biota 424
Using the Data File Password Link 425
   Activating the Data File Password Link 426
   Changing the Data File Password 428
   Deactivating the Data File Password Link 429
   Opening a Password-Protected Data File with a New Copy or New Version of Biota 430
   Using the Data File Link with Backup Files 431
If You Need High Security 432

PART 6 Exporting and Importing Data 433

Chapter 24: Exporting Data 433

Using the Export Editor 434
   Key Fields 435
   Field Types and Field Lengths 435
   Exporting by Tables and Fields: Step by Step 437
Exporting Notes 443
   Exporting Notes with the Export Editor 443
   Exporting Notes Records Linked to Records in the Current Parent-Table Record Set 443
   Exporting All Notes Records for a Notes Table 444

Exporting Notes Records Based on Their Own Content 445
Exporting Notes Records Based on the Content of Parent Records 445
Exporting Notes Records Using the Export Notes Tool 446

Exporting Auxiliary Fields 448

Exporting Images 449
Options for Exporting Images 449
Using the Export Images Tool: Step by Step 450
How Biota Assigns Disk File Names to Exported Image Files 450

Exporting Taxonomic Flatfiles 451
Using the Export Taxonomic Flatfile Tool: Step by Step 452

Exporting Specimen Flatfiles 454
Using the Export Specimen Flatfile Tool: Step by Step 454

Exporting Custom Flatfiles 457
When to Use the Custom Flatfile Tool 457
Using the Export Custom Flatfile Tool: Step by Step 458
Examples of Custom Flatfiles Exported by Biota 462

Exporting Specimens Examined Lists for Publications 463
What the Specimens Examined Tool Exports 463
Exporting a Specimens Examined List: Step by Step 465
Examples of Specimens Examined Lists Exported by Biota 468

Exporting Collections-by-Species Incidence or Abundance Tables 469
Exporting Collections-by-Species Tables: Step by Step 470
An Example of a Collections-by-Species Table Exported by Biota 472

Chapter 25: Biota and the Internet 473

Exporting Web Pages Automatically from Biota Data Files 473
Choosing a Strategy for Creating Web Pages 474
Creating Web Pages: Step by Step 475

Other Internet Options 483
Dynamic Access to Biota Data Files 483
4D Client-Server Access to Biota Data Files 484

Chapter 26: Importing Data 485

Import Step 1: Match Fields 485
Import Step 2: Prepare the Text Files 486
The Column Heading Row Option 486
Automatic Stripping of Initial and Terminal Space Characters 487

Key Fields 487
Field Types and Field Lengths 488

Import Step 3: Set Up and Launch the Import or the Update 490

Displays and Error Messages during Record Importing or Updating 495
The Import Progress Indicator 495
Importing New Records: Undoing an Aborted Import 495
Updating Existing Records: Redoing an Aborted Import 496
Too Many Fields Error 497
Missing Key Error: Importing New Records 497
Missing Key Error: Updating Existing Records 498
Duplicate Key Error: Importing New Records 498
Update Key Match Error: Updating Existing Records 499
Blank Key Error 499
Field Length Error 500
Mixed Characters Error 500
Date Format Error 500
Invalid Date Error 501
Boolean Field Error 501

Importing Data to Notes Tables or to the Determination History Table 502
No Unique Key Update Error 502
Unknown Key Error 503

Importing New Records and Updating Existing Records in Auxiliary Fields Tables 503
Importing or Updating Records in Field Name Tables 504
Importing or Updating Records in Field Value Tables 504
Auxiliary Field Setup Error 505
Auxiliary Field Blank Value Error 505
Duplicate Auxiliary Field Value Key Error: Importing New Records 505
Auxiliary Field Value Key Match Error: Updating Existing Records 506

Importing Records for the Lists Table 506

A Strategy for Preparing Text Files to Import into Hierarchically Linked Tables 507
Strategy: Preparing Text Files for Import from an Existing Flatfile 507
Step by Step: Preparing Text Files for Import from a Taxonomic Flatfile 508
Step by Step: Preparing Text Files for Import from a Specimen Flatfile 512

Appendix A Biota Structure 521

Appendix B The Lists Table 524
Record Structure of the Lists Table 524
Field Structure and Field Lengths in the Lists Table 525

Appendix C Setting Preferences 526

Appendix D Troubleshooting and Support 527
Troubleshooting 527
  Problems with Data Files 527
  Problems with Passwords 529
  Problems with Biota Windows and Records 530
  Problems with Web Pages 531
  English and Spanish Screens and Dialogues 531
  Other Problems 531
  Error Messages 532
Biota Support 535
  "How To" Questions 535
  Visit the Biota Web Site for Help and Information 535
  Support by Electronic Mail 535
  Bug Reports 535
  Suggestions for Improving Biota 537

Appendix E Using Biota4D with 4D Server or 4th Dimension 538
About Biota4D 538
Running Biota4D under 4D Server 539
  4D Server: Important Caveats and Recommendations 539
Running Biota4D under 4th Dimension (Single User) 540
Sources and Prices for 4D Server and 4th Dimension 540

Appendix F Data File Backup, Recovery, Compacting, and Segmenting 541
Backup Strategies for Data Files 541
Checking and Recovering Damaged Data Files 541
Compacting Data Files by Using 4D Tools 543
Segmenting Data Files 545
  Segmenting a New Data File 545
  Adding a New Segment to an Existing Data File 548
  Deleting Existing Segments or Splitting an Existing Data File into Segments 549

Appendix G Biota Data File Conversion: Macintosh and Windows 550
Data and Resources 550
Data File Conversion: Step by Step 551

Appendix H Barcodes 552
Barcodes 552
Barcode Scanners 552

Appendix I Biota Menus: Quick Reference 554
Access Privileges for Menu Items 554
The Menu Bar 554
The File Menu 554
The Edit Menu 555
The Input Menu 555
The Series Menu 556
The Tree Menu 556
The Find Menu 557
The Display Menu 559
The Labels Menu 559
The Import/Export (Im/Export) Menu 560
The Loans Menu 561
The Special Menu 562

Index 564

Preface

How Biota Began

Biota was originally developed to manage biodiversity information for the Arthropods of La Selva (ALAS) project at La Selva Biological Station in Costa Rica. A collaboration between the Organization for Tropical Studies (OTS), the Instituto Nacional de Biodiversidad (INBio), and two dozen systematist collaborators, Project ALAS is a quantitative, ecologically-structured inventory of rainforest insects and arachnids, a special component of INBio's country-wide general biological inventory (http://viceroy.eeb.uconn.edu/ALAS/ALAS.html).

Project ALAS—and the development of Biota—began in early 1992. By mid-1996, the database had been used intensively by ALAS personnel for more than four years, with continual addition of functionality, refinement of structure, and improvement of the user interface in response to the daily needs of the project. During that period, records for nearly 80,000 specimens, with associated collection and taxonomic information, were entered into the database, with no data-entry backlog. Trained local parataxonomists were primarily responsible for data entry.

Biota Today

From the comments and enthusiasm of visitors to the ALAS laboratory in Costa Rica and from the response of colleagues elsewhere (including NSF program directors), it became apparent that, if Biota could be made sufficiently flexible, it might meet the needs of a wide variety of individual ecologists, systematists, and conservation biologists for an inexpensive but powerful toolkit for managing all kinds of biodiversity information for their own research.

Moreover, many field stations, reserves, museums, and herbaria now rely on either antiquated or "home-made" database systems of limited usability and scalability, but cannot afford the cost of developing custom software or of licensing existing high-end specimen management systems.

Biota has now been made quite general in its design, with many ways to customize entry, archival, and analysis of biodiversity data and images. Building on the experience of Project ALAS and dozens of beta testers, Biota has been broadened to accommodate the needs of a wide variety of projects in ecology, biogeography, systematics, and collections management. Many key features of Biota were suggested (or demanded) by users based on their own experience and needs.

Acknowledgments

Project ALAS and the development of Biota would not have been possible without grants from the National Science Foundation (BSR 90-25024, including a component contribution from the Agency for International Development, and DEB 94-01069) and donations from Apple Computer's EarthGrants Program, ACI, Pacer Software, and the Japanese Television Workshop. Through my professorial salary, the University of Connecticut supported me throughout the development of Biota. The personal encouragement and support of Donald Stone and Deborah Clark of the Organization for Tropical Studies, as well as institutional support from OTS, have been crucial to the success of ALAS and Biota.

Although the technical design and programming of Biota are my own work, several of the application's most innovative and useful features are the direct result of the suggestions and taxonomic experience of Jack Longino, my principal collaborator in Project ALAS.

Felipe Oñoro, Werner Bohl, and Gilbert Solís of INBio also made helpful suggestions. Danilo Brenes, Jonathan Coddington, Carolina Godoy, Henry Hespenheide, Nelci Oconotrillo, Maylin Paniagua, Angel Solís, Ronald Vargas, and other members of the ALAS team put Biota to the acid test of daily use and contributed ideas for improvement.

Many Biota beta testers contributed suggestions, bug reports, enthusiasm, and occasional despair, but special thanks are due to Robin Chazdon, Paul Cislo, Juan Dupuy, Brian Farrell, Brian Fisher, Michael Kaspari, Stuart McKamey, Piotr Naskrecki, Tila Perez, Virginia Scott, Derek Sikes, and Fred Stehr. David Maddison and Larry Gall made crucial suggestions after reviewing an early version.

Michelle Craven Nicholas of ACI, the makers of 4th Dimension, provided invaluable encouragement and advice on program design in the early stages. Kevin Shay helped out in key ways later on. Pat Kuras of Apple Computer and Randy Downer of Applied Biomathematics gave essential assistance with questions of hardware and networking.

The publication costs of Biota, including license fees for the 4th Dimension engine, are supported by your purchase of the program and manual.

This book would not be in your hands were it not for Andy Sinauer's steadfast confidence in the Biota project and his infinite patience with many delays. ("The seasons, they come and go…") The Sinauer team has made it possible to produce a truly usable and elegant manual for Biota that stands ready to help you efficiently when you need it. Kathaleen Emerson, Roberta Lewis, Jeff Johnson, and Christopher Small at Sinauer Associates have each made essential contributions.

Two gifted artist-biologists have added their own touch to Biota. Shahid Naeem gave permission to use his spectacular drawing for the book cover. Piotr Naskrecki designed the icons and improved on my Biota wasp logo. The logo is based on a 19th Century lithograph of an ichneumonid wasp of the genus *Ophion*, from A. S. Packard's *A Guide to the Study of Insects* (1868). Thanks to Mark O'Brien of Entomation for the scanned lithograph, and apologies to Leonardo.

To all I am grateful, but to no one more than my wife, Robin Chazdon, for her gifts of steadfast support and understanding.

ROBERT K. COLWELL

Before You Begin

Biota Documentation

Although this book provides comprehensive coverage of the tools, features, capabilities, and limitations of Biota, there are two other important sources of up-to-date information.

Read the ReadMe File

Be sure to check the ReadMe file, which you will find on one of the Biota diskettes, for last-minute updates or corrections to this book.

Visit the Biota Website

Post-release updates to the documentation, suggestions, warnings, workarounds for problems, and news of program updates will be posted at the Biota Website, http://viceroy.eeb.uconn.edu/biota.

How to Use This Book

This book is designed to be used as both a guide for getting started and as a reference, later on. The more of it you read, the more efficient and productive your work with Biota is likely to become.

How much you read before starting to use Biota depends upon your own learning style and on your level of confidence with relational databases.

- **If you are have some experience with relational databases,** you might begin after reading the very brief "Quick Start" (Chapter 4), scanning the titles in "Overview: Biota Tools and Features" (Chapter 5), and taking a careful look at the diagrams of Biota's structure in Appendix A.

- **If you want a condensed, but thorough** *overview* of Biota's capabilities, read "Key Concepts" (Chapter 2). "Quick Start" (Chapter 4), and "Overview: Biota Tools and Features" (Chapter 5).

- **If you are new to relational databases, or simply prefer to take new things step by step**, start by reading Chapter 1 ("Biodiversity Data and Relational Databases") and Chapter 2 ("Key Concepts). Then work through the tutorial (Chapter 3). Finally, read Chapters 6–11 for in-depth information on data entry and information retrieval.
- **If you want a short tour of how Biota behaves** with a complex data file, work through "A Brief Tutorial With the Demo Database" (Chapter 3)—or just start up Biota with the Demo Data file and explore on your own.

Chapters 12 through 26 and the Appendices provide detailed help with Biota's special tools and features. These chapters are intended to be consulted as you need them, but paging through them quickly before you begin in earnest may pay off later, simply in knowing what is there.

Typographic and Terminological Conventions

Throughout this book, the following typographic conventions are used.

Convention	Examples	Type of Information
Name in square brackets	[Specimen], [Order]	Names of Biota tables (see Appendices A and B)
Name in square brackets followed by a name with no brackets	[Species] Author, [Order] Subclass	Names of Biota fields (see Appendices A and B)
SMALL CAPITALS	TAB, COMMAND, SHIFT	Names of keys on the keyboard (Exception: lowercase letter keys appear in lowercase)
BOLDFACED SMALL CAPITALS with top and bottom rules	**NOTE:** If the name of the Data File is blank… **WARNING:** Because Record Code fields are alphanumeric…	Cautionary or explanatory notes and warnings of special importance

Here are a few instructional terms used frequently in this book, with examples and definitions.

Term	Example	Meaning
press	*Press* the TAB key to move to the next field.	*Press* and release a key on the keyboard (*not* a mouse action with an object on the screen).

press and hold	*Press and hold* the OPTION key while clicking the OK button.	*Press and hold* a key on the keyboard while performing some other action (pressing another key or performing some action with the mouse).
click	*Click* the Accept button to accept the record.	*Click* a button or other object on the screen using the mouse (*not* a key on the keyboard).
drag	To add a user to a group, *drag* the name of the user to the Groups panel.	Point with the mouse, and then hold down the mouse button as you move the object with the mouse.

Hardware and Software Requirements

Because processors, other hardware options, and operating systems are continually changing, be sure to supplement what you read in this section by consulting the ReadMe file on the Biota diskettes you received with this book, and check the Biota Website (http://viceroy.eeb.uconn.edu/biota) for the latest news.

Biota Versions and Flavors

The *version* of Biota you have refers to the "edition" of the source code from which the application was compiled. You can find out what version you have by selecting About Biota from the Find menu when Biota is running.

The *flavor* of Biota refers to the operating system (Macintosh or Windows) and processor (e.g., PowerMacintosh, Macintosh, 486, Pentium, etc.) for which Biota has been compiled. For a given version, all Biota flavors are compiled from the same, multiplatform source code. You can find out what flavor you have by looking at the distribution diskettes.

NOTE: At the time of this writing (October, 1996), Biota is available in three flavors, BiotaApp for PowerMacintosh, BiotaApp for Macintosh, and Biota4D for Macintosh. But watch the Biota Web site for the announcement of the release of BiotaApp for Windows and Biota4D for Windows.

BiotaApp for all platforms and processors is identical in functionality. BiotaApp and Biota4D are virtually identical, but differ in one important respect.

- **BiotaApp.** BiotaApp is a stand-alone ("double-clickable"), single-user application.

There are two flavors of BiotaApp for the Macintosh Operating System: BiotaApp for PowerMacintosh computers (in native 32-bit PPC code) and BiotaApp for Macintosh computers (68030 or 68040 processor; the oldest Macintoshes, with the 68020 chip, are really too slow to be practical and are unlikely to have enough RAM). The Macintosh version of BiotaApp also runs well on PowerMacintosh computers, although somewhat more slowly than the native PowerMacintosh version.

- **Biota4D**. Biota4D is intended primarily for multiuser environments. Biota4D runs in true client/server mode under 4D Server on either a PowerMacintosh or a Macintosh computer. See Appendix E for details.

  If you have 4th Dimension or 4D First, you can also run Biota4D in single-user mode, on either a PowerMacintosh or a Macintosh computer. See Appendix E for details.

  You must purchase 4D Server or 4th Dimension separately from a commercial vendor to run Biota4D.

Operating Systems

On the Macintosh platform, Biota expects Macintosh Operating System 7 or higher. QuickTime is required for images, and Window Shade is strongly recommended for working with multiple windows. (QuickTime and Window Shade are standard components of Macintosh System 7.5 and above.)

When released for PC machines, Biota will require Windows 3.1, Windows 95, or later versions. Biota for Windows is compiled in true 32-bit code.

Hardware Requirements

- **RAM.** The more RAM you can give Biota, the more windows you can have open at the same time, and thus the more tasks it can manage simultaneously. With a minimum of about 4MB devoted to Biota, you can have two or three windows (processes) open at once on a Macintosh. For a PowerMacintosh, 5–6 MB is the equivalent (PPC code is less compact). In the Macintosh Operating System, you can set the amount of RAM committed to Biota (or any other application) in the Get Info window (from the File menu in the Finder).

  Check the ReadMe file for information on RAM requirements in the Windows operating system.

- **Screen Size.** The largest fixed-size layouts (windows) in Biota are 640×360 pixels. (Output screens expand vertically to the size of your monitor.) All Macintosh and PC screens in current production are at least this large. Macintosh Plus, SE, and SE30 screens (512×342) are too small for many of Biota's layouts and are thus not recommended.

Installing Biota on Your Computer

For installation instructions, see the ReadMe file on one of the Biota distribution diskettes.

Should You Cite Biota?

Should you cite Biota when you publish? Viewed simply as a record-keeping tool, like a set of note cards or a spreadsheet application, there is no need to cite Biota. On the other hand, if Biota's tools have made possible—or feasible—results that would not have otherwise been part of your published work, you may want to acknowledge Biota's contribution.

If you find that the ideas and strategies for biodiversity data management discussed in this book have influenced your research, you may want to cite this book.

Of course, if you create software tools that have been shaped or inspired by Biota's approach or user interface, it would be appropriate to cite Biota.

The following citation can be used for both the program and this book:

Colwell, R. K. 1996. *Biota: The Biodiversity Database Manager*. Sinauer Associates, Sunderland, Massachusetts.

PART 1 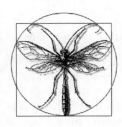 Introducing Biota

Chapter 1 Biodiversity Data and Relational Databases

Biodiversity Data

Consider an ecologist who needs to record data for a quadrat-based plant survey of a large area being considered for protection as a reserve. Or, imagine a systematist or a biogeographer who must keep track of the classification and geographical origin of a large number of specimens.

In each of these examples (and many others in ecology, conservation biology, systematics, biogeography, and evolutionary biology), the biologist needs to record not only taxonomic information (Species, Genus, Family, and so on) for each specimen or group of specimens, but also collection and location data (collector or observer, collection or observation date, method, site characteristics, map coordinates, and names for localities and political units). This book will refer to information with these characteristics as *biodiversity data*.

Why Use a Relational Database?

Each individual organism in a biodiversity dataset has two allegiances. It belongs to a particular species and it belongs to a particular place and time (a particular observation or collecting event). The organism shares with all other members of its species—wherever they are—precisely the same taxonomic data, all the way up the Linnean hierarchy. At the same time, it shares with all organisms from the same collecting event—whatever their species—the same time and place data.

Flatfiles

To record all these data for each specimen in a single column-by-row table requires repeating all the taxonomic data for each member of the same species and all the collection data for each specimen from a given collecting event. This kind of table is called a *flatfile* (or a simple spreadsheet). As anyone who has made a large one knows, flatfiles for floristic, faunistic, and collections data are not only clumsy and error-prone, but highly repetitive and therefore wasteful of skilled human effort—the most precious scientific resource of all.

The Relational Model

A powerful alternative to a flatfile is a *relational database*, a design for storing and retrieving data in *linked tables* that reduces repetitive input and redundant storage of data to an absolute minimum. Each row of a *table* represents a *record* (sometimes called an instance or a case). Each column represents a *field* (or attribute).

For hierarchical data, the familiar format of an indented text table shares the property of non-repetitive entries with a relational database. In the indented taxonomic table below, each taxon name appears only once, regardless of taxonomic rank.

Family	Genus	Specific Name
Campanulaceae	*Centropogon*	*caoutchouc*
		erianthus
	Lobelia	*laxiflora*
		salicifolia
	Siphocampylus	*ecuadoriensis*
		sanguineus
		scandens
Ericaceae	*Anthopterus*	*verticillatus*
	Cavendishia	*forreroi*
		gilgiana
		leucantha
		lindauiana
	Ceratostema	*nodosum*
		peruvianum
		reginaldi
	Macleania	*bullata*
		cf. ericae
		coccoloboides
		glabra

In fact, translating an indented table, like the one above, to the relational equivalent is straightforward. Each column of the indented table forms the basis for a separate *relational table*. In this example, three separate tables (for specific names, genera, and families) are required.

Linking Fields

In the indented format, the physical layout of the table indicates the generic membership of each species and the familial membership of each genus. In the separate tables of the relational equivalent, illustrated below, the generic membership of each species is recorded by adding a *linking field*—the Genus column, to the Specific Name table. Likewise, the familial membership of each genus is indicated in a linking field—the Family column in the Genus table.

Specific Name Table

Specific Name	Genus
caoutchouc	*Centropogon*
erianthus	*Centropogon*
laxiflora	*Lobelia*
salicifolia	*Lobelia*
ecuadoriensis	*Siphocampylus*
sanguineus	*Siphocampylus*
scandens	*Siphocampylus*
verticillatus	*Anthopterus*
forreroi	*Cavendishia*
gilgiana	*Cavendishia*
leucantha	*Cavendishia*
lindauiana	*Cavendishia*
nodosum	*Ceratostema*
peruvianum	*Ceratostema*
reginaldi	*Ceratostema*
bullata	*Macleania*
cf. ericae	*Macleania*
coccoloboides	*Macleania*
glabra	*Macleania*

Genus Table

Genus	Family
Centropogon	Campanulaceae
Lobelia	Campanulaceae
Siphocampylus	Campanulaceae
Anthopterus	Ericaceae
Cavendishia	Ericaceae
Ceratostema	Ericaceae
Macleania	Ericaceae

Family Table

Family
Campanulaceae
Ericaceae

Although the taxonomic example above used a familiar hierarchy to illustrate the relational representation of data, the relational model can accommodate any logical structure, not just hierarchies.

Biota's structure, discussed in detail in the next chapter and shown in Appendix A, takes advantage of the relational model to represent and manage data for a variety of non-hierarchical logical relationship, as well as for the taxonomic and geographic hierarchies.

Efficiency of Data Input and Updating

Biota organizes biodiversity information according to the relational data model. For each class of information, such as the names and other attributes of species, genera, localities, or personnel connected with a dataset, Biota records information in a separate relational table, with just one record for each genus, locality, or person.

Because of the repeated values in linking fields, it might appear that the relational design is not much of an improvement over a flatfile, in terms of avoiding data entry error and the burden of updating records. In fact, an efficient interface for a relational design allows one-time entry for each distinct value and automatic updating for linking fields, if necessary.

In a new Species record in Biota, for example, the specific name and author of a species need be entered exactly once. If a correction or update is required later, the correction is made in just one record.

Using Biota's lookup and wildcard entry tools, you can quickly enter the linking field values for new records (for example, new Species records), when the related parent record (a Genus record, in this example) is already in your database (Chapter 10). In short, none of the repetitive data that are characteristic of flatfiles ever need be entered more than once.

Efficiency of Relational Searches

Relational databases are much more quickly searched and sorted than equivalent large flatfiles, especially for complex queries. Suppose you want a list of the Collection records for all 10,000 individual ant specimens in a Biota Data File of 100,000 records. Using Biota's Lower Taxa for Higher Taxa tool (pp. 201–205), the ant Specimen records can be found and displayed in a few seconds, simply by constructing the query: "Find All Specimens of the Family Formicidae." A secondary query, "Find all Collections for the Specimen Record Set" (using the Places for Specimens or Species tool, pp. 208–210) then displays the desired Collection records in another few seconds.

The speed of the search is entirely a result of the relational structure of the data and the use of powerful relational commands in the programming language (4th Dimension, in the case of Biota). In contrast, a flatfile search tool would have to look at the Family column of each of the 80,0000 specimen records to find those listed as Formicidae (ants). Once the ant records were found, you would need to eliminate—manually—all redundant collection data for specimens derived from the same collecting events.

Chapter 2 Key Concepts

Tables Biota is structured around twelve *Core tables* of data (Appendix A). In addition to a table for each of the seven obligatory levels of the *taxonomic* hierarchy (Species through Kingdom) and the Specimen table, there are two tables in the *place* hierarchy—Collection and Locality, plus special tables to record specimen Loans and Personnel data. *You need to understand what each Core table is used for, since all Biota records are entered and displayed by Core tables.*

In addition to the 12 Core tables, Biota's relational structure encompasses more than a dozen *Peripheral tables* (Appendix A). You will generally not need to understand what the Peripheral tables record and how they are linked to each other or to the Core tables, unless you wish to import data into Peripheral tables or export data directly from them. Peripheral tables keep track of Notes (the Specimen Notes, Species Notes, Collection Notes, Locality Notes and Loan Notes tables; pp. 179–183), specimen determination histories (the Det History table, Chapter 19), images (the Image Archive table, Chapter 18), personnel groups (the Group table, pp. 174–177), Auxiliary Field names and entries (four Field Name and four Field Value tables, Chapter 15), Core Field Aliases (the Lists table, pp. 323–326, and Appendix B), and input choice lists for fields (the Lists table, pp. 315–322, and Appendix B).

Biota's tables are internal components of the database and thus do not appear as ordinary files in your folders or directories.

Fields and Records

Each Biota table consists of a unique set of *fields* that represent data elements (attributes) appropriate to that table. These fields are listed for each table in Appendix A.

Most of Biota's input (data entry) screens, output (display) screens, and printed reports and labels combine fields from several tables at once. The Specimens Report, for example, lists specimens organized by the taxonomic hierarchy, with collection and locality data for each specimen—it thus includes fields from 10 different tables.

The information in each table is organized by *records*. Each record in a table may include entries for each field in that table. For example, in the Specimen table, each record represents an individual specimen (or lot), with entries for that specimen's unique Specimen Code, the Collection Code for the sample or collection site from which the specimen came, the Species Code for the species to which the specimen has been assigned or determined, the name of the person who made that assignment (Determined By), and so on through the fields listed for the Specimen table in Appendix A.

You can think of an individual database table as a spreadsheet, with fields as columns and records as rows—indeed, records and fields appear in this format in most Biota output display screens.

Key Fields

In most relational database tables (including all of Biota's Core tables) each record must be uniquely identifiable and retrievable. For this reason, one field (in some tables a pair of fields) in each such table is designated as the *Key field*—or simply the *Key*. Every record in such a table must have a unique alphanumeric value for the Key field (or a unique combination of values for a *composite* key). In Biota, the Key field for the Locality, Collection, Specimen, Species, and Loans tables is a unique alphanumeric *Record Code* for each record. (You can define Record Codes yourself, or ask Biota to supply them automatically). Chapter 7 discusses Record Codes in depth.

For taxon tables above the species level, the taxon name itself is the Key field, relying on the uniqueness criterion of the Rules of Nomenclature. Names of genera, however, may sometimes be ambiguous. If you need to record generic junior synonyms, or in the rare cases in which genera in two or more Kingdoms share the same genus name and are listed in the same database, you will need to add a distinguishing suffix to generic names. This approach and Biota's ability to exclude appended special characters in generic names on labels are discussed on pp. 169–170.)

The Key field for the Personnel table is a user-defined, unique "Short Name" (initials and last name for a person, for example, or an acronym for an institution). The uses of Short Names are discussed on pp. 173–174.

Duplicate Key Errors

If you enter a value in a Key field that duplicates the Key field entry for any existing record in the same table in your database, Biota will post an error message like the one shown here. You will have to choose a different value for the Key, or change the Key in the existing record.

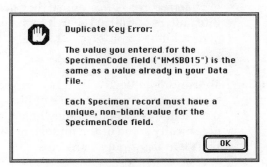

Linking Fields: Parent and Child Records

Relational links (or relationships) join tables through *linking fields*. These links give a relational database its power and efficiency. Appendix A shows the relational links between Biota's data tables.

In Biota, as pointed out in Chapter 3, the Specimen table lies at the base of two hierarchies, or paths of *many-to-one* or relationships.

In the place hierarchy, the Specimen table is linked to the Collection table through the Collection Code linking field. (There may be *many* specimens in *one* collection.) In turn, the Collection table is linked to the Locality table through the Locality Code linking field. (There may be *many* collections from *one* locality.)

In the taxonomic hierarchy, the Specimen table is linked to the Species table through the Species Code linking field. (There may be *many* specimens of *one* species). The Species table, in turn, is linked to the Genus table (there may be *many* species in *one* genus), and so on up the Linnean hierarchy.

In menus, screens, and dialogs, and in most contexts in this manual, Biota does not use the formal "many-to-one" or "one-to-many" terminology. Instead, a "parent-child" metaphor is used. For example, a particular Family record is the *parent* record for all the Genus records for all genera in that family—the *child* records. Of course, in many cases a particular record is both parent and child; a Genus may be the child record of a Family record, and at the same time parent record to one or more Species (child) records. (The parent-child metaphor works well here, since each of us may be both child and parent, as our own parents undeniably were.)

The metaphor is carried further in the concepts of *orphan* records (pp. 216–218)—those that are not linked to any record in the parent table and *childless* records (pp. 215–216)—those not linked to any record in the child table, for a particular database relation. For example, a Genus record not linked to any Family record is an orphan record. A Genus record not linked to any Species record is a childless record. Orphan records, especially, and childless records, to a lesser extent, tend to be "invisible" in relational searches. (In relational database terminology, orphan records compromise the *referential integrity* of the database.) Biota offers optimized search tools for finding and displaying orphan and childless records (pp. 215–218).

Data Files

To use Biota effectively, it is essential to understand the distinction between four representations of Biota data records, discussed and contrasted in this section and the three that follow. The figure below shows graphically some of the important distinctions among the four ways Biota handles groups of records.

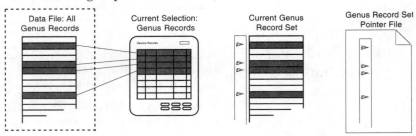

The actual records for a table are recorded in a Biota *Data File*, in an internal format not directly accessible to the user. The figure above illustrates a Biota Data File in the dashed rectangle, with the Genus table as an example. Biota Data Files are separate from the application file itself (BiotaApp or Biota4D).

A Biota Data File can be selected, opened, or created only while Biota is being launched.

To close the active Data File and open another, or to create a new, empty Data File, you must quit Biota and launch it again.

Opening Biota Data Files

To choose among existing Biota Data Files, follow these steps.

1. **Press and hold the OPTION key (Macintosh) or ALT key (Windows) while launching Biota.** If you have implemented the user password system (see Chapter 23), you will need to press and hold the OPTION key (Macintosh) or ALT key (Windows) while clicking the Connect button in the Password screen, shown here.

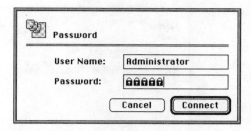

The Open File window for your operating system will appear (the Macintosh window is illustrated below).

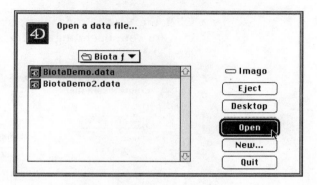

2. **Find the Data File you want to open, select it, and click the Open button.**

Creating an Empty Biota Data File

To create a new, empty Biota Data File, follow these steps.

1. **Press and hold the OPTION key (Macintosh) or ALT key (Windows) while launching Biota.** If you have implemented the user password system (see Chapter 23), you will need to press and hold the OPTION key (Macintosh) or ALT key (Windows) while clicking the Connect button in the Password screen, shown here.

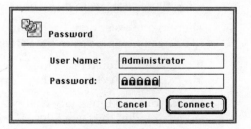

The Open File window for your operating system will appear (Macintosh windows are illustrated below and in the next step).

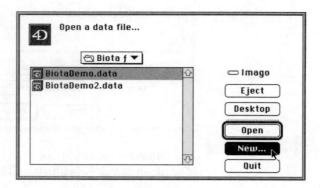

2. Click the New button (above; *not* the Open button). A Save File window for your operating system appears, with the instruction "Create a Data File."

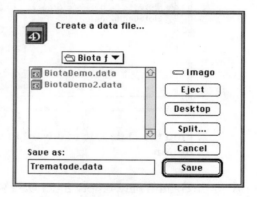

3. Name and place the file then click the Save button. Biota will take a few moments to create the Data File, so be patient.

Saving Changes in a Biota Data File

Biota uses temporary buffers in RAM to keep track of any changes you make in the active Data File. Every three minutes, Biota automatically saves these changes to disk and clears the buffers.

There is one exception: Each group of new records you create or modify using commands from the Series menu is saved immediately.

If you want to force Biota to save changes immediately, press the keys ⌘(COMMAND)+W (Macintosh) or CTRL+W (Windows) at any time.

When you quit Biota, any changes made since the last automatic save are recorded in the Data File.

The small window shown here appears in the corner of the screen during a save operation. Other ongoing operations may be momentarily slowed, but are not halted.

NOTE TO MACINTOSH USERS: The key combination ⌘(COMMAND)+W, normally used for closing windows in Macintosh applications, is reserved by 4th Dimension for saving data to disk from the buffers. To close a Biota window from the keyboard, press OPTION+W instead.

To Find Out Which Data File Is Currently Open

Biota can open only one Data File at a time. If you have enough memory available, however, you can open more than one copy of BiotaApp simultaneously, with different Data Files.

You can find out which Data File is open by selecting About the Current Data File from the File menu. The information window that appears also tells you how many records there are in each Biota table in the current Data File.

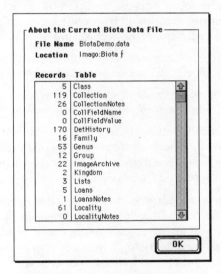

NOTE: If the name of the Data File is blank in the information screen, or you get an error comment, you need to place the Mac4DX folder, which came with your copy of Biota, in the same folder with Biota4D or BiotaApp (Macintosh); or the Win4DX directory in the same directory with Biota4D or BiotaApp (Windows). These folders contain special code segments that allow Biota to communicate with the operating system.

Selections

Because a database may become very large, displaying *all* the records in a table every time you want to look up a record, correct mistakes, or view a set of recent entries after entering them would soon become extremely cumbersome and time-consuming. To solve this problem, Biota displays in a window (or prints) only those records in a *Selection* for a particular table—the second way Biota handles records (illustrated below in the dashed rectangle).

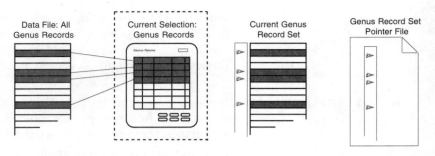

A Selection can include all of the records for a table or any subset of them. If you select all records in the Species table for the Family Ericaceae, for example, all the Species records for ericads will be placed in a Selection and displayed in the Species output screen. If a Selection is empty, which will happen (for example) if a search fails to find a match for your query, no records are displayed.

In short, a Selection is simply a "list" of selected records from the database, which may be sorted, displayed on the screen, or printed.

Placing records in a Selection or removing records from a Selection has absolutely no effect on their presence the Data File itself.

During data entry, as each record is accepted, it is added to the Data File itself. Once entered, each record remains in the Data File until and unless it is specifically deleted, regardless of how often it is included in or removed from a Selection.

You can retrieve and display several Selections for the same table simultaneously in different windows—as many as memory allows. Because simultaneous Selections are entirely independent from one another, they can include distinct, overlapping, or even identical sets of records. (You may edit an individual record in only one input screen at a time, however; see p. 533 on *record locking*).

Record Sets

The third way you can represent and work with records is by using *Record Sets*. You can tell Biota to "remember" a Selection of records from a particular table by establishing it as a Record Set for the table (illustrated below in the dashed rectangle).

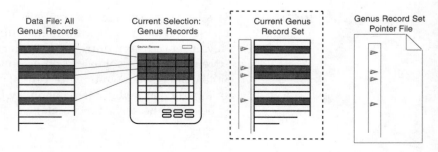

Each of Biota's seven Core tables, plus the Determination History table (Appendix A) has its own *current* Record Set, at all times. Unlike Selections for a particular table, of which many can exist simultaneously in different windows, there is always only one current Record Set per table. (You can save and retrieve as many Record Sets per table as you wish, however. See the next section.) When you launch Biota, all current Record Sets are empty and remain empty until you retrieve or create records to place in them.

The Record Set Options Screen

Understanding the Record Set options screen is the key to working effectively with Record Sets in Biota. The options screen (illustrated below) appears whenever you click the Done button in an output (record listing) screen:

- **after finding and retrieving records** from the Data File using any of Biota's search tools (Chapter 11),
- **after adding or deleting records** (Chapter 9),
- **or after changing the Selection** by using the New Selection button (Chapter 9).

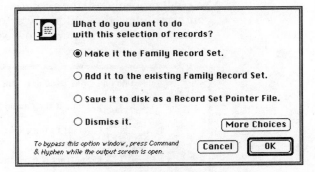

The Record Set options screen offers the following primary alternatives:

- **Make it the [*table name*]Record Set.** If you choose this option, the current Selection of records is declared the current Record Set for the table.
- **Add it to the existing [*table name*]Record Set.** This option allows you to combine the Selection with the current Record Set for the table. If the current Record Set is empty, this option and the More Choices button are disabled (dimmed) and a message appears reading: "The [table name] Record Set is empty," as illustrated below.

> ○ Add it to the existing Family Record Set.
>   (The Family Record Set is empty.)

- **Save it to disk as a Record Set Pointer File.** This option saves a disk file that allows you to reload the Selection quickly at a later time. See "Saving a Record Set Pointer File: First Method," pp. 16-17.

- **Dismiss it.** Choosing this option dismisses the Selection of records without affecting the current Record Set for the table.

NOTE: You can achieve the same result as the "Dismiss it" option by bypassing the Record Set options screen entirely. To do so, press the COMMAND key plus the HYPHEN key (Macintosh) or the CTRL key plus the HYPHEN key (Windows), simultaneously, to dismiss any output (record listing) screen, instead of clicking the Done button. (You can also use this keyboard shortcut to dismiss most input screens, instead of clicking the Cancel button. The HYPHEN key is the key next to the ZERO key; do not confuse it with the MINUS key on the numeric keypad.)

- **The More Choices button.** If the current Record Set for the table is not empty, the More Choices button in the Record Set options screen is enabled. (If the current Record Set is empty, the button is dimmed.) If you click the More Choices button when it is enabled, the Record Set options screen expands to include not only the four original options (illustrated above) but three additional options that allow other operations on Record Sets in relation to the current Selection.

You will probably use the additional options infrequently, but when you need them, they offer powerful solutions to handling sets of records. The Venn diagrams show each set operation graphically, on the left side of the screen.

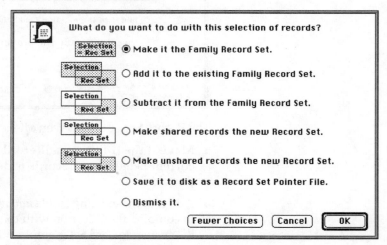

- **Make it the [*table name*]Record Set.** See above (p. 13). The Selection *itself* becomes the new current Record Set for the table (the *assignment* set operation).

- **Add it to the existing [*table name*]Record Set.** See above (p. 13). The *union* of the Selection and the Record Set becomes the new current Record Set for the table.

- **Subtract it from the existing [*table name*]Record Set.** The set of all records in the current Record Set that are *not* in the Selection (the *difference* between the Record Set and Selection) becomes the new current Record Set for the table.

  NOTE: To subtract the current Record set from the Selection, instead, save the current Record Set as a Record Set Pointer File (pp. 17-18), make the Selection the new current Record Set, then load the saved Record Set Pointer File (pp. 18-19) and use the subtract operation.

- **Make shared records the new Record Set.** The *intersection* of the Selection and the Record Set becomes the new current Record Set for the table.

- **Make unshared records the new Record Set.** The set of all records in *either* the Selection or the current Record Set, *but not both*, becomes the new current Record Set for the table.

- **Save it to disk as a Record Set Pointer File.** See above (p. 13).

- **Dismiss it.** See above (p. 14).

How Biota Uses Record Sets

Biota uses Record Sets to define groups of records for several kinds of special tasks. For example, the Search Editor can, as an option, search within a Record Set instead of searching through all records in a table (pp. 192-200).

Four search tools find records in one table based on the current Record Set for another table: Lower Taxa for Higher Taxa, Higher Taxa for Lower Taxa, Places for Specimens Or Species, and Specimens or Species for Places (pp. 200-213).

All of Biota's Export tools can (some of them must) use the appropriate Record Set as the basis for export (Chapters 24 and 25), and all Labels tools print labels for a Record Set (Chapter 13).

A new Specimen Loan can be made for the current Specimen Record Set, or the current Specimen Record Set can be recorded as a returned loan (Chapter 17).

The Find and Replace tool can be used on Record Sets, as an option (pp. 220-225).

To Display a Record Set

As their only function, commands in the Display menu display the current Record Set for each of the Core tables in the standard output screen for each table, with all the usual tools for output and input screens available (Chapter 9). If the current Record Set is empty, Biota informs you that the Record Set you asked to display is currently empty, and nothing is displayed.

NOTE: The command for displaying the current Record Set for the Det History (Specimen Determination History) table is in the Special menu (Chapter 19). If you have activated the user password system, only a user in with Administration access privileges can use this command (Chapter 23).

To Empty All Record Sets

You can empty all current Record Sets by selecting Empty Record Sets from the Special menu. Record Sets always start out empty each time you launch Biota.

Record Set Pointer Files

Record Set Pointer Files, the fourth way to represent Biota records, allow you to save the results of searches, input sessions, or import operations for later use, even between Biota sessions. A Record Set Pointer File is illustrated below in the dashed rectangle.

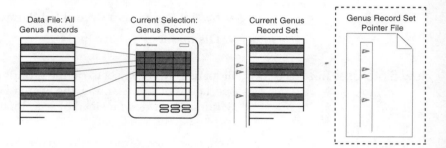

Internally, Record Sets themselves are simply lists of "pointers" to records, not the records themselves. (Record pointers are a very compact and efficient way to keep track of Selections of records.) A Record Set Pointer File is simply a disk file (which you name as you wish) containing a set of record pointers, which can later be loaded from disk to recover a Selection of records for a particular table.

You can create as many Record Set Pointer Files as you wish for each Core table. They take up virtually no disk space, and they are saved or loaded virtually instantaneously.

Record Set Pointer Files are not readable or useable with other applications. (If you want to create a text file from a set of records, use the Export by Tables and Fields tool, pp. 434–443, or another export tool from Chapter 24.)

Saving a Record Set Pointer File: First Method

You use this method to create a Record Set Pointer File for a selection of records in an out put screen when you close the output screen. If a Record Set already exists for the table, creating a Record Set Pointer File with this method leaves the existing Record Set unchanged.

1. **Choose "Save as a Record Set Pointer File"** in the Record Set option window that appears when you close an output screen (pp. 13–15).

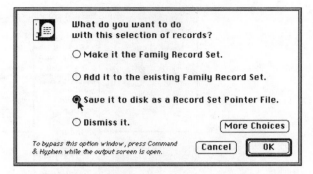

2. **Click the OK button** in the option window. A message appears reminding you to name the disk file appropriately.

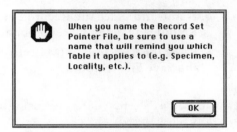

3. **Click the OK button in the reminder message.** The Save File window for your operating system appears.
4. **Name and place the Record Set Pointer File.**
5. **Click the Save button (Macintosh) or the OK button (Windows) in the Save File window.**

Saving a Record Set Pointer File: Second Method

You can use this method at any time to create a Record Set Pointer File for the current Record Set of a particular table. No records need be displayed.

1. **Choose Save Record Set Pointer File from the File menu.** A table selection window appears.

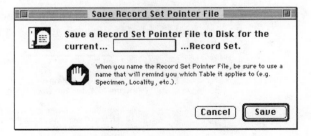

2. **Choose the table name from the popup list** in the table selection window.

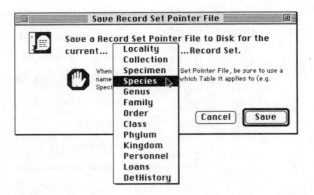

3. The Save File window for your operating system appears.

4. **Name and place the Record Set Pointer File.** Be sure to name the Record Set Pointer File appropriately.

5. **Click the Save button (Macintosh) or the OK button (Windows) in the Save File window.**

Loading a Record Set Pointer File

To load a Selection of records using a saved Record Set Pointer File, follow these steps.

1. **From the File menu, choose Load Record Set Pointer File.** A table selection window appears.

2. **From the popup list in the selection window (above), select the Biota table for which you intend to retrieve records.**

3. **Click the Load** button when ready (above). The file navigation window appears

4. **Find the Record Set Pointer File** in your folder or directory structure and open it.

WARNING: Be certain that the file you choose applies to the Biota table you have selected in step 2. If not, you will have unpredictable results, or no results.

The records will be displayed in the output screen for the appropriate table.

Loaded Record Set Pointer Files and the Current Record Set

Records retrieved using a saved Record Set Pointer File are not automatically designated as the current Record Set when they are displayed. If you want to make them the current Record Set, or add them to the current Record Set for the table, you must do it in the usual way, using the Record Set Option window that is displayed when you dismiss the output screen (see "The Record Set Options Screen," pp. 13–15, earlier in this chapter).

Why does Biota not declare a retrieved selection the current Record Set automatically? Were that the case, you would be unable to combine selections saved as separate Record Set Pointer Files to create a pooled current Record Set, nor could you pool records from a new search with a saved Record Set.

Chapter 3 A Brief Tutorial with the Demo Database

It is difficult to see Biota in action with a new, empty Data File. This chapter takes you on a quick tour of Biota's basic capabilities using the Demo Data File that you received on one of the Biota distribution diskettes. More advanced and less frequently used features are covered in the chapters that follow.

Don't be intimidated by the number of pages in this chapter. They go by very quickly, and the chapter is broken up into separate lessons so that you need not do the entire tutorial at one sitting. Although each lesson stands alone, later lessons assume you have done the earlier ones first.

Before starting the tutorial tour, please read Chapter 1, "Biodiversity Data and Relational Databases," and Chapter 2, "Key Concepts."

Installing Biota and Opening the Demo Data File

If you have not yet installed Biota, please install it, following the instructions in the ReadMe file on one of the Biota distribution diskettes.

1. Launch Biota in Data File finding mode.

- *Macintosh*: **Press and hold the OPTION key** while you launch Biota by either of the following two methods.
  - ◊ *Either:* **Double-click the BiotaApp (or Biota4D) icon or application name.**
  - ◊ *Or:* **Select the Biota icon or application name, then choose Open from the File menu** while in the Desktop environment.
- *Windows*: **Press and hold the ALT key while you launch Biota by selecting the BiotaApp (or Biota4D) icon or application name and choosing Open from the File menu** in the directory window.

(Double-clicking an icon while pressing ALT has a different function in the Windows operating system.)

NOTE: If you have implemented the password protection system (see pp. 409–415), you will need to press and hold the OPTION key (Macintosh) or ALT key (Windows) while clicking the Connect button in the Password dialog.

2. **When the file navigation window appears, find and open the Biota Demo Data File.**

    - *Macintosh*: The file is named BiotaDemo.data. Find it and open it when the "Open a data file" window appears.
    - *Windows*: The Demo Data File is named BiotaDemo.4dd, shown as BiotaDemo in the "Open which data file" window.

   Once you have opened a particular Biota Data File using this method, the same Data File will be opened automatically next time you launch Biota—*without* having to use the OPTION/ALT key technique.

   The Biota startup screen will appear briefly while the program sets up. Then you will see this menu bar (the Macintosh menu bar is shown, but the Windows one is very similar).

   | File | Edit | Input | Series | Tree | Find | Display | Labels | Im/Export | Loans | Special |

3. **From the File menu, choose "About the Current Data File"** to confirm that the Demo Data File has been loaded, A screen like the one here should appear, identifying the Data File, specifying its location on your hard disk, and listing the number of records it currently contains for each Biota table. (You will have to scroll the list to see all tables.)

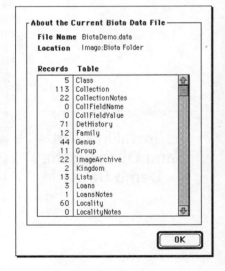

Note: If you get an error message, you need to place the Mac4DX folder (Macintosh) or the Win4DX folder (Windows) in the same folder (directory) with Biota. See the installation instructions in the ReadMe disk file if necessary.

What's in the Demo Data File?

The Demo Data File began life as an actual research dataset from a project on the hummingbird flower mites associated with tropical plants of the families Ericaceae and Campanulaceae. The mites were collected from host flowers preserved in fluid, collected by James Luteyn of the New York Botanical Garden. (You will see Luteyn's name in many places in the file.)

The file thus includes records for both plant and mite taxa and specimens and demonstrates the "host–guest" capability of Biota (Chapter 21).

For this tutorial, in the hope of creating some simple examples using species familiar to all biologists, a small set of fictitious Specimen records for Darwin's finches was added, along with appropriate higher taxon, Collection, and Locality records.

In addition to mites, their host plants, and fictitious finch specimens, a few records of ants from the Project ALAS database at La Selva Biological Station in Costa Rica (for which Biota was originally developed, see the Preface) have been added to demonstrate Biota's image capability.

NOTE: Any unpublished species names that may appear in the Demo Data File or in this book are included for illustration only, and are not intended nor should they be interpreted as formally published descriptions.

Lesson 1. Displaying and Editing Existing Records

Biota displays groups of records in *output screens*. In an output screen, you can sort the records displayed, print a list of them, select some of them to delete from the Data File, or select some of them to retain as a Subselection and dismiss the rest from the display. When you leave an output screen, you can declare the Selection of records last displayed in the output screen to be a *Record Set*.

An individual record can be displayed and edited in an *input screen*. In an input screen, you can change the content of the record, attach Notes or Auxiliary Field Values, use the record as a template for a new record, or delete the record. The input screen, as the name suggests, is also used when you create new records from scratch, as you will do in a later lesson of the tutorial.

Step by Step

1. **From the Find menu, choose All Species.** All Species records in the Demo Data File will be listed in the Species output screen.

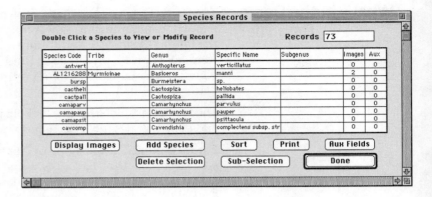

(The output screen expands vertically to fit your monitor, although you can adjust the size of the window manually. The illustrations use a window that has been adjusted in size vertically to save space on the page.)

Notice that the Species records are sorted, by default, by Genus and Specific Name (you can change the default in the Preferences screen; p. 101).

2. **To sort the records on a different field in the output screen, click the Sort button at the bottom of the display,** as shown below.

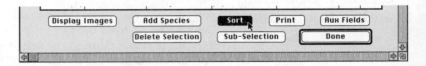

The Sort Editor appears, with the title "Sort Species."

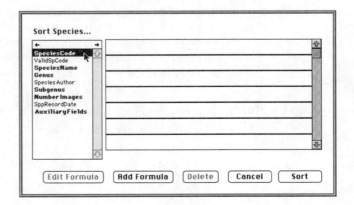

3. **Click Species Code in the field list panel** (illustrated above) to sort the records on the Species Code field (the Key field for the Species table, with a unique value for each record). Species Code is added to the sort criteria (below).

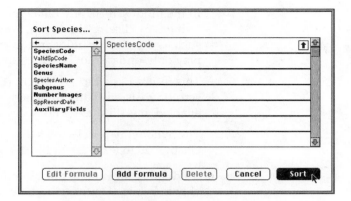

You can add additional criteria—including fields from related tables—or change the direction of the sort; see pp. 131–139.

4. **In the Sort Species window, click the Sort button** (as illustrated above). The output screen reappears, with the records sorted by Species Code.

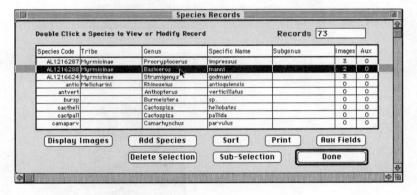

5. **To display an individual Species record, double-click its record in the output screen.** Try it with *Basiceros manni*, as shown above. The full record appears in the Species *input screen*.

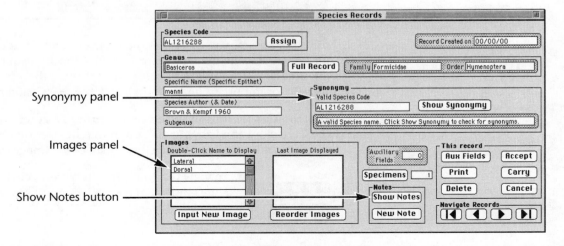

In the input screen, any data entry area can be edited. You save any changes by clicking the Accept button. (Information in stippled areas, such as the green Family area in the Species input screen, is for display only, and cannot be edited in this screen.)

This record has attached Images (there are Image Names in the inset scrolling display) and attached Notes (the Show Notes button is enabled; it is dimmed if no Notes are attached).

The Synonymy panel shows that *Basiceros manni* is a valid name. See Chapter 20 to learn about Biota's Synonymy system.

6. **To display an Image, double-click its name in the Image list,** as shown below. (Double-click Lateral.)

The Image appears in the Image input screen.

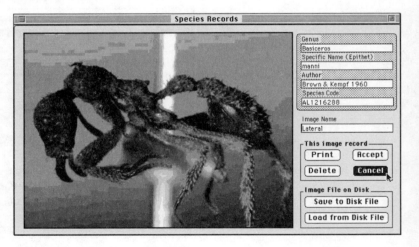

From this screen, you can print the Image or save it as a disk file. You can create a new Image record for this Species using the Input New Image button in the Species input screen, then paste a new Image in, or import an Image from a disk file. See Chapter 18.

7. **Click the Cancel button in the Image input screen to dismiss it.** (The Cancel button is being clicked in the illustration above.) The Species record reappears, with a thumbnail of the Image displayed.

8. **To view the Notes for this record, click the Show Notes button** in the Species input screen.

The Species Notes window appears.

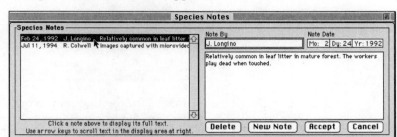

9. **In the Species Notes window, click a Note record in the scrolling display on the left** to display it in full in the right hand panel. The first of two notes has been displayed in the illustration above.

   In this same window, you can create new Notes for this Species (using the New Note button in the display or the New Note button in the Species input screen) or delete the Note displayed in the right-hand panel of the Notes window (by using the Delete button in the Notes window). See pp. 179–183.

10. **Dismiss the Species Notes window by clicking its Cancel button,** illustrated at the right (not the Cancel button in the Species input screen).

11. **To move to the next Species record** in the Selection of records that you displayed and sorted in the output screen, click the "forward" navigation button.

    The record for the ant *Strumigenys godmani* appears. The navigation buttons take you to the first, previous, next, and last record in the current selection, respectively, as sorted in the output screen. There are keyboard equivalents for these buttons and for several others; see pp. 91–92.

12. **Dismiss the Species input screen by clicking its Cancel button, illustrated at the right.**

    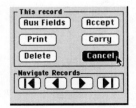

    The Species output screen reappears, with the records sorted in the same sorted order they were when you left it.

13. **In the output screen, click the Display Images button.**

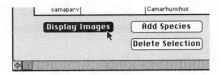

Images for the first few records appear in the Image output screen. Each row of Images belongs to one Species record. (The name of the species, its Species Code, and it Family appear above its Image row.) A row of blank frames appears if no Images are linked to a Species record; if fewer than four Images are linked, the remaining frames are blank. If the Species records are still sorted by Species Code, the first three records will be ants with Images, of which two are shown here.

Note: The Display Images button exists only in the Species output screen—not in the output screen for any other table—because in Biota, Image Archive records are linked to Species records.

14. **Click the Done button in the Image output screen to dismiss it.** The Species output screen reappears.

15. **Select a set of records in the Species output screen.** Select any records you wish, but be sure to include the Galápagos finch *Camarhynchus parvulus* (Species Code "camaparv"), which is needed for a later step.

   ♦ **To select consecutive records**, click the first record in the group then hold down the SHIFT key while you click the last record in the group.

   ♦ **To select nonconsecutive records,** hold down the COMMAND (⌘) key (Macintosh) or the CTRL key (Windows) while clicking the individual records.

16. **Click the Sub-Selection** button to dismiss all unselected records.

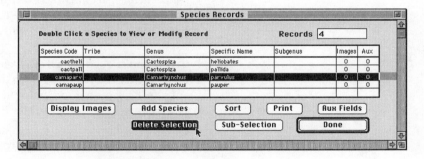

The Species output screen is now shows only the Species records you selected (illustrated in the next step).

Dismissing records from a Selection (or including them in one) does not affect the records in the Data File in any way. You can prove it to yourself by choosing All Species from the File menu again, leaving the current window open. The same records can simultaneously be part of any number of different Selections or Record Sets (Chapter 2).

17. **In the Species output screen, select the record for *Camarhynchus parvulus*** (Species Code "camaparv").

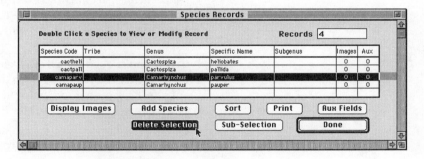

18. **Click the Delete Selection button in the Species output screen** (illustrated above). A warning message appears.

19. **Click the OK button in the warning window shown above.** (The deletion will be canceled in the next warning screen). If a record you propose to delete has any linked child records in a Core table (see Chapter 2 for definitions), a second warning message appears, like the one below. (If you have asked to delete *more* than one record, and at least one of those records is linked to child records, a somewhat different screen appears; see pp. 129–131.)

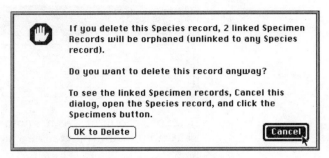

In this example, there are records for two Specimens of *Camarhynchus parvulus* in the database, which will no longer be linked to any Species record (and thus no longer linked to a Genus record, Family record, or higher ranks) if you confirm the deletion.

20. **Click the Cancel button in the warning screen above** to avert the pending deletion. The Species output screen reappears, with the *Camarhynchus parvulus* record intact. As the warning screen above describes, there is an easy way to take a look at linked Specimen records for *Camarhynchus parvulus*, which you will do next.

NOTE: If you clicked the OK to Delete button in error, the Demo Data File can always be reloaded from the distribution diskette if you later want a full copy—a good example of why you should always make backup copies of your own Data Files.

21. **Double-click *Camarhynchus parvulus* in the Species output screen.** The record appears in the Species input screen (as in step 5, above).

22. **Click the Specimens button in the Species input screen** (illustrated below). Notice that the small display field next to the Specimens button shows the number of linked Specimen records. This button is always enabled for use if any child records exist—not just when you have threatened to delete a record. If there are no child records for the Species record displayed, the button is disabled and the display field reads "0."

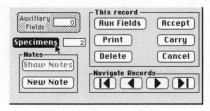

NOTE: The input screen for each Core table has an analogous button to display linked child records. The name of the button is always the name of the child table (in plural form); generically, these are called *Child Records* buttons (pp. 125–126).

The two Specimen records for *Camarhynchus parvulus* appear in the Specimen output screen in a new window. The Species input screen remains open in its own window, although it may be hidden.

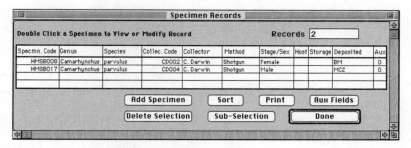

NOTE: You can use Window Shade (Macintosh System 7.5 or later) or Minimize (Windows) to help you work with multiple Biota windows.

23. **Double-click one of the records in the Specimen output screen.** The record appears in the Specimen input screen.

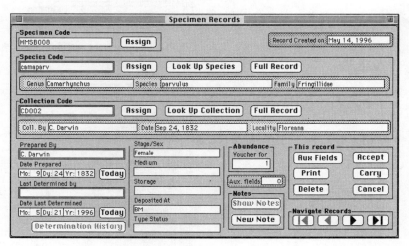

The Specimen input screen shows not only all Specimen fields, but essential information from the related Species and Collections records. You will see more of this screen later in the tutorial.

This is the record for a fictitious specimen that exists only in the Demo Data File, although the collector and collection dates are plausible, based on *The Voyage of the Beagle*.

24. **Click the Cancel button in the Specimen input screen to dismiss it.** The Specimen output screen reappears.
25. **Click the Done button in the Specimen output screen.** The Record Set option screen appears.

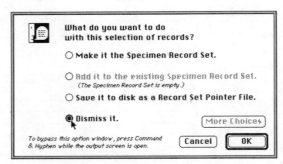

26. **Click the "Dismiss it" button, then the OK button in the Record Set option screen (above).** See pp. 13–15 for a discussion of the other options offered. You will create a Record Set in the next lesson.
27. **Return to the record for *Camarhynchus parvulus* displayed in the Species input screen.**
28. **Click the Full Record button in the Genus panel of the Species input screen.**

The full Genus record for *Camarhynchus* appears in the Genus input screen.

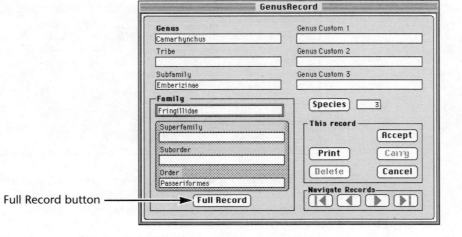

Full Record button

If you want to see the full record for Fringillidae, the linked Family record, you can click the Full Record button in the Family display area of the Genus screen, above, then Cancel the Family input screen.

The input screen for each Core table (except for Kingdom and Locality) offers a Full Record button. These buttons make it easy to display information on parent records—all the way to top of the taxonomic and place hierarchies (see pp. 124–125).

29. **Click the Cancel button in the Genus input screen.** The Species input screen reappears.

30. **Click the Cancel button in the Species input screen.** The Species output screen reappears.

31. **Click the Done button in the Species input screen.** The Record Code option window appears.

Notice the small note in the lower-left corner: *To bypass this option window, press* COMMAND *and* HYPHEN *while the output screen is open.* (This is the Macintosh screen. The Windows version substitutes CONTROL for COMMAND.)

You may not always want to declare the current Selection of records to be a Record Set, add it to the existing Record Set, or save it to disk as a Record Set Pointer File. (See Chapter 2 for the distinctions.) In the next two steps, you will try the keyboard bypass explained in the small note, which has the same effect as the "Dismiss it" option in the screen above.

32. **Click the Cancel button in the Record Set Option window** (above). The Species output screen reappears.

33. **This time, dismiss the Species output screen from the keyboard,** bypassing the Record Set option window:

    • *Macintosh*: **Hold down the** COMMAND **key (⌘) and press the** HYPHEN **key** (above the P key on the keyboard, not the minus key in the numeric keypad).

    • *Windows*: **Hold down the** CTRL **key and press the** HYPHEN **key** (above the P key on the keyboard, not the minus key in the numeric keypad).

> **NOTE:** *You can use this keyboard shortcut to dismiss virtually any output or input screen in Biota*. In the case of an input screen, it is the equivalent of clicking the Cancel button in the screen. There are several useful keyboard equivalents for onscreen buttons, as well; see pp. 91–92.

Overview

In this lesson, you entered the Biota structure through the Species table by choosing All Species from the Find menu. (The structure is shown in full in Appendix A and discussed in Chapter 2.) Once "inside," you learned how to:

- **sort records in the output screen.**
- **delete records from the output screen.**
- **make a Subselection of records in the output screen.**
- **display an individual record in the input screen.**
- **move *down* a hierarchy** using Child Records buttons (the Specimens button in the Species input screen).
- **move *up* a hierarchy** using Full Record buttons in input screens.

Each of the above techniques applies to output and input screens throughout Biota's structure—regardless of how and where you enter the structure, which will depend on the task at hand.

You also learned how to:

- **display Notes.**
- **display Images** from the Species input and output screens.

The Notes techniques apply to all tables with Notes—Species, Specimen, Collection, Locality, and Loans. Only the Species table has associated Images.

For complete information on these topics, refer to Chapters 1, 2, and 9.

Lesson 2. Creating New Records One Table at a Time

This lesson shows you how to new create records for one table at a time. In Lesson 3, you will learn how create records in related tables "on the fly," as they are needed.

The input screen for each table is the tool you use to create and enter data in new records for that table. As much as possible, all of Biota's input screens share the same features. These common elements are discussed in Chapter 6, "Entering Data," and Chapter 9, "Working with Records in Output and Input Screens."

Step by Step

1. **From the Input menu, choose Specimen.** The Specimen input screen appears, with the insertion point (blinking cursor) in the Specimen code *entry area*. You are going to add new records to the Demo Data

File for additional fictitious specimens from Darwin's visit to the Galápagos Islands.

2. **Enter *HMSB018* in the Specimen Code entry area and press the TAB key.**

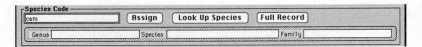

The insertion point moves to the Species Code entry area. Let's say this is a specimen of the Small Tree Finch, *Camarhynchus parvulus*, collected on Santiago Island (James Island, in Darwin's time) on October 9, 1832, when the *Beagle* was anchored there.

NOTE: If you get an error message indicating that this record already exists, and you are certain you entered *HMSB018*, it means that someone has already done the tutorial using this copy of the Demo Data File. Either quit Biota and install a fresh copy of the Demo Data File from the distribution diskette (see pp. 21–22 to find the new copy), or delete the existing record (select All Specimens from the Find menu, find it and delete it).

The next step is to link this Specimen record with the Species record for *Camarhynchus parvulus*, which is done by entering the Species Code for *Camarhynchus parvulus*. Suppose you recall that the Species Code starts with "cam," but you are not sure of the rest of it.

3. **Enter "cam" in the Species Code entry area and press tab.**

Biota presents a list of all Species records in which the Species Code begins with "cam." (This option for making an entry in a *linking field* is called the *wildcard* method.)

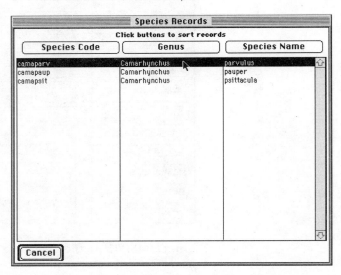

4. **Click the line in the Species Records list (above) for *Camarhynchus parvulus*.** The Species Code is automatically entered in the Specimen record, with additional information displayed about the species in the blue Species display panel.

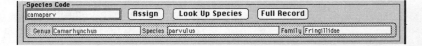

The insertion point moves on to the Collection Code entry area. The next step is to link this Specimen record with the Collection record for Darwin's specimens from Santiago Island on October 9, 1832.

You do not know whether such a Collection record even exists yet, nor its Collection Code if it does. This time, to seek the Collection record, you will use a Look Up button instead of the wildcard method you used to find the Species record. (Notice that there is also a Look Up button for Species Codes, which you will use later.)

5. **Click the Look Up Collection button in the Specimen input screen.**

The Look Up Collection option window appears.

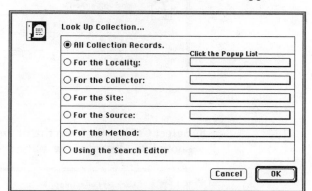

This option window helps you narrow the search for existing Collection records—although the first option in the screen (the default) displays them all if you prefer. (If there are hundreds or thousands of Collection records in database, finding a record among those displayed can take a long time.)

Each of the next five option buttons in the Look Up Collection screen limits the search by a different attribute (field) of the Collection records (Locality, Collector, Site, Source, or Method). If you choose the last option in the list, Biota opens the Search Editor (pp. 192–200) to allow searches not covered by the other options.

6. **Click the button labeled "For the Collector" in the Look up Collection window,** since you know the record you are looking for will have Darwin listed as the Collector.

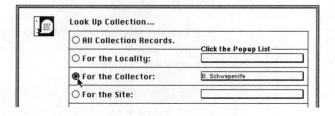

When you click the button, Biota creates a list of all collectors represented in the database. The first entry from the list ("B. Schwepenife" in this case) is displayed in the popup area to the right of the "For the Collector" button to let you know the list has been completed. The rest of the list is hidden in the popup list.

7. **Click and hold the popup list to the right of the "For the Collector" button** to display the list of collectors' names.

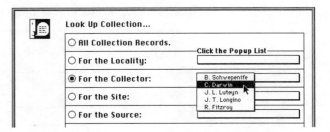

8. **Select C. Darwin from the popup list** (illustrated above). Biota will show the selected value in the popup area (shown below).

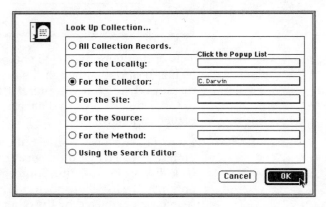

You can try out the other button and popup lists if you want. No selection is final until you click the OK button in the Look Up Collection option window. If you do this, be sure the "For the Collector" button is selected and C. Darwin has been selected in the popup before you do the next step.

9. **Click the OK button in the Look Up Collection option window.** The Look Up Collection display screen will appear, listing all Collection records for C. Darwin.

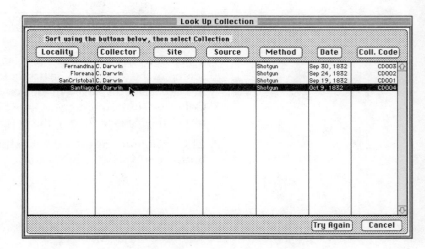

The fourth record is the one you want: Santiago Island on October 9, 1832.

If there were many records in the display screen, you might want to use the sort buttons at the top of each column to sort the records by the values in that column. Try the Date button now.

10. **Click the record for Santiago Island on October 9, 1832.** The Look Up Collection window is dismissed. The Collection Code for the record you selected is entered in the Specimen record, with additional information displayed about the Collection in the yellow Collection display panel.

Note: If you want to check the full Collection record, click the Full Record button in the Collection Code panel, above, then return to the Specimen input screen.

The insertion point moves on to the Prepared By entry area. The next step is to record the name of the person who prepared the finch specimen. We will assume that Darwin himself prepared the specimen on the same day he collected it. (He probably would have.)

The Prepared By field of the Specimen table is linked to the Personnel table (notice the double-bordered entry area, which always indicates a link to a parent table). You could simply enter "C. Darwin" to link the Prepared By field of this Specimen record to Darwin's record in the Personnel table, but you will try a different approach this time, as an illustration of the procedure.

11. **In the Prepared By entry area, enter the "at" character @ and press the TAB key.**

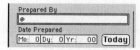

All records from the Personnel table are displayed in a choice window.

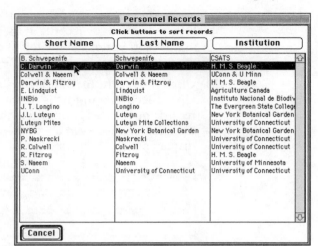

This trick works for any linking field (double-bordered entry area). The @ character is Biota's *wildcard character*.

12. **Click the record for Darwin.** The record is entered in the Prepared By entry of the Specimen record.

    The next step is to enter the Date Collected.

13. **Enter 10 in the Mo entry area, 09 in the Dy area, and 1832 in the Yr area** using the TAB key to enter each value.

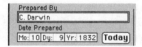

You make all date entries in Biota in this format, although collection and specimen labels are printed or exported using the international date format (9 October 1832, in this example).

You can also enter a *partial date* (month and year only, or just the year) in the Prepared By or Determined By fields of Specimen records, or in the Collected By field of Collection records. See pp. 111–114 for complete details on dates in Biota.

The Today button enters the date from your computer's internal clock.

14. **Press the TAB key repeatedly to skip through the Last Determined By and Date Last Determined areas.** Data entry for these fields is identical to the Prepared By and Date Prepared fields. (Try it out if you want.)

    When the insertion point moves into the Stage/Sex entry area, an *Entry Choice List* window appears with the title "Choices for Stage/Sex."

Entry Choice Lists are under your control. You can enable or disable them and change the composition or the order of the items displayed. See pp. 315–322 for details.

Note: As distributed, the Demo Data File has this List enabled. If the List does not appear, however, a previous user of the Demo Data File may have disabled it. *When you are done with this lesson,* choose Entry Choice Lists from the Special menu and enable the List if you like.

15. **Click Male in the Choices for Stage/Sex Entry Choice List.** The List disappears and Male is entered in the record.

Note: To override a Choice List and make a manual entry, press TAB when the List appears.

Take a look at the remaining fields in the Specimen record. (See pp. 143–152 for details on any field.) If you want to make any further entries in the record, go ahead and do so now before proceeding to the next step.

16. **When the record is ready, click the Carry button in the Specimen input screen.**

Two things happen when you click the Carry button. The record you have been working on (HMSB018) is accepted, and Biota displays a *new* Specimen record with all information *except for the Specimen Code* "carried over" (copied in) from the record you just completed.

17. **Enter *HMSB019* in the Specimen Code entry area of the new record and press the TAB key.**

    Using the Carry button saves time when you are entering a series of records that share information. (You can also click the Carry button to use a saved record as a template for a new one; see pp. 127–128.)

    For the new record HMS019, assume that Darwin also collected a female *Camarhynchus parvulus* on the same day and in the same tree as the male whose record you just completed. Thus only the Stage/Sex field needs to be changed.

18. **With the mouse, click in the Stage/Sex entry area.** The Entry Choice List appears, as in step 14 above.
19. **Click Female in the Entry Choice List.** The value is entered in the record.
20. **Click the Accept button in the Specimen input screen.**

    The record is accepted into the Demo Data File and a new, blank Specimen record appears.

21. **Click the Cancel button in the Specimen input screen** to dismiss the blank record.

    The two records you created appear in the Specimen output screen.

22. **Click the Done button in the Specimen output screen.** The Record Code Option window appears.

23. **Click the OK button in the Record Code Option screen** to accept the default option, "Make it the Specimen Record Set." The Specimen output screen is dismissed and the two records you created are declared the current Specimen Record Set.

    NOTE: For double-bordered buttons, like the OK button in the Record Set option window or the Done button in record output screen, you can always use the RETURN key (Macintosh) or ENTER key (Windows) to execute the button action, instead of clicking the button with the mouse.

24. **From the Display menu, select Specimens.** The Specimen output screen appears with the two records you created. Once the records are displayed, you can use all the operations for working with records that you have learned.

    Choosing a command for a Biota table from the Display menu always displays the records in the current Record Set for that table. If the current Record Set is empty, Biota posts a message.

25. **Click the Done button in the Specimen output screen.** The screen is dismissed *without* displaying the Record Set Option window, since these records have already been declared the current Specimen Record Set.

    If you had changed the Selection of records after displaying the Record Set (by using the Subselection button, the Add Specimen button, or by deleting one or more of the displayed records from the database), the Record Set Option window would have appeared.

Overview

In this lesson, you learned how to:

- **create a Specimen record and link it to a Species Record, to a Collection Record, and to a Personnel record.** The very same techniques apply to creating records for each of Biota's other tables and to linking them to existing parent records.

- **use the wildcard entry method** for entering data in linking fields. You used this method to enter the Species Code based on its first few characters, and, later, to display all records from the Personnel table using the @ character.

- **use a Look Up button** to find the right link to a parent table. You used the Look Up Collection button to find an existing Collection record.

    There is also a Look Up Species button in the Specimen input screen for finding an existing Species record. When you click it, Biota asks for the Genus of the Species you are looking for, then displays all Species records for that Genus in the selection window. You will use this tool in the next lesson.

    In the Collection input screen, a Look Up Locality button for linking

Collection records to Locality records works just like the Look Up Collection button you learned about in this lesson.

- **use an Entry Choice List** to enter data in a field. You can set Entry Choice Lists for many fields in Biota (pp. 315–322), or have Biota enter a particular value in a field automatically (not demonstrated in the lesson; see pp. 313–315).
- **use the Carry button** to accept a record, then use it as a template for the next new record.
- **use the Accept button** to accept a new, single record.
- **create and display a Record Set.**

For a comprehensive treatment of special features of particular input screens consult Chapter 7, "Record Codes," Chapter 8, "Special Data Types," and Chapter 10, "Input—Table by Table."

Lesson 3. Creating New Records "On the Fly"

In Lesson 2, you learned to link new records to related records in a parent table. The Species, Collection, and Personnel records to which you linked the new Specimen records were, conveniently, already in the Data File. Often, of course, this will not be the case.

When entering a new record or editing an existing one, it is often necessary to create a new record in a parent table. This lesson teaches you how to use Biota's ability to do this "on the fly" as you create or edit the primary record.

Step by Step

1. **From the Input menu, choose Specimen.** The Specimen input screen appears. You are going to add a new record to the Demo Data File for yet another fictitious specimen from Darwin's visit to the Galápagos Islands.

2. **Enter *HMSB020* in the Specimen Code entry area and press the TAB key.** The insertion point moves to the Species Code entry area.

   NOTE: If you get an error message indicating that this record already exists, see the Note at step 2 in Lesson 1.

   Let's say this is a specimen of the Charles Mockingbird, *Nesomimus trifasciatus*, collected by Darwin on Floreana Island (Charles Island, in Darwin's time) on September 24, 1832.

3. **To find out if there is already a record for this Species in the database, click the Look Up Species button** in the Species Code panel of the Specimen input screen.

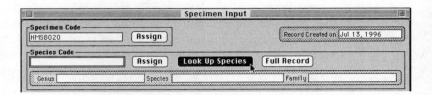

The Look Up Species selection window appears.

You use this tool by entering a Genus name—or the first part of a Genus name using the wildcard method—in the entry area at the top of the window, then press TAB. Biota then looks in the Genus table for matching entries.

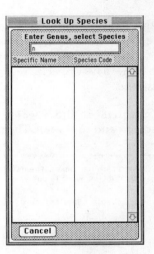

- **If Biota finds only one match** among Genus records, all Species records linked to that Genus record are displayed in the lower panel. You click the one you are looking for to enter it in the Specimen record.

- **If Biota finds more than one match** among Genus records, the list of candidates is displayed, as you will see in the next step.

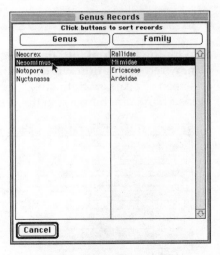

4. **Enter the letter "n" in the Genus entry area of the Look Up Species window and press TAB.** A list of four genera that begin with the letter "n" is displayed, including *Nesomimus*.

5. **Click the entry for *Nesomimus* in the list of genera** (see above). The Look Up Species window now shows three Species records for *Nesomimus* species.

If the species you were looking for were here, you would click its name to enter it in the Specimen record. But, clearly, *trifasciatus* is not yet in the database.

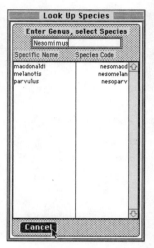

6. **Click the Cancel button in the Look Up Species window** (right) since the record you seek is not listed. You will need to create a new Species record for *N. trifasciatus*.

7. **Type "nesotrif" in the Species Code entry area of the Specimen input screen and press TAB.** The new record option window appears.

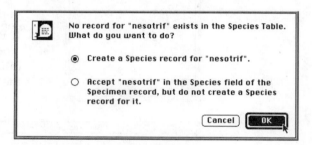

8. **With the first (default) option selected, click the OK button in the new records option window** (illustrated above). The Species input screen appears, with the new Species Code already entered and the insertion point blinking in the Genus entry area.

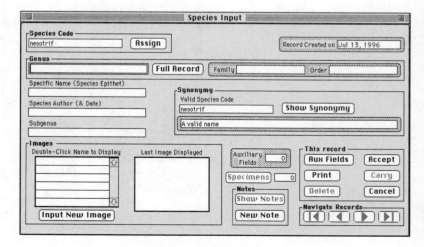

9. **In the Genus entry area, type "nes" and press TAB.** The wildcard entry system completes the entry for *Nesomimus*, since no other Genus name in the database starts with *nes*.

   Note: If the genus record had not existed, you could create a new one "on the fly," just as you are in the process of doing for a new Species record. In fact, you could create new records all the way to Kingdom this way, Accepting each as you work your way back down.

10. **In the Specific Name entry area, type "trifasciatus" and press TAB.**

11. **Click the Accept button in the Species input screen.** The Species input screen is dismissed and you return to the Specimen record in progress.

    Note: If you Cancel the Specimen record now, instead of Accepting it, the Species record you just created will nonetheless remain in the database.

12. **Click the Look up Collection button in the Specimen input screen** to find and enter the Collection Code for the record of Darwin's collections on Floreana Island for September 24, 1832. If you need help, refer to steps 5–10 in Lesson 2.

13. **In the Stage/Sex input area, record the specimen as Female.**

14. **Accept the Specimen record.**

15. **Click the Done button in the Specimen output screen and dismiss the Record Set option screen.**

Overview

This lesson covered the creation of new records and how to link them to parent records that you create "on the fly." In this lesson, you learned how to:

- **find out if a parent record already exists.** You did this for a Species record from the Specimen input screen, using the Look Up Species tool.

  For higher taxa (e.g., to see whether a Family records exists when you need to link a new Genus record), just enter the first few letters of the name you seek (or @ for all records) and press TAB. Biota will complete the entry if there is just one match, list all candidates if there are several matches, or post a message if nothing matches.

- **create a new parent record when you need one.** Just enter the new name. Biota will give you the option of displaying the parent record input screen.

For a comprehensive treatment of special features in particular input screens consult Chapter 7, "Record Codes," Chapter 8, "Special Data Types," and Chapter 10, "Input—Table by Table."

Lesson 4. Finding Records with the Tree Hierarchy

In Lesson 1, you learned how to display all the records for a Biota table by using the All... commands from the File menu. In a large database, however, displaying all records is not a very efficient way to find a record.

In this lesson and the next, you will learn several ways to find records using other tools from the Find and Tree menus. Chapter 11, "Finding and Updating Records," provides a comprehensive reference for Biota's search and update capabilities, only some of which can be covered in this tutorial.

Step by Step

1. **From the Tree menu, choose Class Down.** All Class records in the Demo Data File, including both plants and animals, appear in a special version of the Class output screen with the title "Class-Down Tree Lookup."

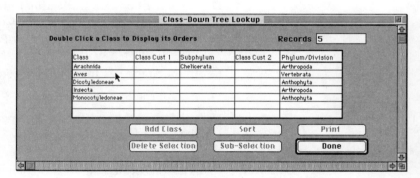

(Notice that all buttons are disabled except the Done button. To use the disabled buttons, you would need to find Classes using a tool from the Find menu instead.)

2. **Double-click the line for Class Aves in the Class-Down Tree Lookup screen (above).** A new screen reading "Orders of the Class Aves" (window entitled "Order Lookup") appears, listing all three Order records for birds that are in the Demo Data File.

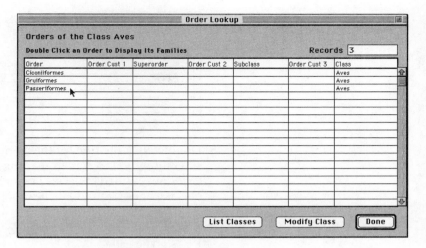

There are other Order records in the database—for plants and arthropods (select All Orders from the Find menu to see them)—but *the Tree menu is a hierarchical lookup system*. Once you get past the entry level (the Class table, in this lesson) Biota displays only the child taxa for each line you double-click.

Note: Recall that double-clicking a record in the *standard* output screen for a table displays the input screen, showing the individual record you clicked—unlike the hierarchical behavior of the Tree menu screens.

3. **Double-click the line for Order Passeriformes in the Order Lookup screen (above).** A screen reading "Families of the Order Passeriformes" (window entitled "Family Lookup") appears, listing the two Family records for passeriform birds present in the Demo Data File.

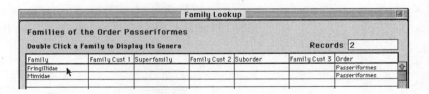

4. **Double-click the line for the family Fringillidae in the Family Lookup screen (above).** A screen reading "Genera of the Family Fringillidae" (window entitled "Genus Lookup") appears, listing the six genera of Darwin's finches (the only fringillid genera in the database).

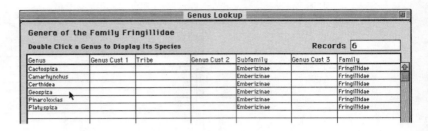

5. **Double-click the line for the genus Geospiza in the Genus Lookup screen (above).** A screen reading "Species of the Genus *Geospiza*" (window entitled "Species Lookup") appears, listing the six species of Galápagos Ground Finches.

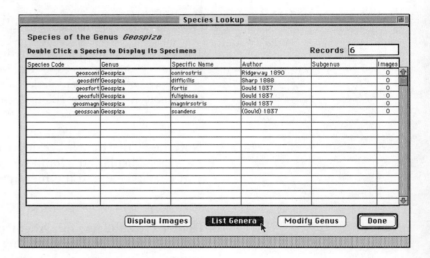

6. **Click the List Genera button** at the bottom of the Species Lookup window (above). The Genus Lookup window reappears, just as it was in step 4.

7. **Click the List the List Families button** at the bottom of the Genus Lookup window.

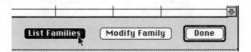

The Family Lookup window reappears, again showing the same passerine families as in step 3. In fact, you can go all the way back up the hierarchy this way, then back down a different branch of the tree if you want.

8. **Navigate the hierarchy until the species of** *Geospiza* **are again displayed** (as in step 5, above).
9. **Click the Modify Genus button** at the bottom of the Species Lookup window.

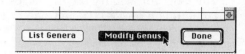

The full Genus record for *Geospiza* appears in the Genus input screen. (You could edit the record, then Accept the changes in this screen.)

The Modify... button at the bottom of each screen of the Tree hierarchy works in the same way: the parent record for the records displayed appears in the input layout.

10. **Click the Cancel button in the Genus input screen** to dismiss it. The display of species of *Geospiza* reappears (as in step 5).
11. **Explore the Tree hierarchy further.** Try double-clicking a Species record to see Specimen records, then double-click a Specimen record to open it in the Specimen input screen. When you click the Cancel button in the Specimen input screen (while in the Tree hierarchy), you can go back up the hierarchy using the "List..." buttons.
12. **Display the species of** *Geospiza* **again, then click the Done button** in the Species Lookup screen to leave the Tree hierarchy. The Record Set option screen appears.
13. **Click the OK button in the Record Set option screen to create a Record Set.** You can create a Record Set in this way for whatever set of taxa was last displayed, at any level. (It does not matter what Record Set you create in this lesson, since it will not be used again.)

Overview   In this lesson, you learned how to:

- **enter the Tree hierarchy.** You entered at the Class level, but you can enter at any point, by using the commands in the lower section of the Tree menu.
- **move down the hierarchy** by double-clicking individual records at any level, all the way to individual Specimen records.
- **display an individual record** at the Species level and above by using a Modify... button.
- **create a Record Set for a group of taxon records** that share the same parent record using the Tree hierarchy.

Lesson 5. Finding Records in One Table Based on a Set of Records in Another Table

In lesson 4, you established a Species Record Set for Species of the genus *Geospiza*, using the Tree hierarchy. In this lesson, you will learn a different way to find a group of records and declare the a Record Set, then *use* that Record Set to find related records in other tables.

The tools you will try out in this lesson (and related tools) are presented in full on pp. 200–213 in Chapter 11.

Step by Step

1. **From the Find menu, choose Lower Taxa for Higher Taxa.** The query setup screen below appears. You will use this tool to find all Species records for the Family Fringillidae (finches) present in the database.

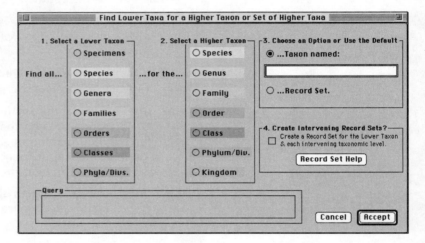

2. **In the list of tables in the left panel (the panel labeled "1. Select a Lower Taxon"), click the Species button.** This is called the *target table*—the table containing the records you seek.

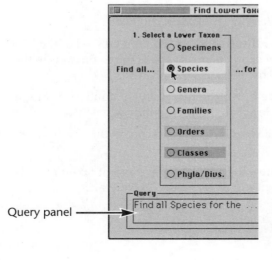

Query panel

The *query panel* displays the beginning of the query at the bottom of the screen.

3. In the list of tables in the right panel (the panel labeled "2. Select a Higher Taxon"), click the Family button—the button for the table to be used in the selection criterion (the *criterion table*).

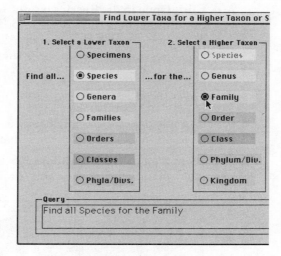

The query panel adds the next phrase to the query.

4. In the entry area beneath the button labeled "...Family named," type "fri" and press the TAB key. (The entry area is in the upper-right corner of the screen.)

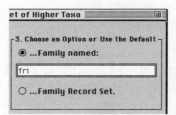

In the panel at the left, notice that, instead of entering a Family name, you could have clicked the lower button in the panel to use a Family record set as a search criterion.

Biota completes the entry so that it reads "Fringillidae," and completes the query to read "Find all Species for the Family Fringillidae."

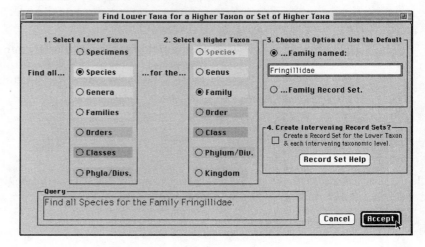

5. **Click the Accept button to launch the search** (above). The 14 Species records for the Family Fringillidae (including species of six different genera) are displayed in the Species output screen.

6. **Click Done in the Species output screen, then OK in the Record Set option screen** to make these records the current Species Record Set. In the next set of steps, you will use this Species Record Set to find all Localities linked to these Species records.

   Notice that the target and criterion tables in the search just completed are not adjacent in the hierarchy (the Genus table lies between the Species and Family tables). The Record Set you just declared could not be made directly using the Tree hierarchy (Lesson 4), which is limited to finding groups of records that share a single parent record.

7. **From the Find menu, choose Places for Specimens or Species.** The query setup screen below appears.

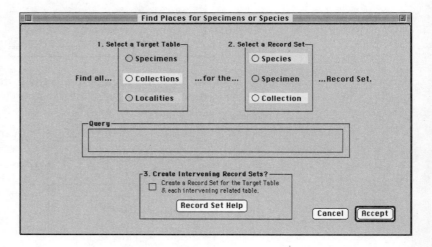

You will use this tool to find all Specimen records linked to the finch Species Record Set, all the Collection records linked to those Specimen records, and all the Locality records linked to those Collection records—all at once.

8. **In the list of tables in the left panel (the panel labeled "1. Select a Target Table"), click the Localities button.**

The query panel displays the beginning of the query.

9. In the list of tables in the right panel (the panel labeled "2. Select a Record Set"), click the Species button.

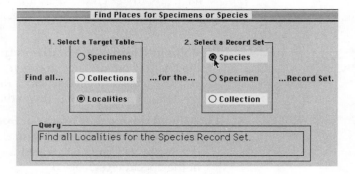

The query panel completes the query, "Find all Localities for the Species Record Set."

10. In the panel labeled "Create Intervening Record Sets?" click the checkbox, to set it.

When you set this option, Biota creates Record Sets automatically for the target table records and for the linking records in any intervening tables—in this case, for the Specimen and Collection tables (see Appendix A).

As you learned earlier in this lesson with the Lower Taxa for Higher Taxa tool (which has an identical checkbox), it is not *necessary* to check this option in order to find the target table records you are searching for, nor in order to create a Record Set for them.

11. Click the Accept button to launch the search.

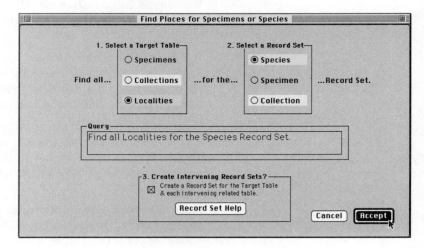

Four Locality records (for four Galápagos Islands) are displayed in the Locality output screen. Because you checked the checkbox, they have already been declared the Locality Record Set. These are the Localities for Collections of Specimens representing Species in the Species Record set.

12. **Without closing the Locality output screen, select Specimens from the Display menu.** The Specimen output screen shows the Record Set you just created for all finch Specimen records (including the ones you created in earlier lessons, if you did not delete them).

13. **Without closing the Locality or Specimen output screens, select Collections from the Display menu.** The Collection output screen shows the Record Set you just created for all Collections for the Specimens in the Specimen screen.

    Notice that Biota can open as many simultaneous windows as your computer's memory can support.

14. **Open a few Locality and Collection records** to see what the input screens look like, then close them with the Cancel button.

15. **Click the Done buttons in all three output screens** to close them.

Overview

In this lesson, you learned how to:

- **find records in a table in the taxonomic hierarchy based on a record or records in a higher table.**

    You may have noticed in the Find menu that there is an "opposite" tool, Higher Taxa for Lower Taxa. It works in the same way in reverse (but requires a lower taxon Record Set to be established first). You could use it, for example, to find all Order records that are linked through intermediate tables to a set of Specimen records.

- **find records in the place hierarchy (Collection and Locality tables), based on a Species Record Set.** You can use the same tool to find records based on other combinations spanning this series of linked tables.

    Again, the Find menu offers an "opposite" tool, Specimens or Species for Places. It works for the same table combinations, but in reverse order. You could use it, for example, to find all Species records linked through intermediate tables to a set of Locality records.

    If you do this based on the Locality set you just created, for example, it will yield not only the finch Species records you started with, but also the Species record for the mockingbird *Nesomimus trifasciatus* that you created in Lesson 3—a nice demonstration of why both tools are needed.

- **create Record Sets automatically for the target table and all intervening tables.** This capability enables you to create linked subsets of your data for several tables at once (for export, for example).

On Your Own—What to Try Next

The lessons in this tutorial have covered many important features of Biota, but there are many others you will eventually want to learn about. Here are a few suggestions for next steps on your own.

- **Learn to use the Search Editor.** From the Find menu, choose Using the Search Editor. Then turn to pp. 192–200 for instructions.
- **Learn to use Record Set Pointer Files**. Read or review Chapter 2, pp. 16–19.
- **Learn how to print or export labels.** From the Table of Contents, choose the section of Chapter 13 that is relevant to your work.
- **Learn how to create and use Auxiliary Fields,** in Chapter 15. This feature enables you to create an unlimited number of special fields that you can name as you like, for the Species, Specimen, Collection, or Locality tables.
- **Learn how to customize Biota** by setting preferences, changing the displayed names of fields, adding Choice Lists and default entries, or setting up Record Code Prefixes. From the Table of Contents, choose the section of Chapter 16 that covers your needs.
- **Scan Chapter 5, "Overview: Biota Tools and Features,"** for additional items that interest you or solve a problem, and turn to the pages cited there.

PART 2 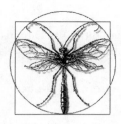 Express Route

Chapter 4 Quick Start

Quick-Start Strategy

If the concept of a relational database is new to you, it would probably be a good idea to read the rest of Chapters 1 and 2 before getting started.

If database concepts are familiar to you, however, or if you detest manuals, you might try the following approach:

1. **If you read nothing else in this Manual before you begin,** spend a few minutes with Appendix A, which diagrams the tables, fields, and relations that define Biota's data structure.

2. **Read the "Idiosyncrasies" section, below,** which highlights differences between the behavior of Biota and most other applications.

3. **Scan Chapter 2, "Key Concepts,"** to make sure you understand how Biota Data Files, Selections of records, Record Sets, and Record Set Pointer Files differ from one another. The diagrams in the chapter may suffice.

4. **Use Chapter 5, "Overview: Biota Tools and Features,"** as a quick guide.

   - Read through the *section titles and subheading titles* of Chapter 5, so you will know what Biota can do.

   - Later, if you run into a problem or remember something from Chapter 5 that you would like to use or to understand better, read the *paragraph* about that feature in Chapter 5, which will often be all you need to know to use it.

   - If you need more details (and a few Biota tools are rather complex), consult the pages referenced at the end of the paragraph about that tool in Chapter 5, for complete instructions.

Idiosyncrasies

The commercial database engine that drives Biota, 4th Dimension, imposes a few idiosyncrasies worth knowing about before you get started.

Opening Biota Data Files

Biota creates Data Files that are separate from the application file itself (BiotaApp or Biota4D).

- *A Biota Data File can be selected, opened, or created only while Biota is being launched.*
- You can copy a Biota Data File to back it up at any time, even when Biota has the file open, as long as no Biota tools are in use.
- To close the active Data File and open another, or to create a new, empty Data File, you must quit Biota and launch it again.
- To choose among existing Biota Data Files or create a new one:
  - *Macintosh*: Press and hold the OPTION key while you launch Biota either by double-clicking the BiotaApp (or Biota4D) icon or application name, or by selecting it then choosing Open from the File menu.
  - *Windows*: Press and hold the ALT key while you launch Biota by selecting the BiotaApp (or Biota4D) icon or application name, then chose Open from the File menu in the directory window.

NOTE: If you have implemented the user password protection system (see Chapter 23), you will need to press and hold the OPTION key (Macintosh) or ALT key (Windows) while clicking the Connect button in the Password dialog.

For complete information on Data Files, see pp. 8-11.

Saving Changes in a Biota Data File

Biota uses temporary buffers in RAM to keep track of any changes you make in the active Data File. Every three minutes, Biota automatically saves these changes to disk and clears the buffers.

There is one exception: Each group of new records you create or modify using commands from the Series menu is saved immediately.

If you want to force Biota to save changes immediately, press the keys ⌘(COMMAND)+W (Macintosh) or CTRL+W (Windows) at any time.

When you quit Biota, any changes made since the last automatic save are recorded in the Data File.

The small window shown here appears in the corner of the screen during a save operation. Other ongoing operations may be momentarily slowed, but are not halted.

> **NOTE TO MACINTOSH USERS:** The key combination ⌘(COMMAND)+W, normally used for closing windows in Macintosh applications, is reserved by 4th Dimension for saving data to disk from the buffers. To close a Biota window from the keyboard, use OPTION+W instead.

Hidden Window Cursor

If the cursor suddenly looks like a page behind another page, it means another window requires attention before you can proceed. (You clicked in an inactive window.) *To return to the active window, press any key on the keyboard.* (The keystroke is not recorded.)

Moving Dialog Windows (Macintosh)

Sometimes, you may wish you could move a dialog window (a window with a simple border) to see another open window. Dialog windows are not movable in most Macintosh applications, but any Biota window, even dialogs, can be moved with the mouse while depressing ⌘(COMMAND)+OPTION.

Ordering of Biota Table and Field Names in Certain Editors

Most of Biota's tools were created from scratch for Biota, but four complex editors are off-the-shelf 4th Dimension tools. In all four (see below), Biota table and field names are listed for selection, *but the names are not listed in alphabetical order.* Instead, unfortunately and unavoidably, they are presented in order of creation during the development of Biota.

In short, do not expect alphabetical listing of tables and fields in the following editors (all other listings of tables and fields in Biota are alphabetical):

1. **The Search Editor.** (Choose "By Using Search Editor" from the Find menu. See pp. 192-200.)

2. **The Sort Editor.** (Click the Sort button on output screens. See pp. 131-139.)

3. **The Quick Report Editor** (pp. 269–278). (Choose "Design and Print a Special Report" after clicking the Print button in output screens, pp. 266-268, or select "Export Custom Flatfile" from the Im/Export menu, pp. 457-462.)

4. **The Custom Label Editor.** (Choose "Design & Print Custom Labels" from the Labels menu. See pp. 261-264.)

If you have defined any Field Name Aliases, a list of current aliases will automatically appear to help you use these editors, since they list all fields by their internal (default) names. All other Biota screens, tools, and reports use the Field Name Aliases you define. See pp. 323-326 for details.

Chapter 5 Overview: Biota Tools and Features

This chapter presents an overview of the tools and features that Biota offers. Often, there are several different ways to accomplish the same task with Biota. General-purpose tools (like the Search Editor, pp. 192–200) are included to cover special cases, but for many purposes Biota offers special tools that have been optimized to do a specific task.

Reading through this section now (or at least scanning the titles) will save you time later when you need to accomplish tasks for which special tools exist. In most cases, page references lead you to a more detailed presentation elsewhere in the Manual, if you need help.

Operational Features
Multitasking

Each Biota tool or screen that you use runs in an independent *process* that automatically shares microprocessor time with all concurrent Biota processes, independent of the operating system. This means you can open as many windows and carry out as many tasks simultaneously as your hardware will allow. For example, in a large database you could launch a time-consuming search (e.g., for all records for specimens of Verbenaceae collected between 1989 and 1991 in Heredia Province, Costa Rica, by a particular collector) and, while it is running, print herbarium labels for a different set of specimens, while entering data for a third set.

Macintosh Biota Flavors

BiotaApp, the stand-alone, single-user application, is available for either PowerMacintosh or Macintosh computers. The PowerMacintosh flavor is true native PPC code. The Macintosh flavor will run on any Macintosh with sufficient RAM, but a 68030 or 68040 processor is recommended. The third flavor for Macintosh, *Biota4D*, is intended primarily for client/server use, running under 4D Server (see below), but can also run in single-user mode under 4th Dimension, on either PowerMacintosh or Macintosh computers. (Biota4D is compiled in "fat code" for PowerMacintosh and Macintosh.) All Macintosh flavors of Biota are true 32-bit native compiled machine code.

Windows Biota Flavors

Although no Windows versions of BiotaApp or Biota4D are offered in the initial Biota release, watch the Biota Web site for the announcement of BiotaApp for Windows and Biota4D for Windows servers. This manual is written for both Macintosh and Windows platforms, although the illustrations are of Macintosh window styles. Keyboard shortcuts and other platform-dependent information are included where necessary.

Server Option with Mixed-Platform Clients (Macintosh and Windows)

Once Biota for Windows is released, you can give multiple users simultaneous access to any Biota Data File in true client-server mode, with a mixture of Macintosh and Windows clients. The server can be either a Macintosh (running an upgraded version of Biota4D for mixed platform clients) or a Windows machine, running Biota4D for Windows (which will be released in a mixed-platform version). The number of simultaneous users depends on the license you purchase for 4D Server from its maker, ACI—and, as a practical matter, on the capability of the server machine. Because 4D Server can use TCP/IP as the communication protocol between server and client machines, clients on either platform can access the server over the Internet, using 4D Client software, which is part of the 4D Server package (see Appendix E).

Biodiversity Information Types
Specimen-Based Data

Because Biota was originally designed for a quantitative, specimen-based, geographically referenced biotic inventory (see the Preface), its data-input and analysis tools have been crafted for maximum efficiency in dealing with individual specimens from new collections. But many of the same tools have proven themselves worthy in facilitating specimen-based work in systematics and collections management for historical collections.

Living Organisms

Although Biota was designed to handle specimen-based data from inventories or collections, it is easy to use for surveys of living individual organisms (for example, forest stands or bird sightings).

"Lot"-Based Data

In "lot"-based collections, groups of specimens share site (collection) data, but specimens are simply enumerated by species within each lot, without giving each specimen a separate identifier. (In some disciplines, a lot is a set of specimens from a single species from a single collecting event.) Lot systems are easily handled by Biota's "voucher" system, by creating just one voucher Specimen record for each species in each lot

(p. 152, step 6). If you later need to create individual specimens records for certain specimens (e.g., a mounted series, or types), Biota lets you use the original voucher Specimen record as a template for the new records, using the Carry button (pp. 127–128).

Site-Based Species Lists with No Specimens

By creating one "pseudospecimen" record for each species/site combination, Biota can accommodate species-based data (faunal or floral lists) for which no specific specimens are entered in the database. In this case, the pseudospecimens link taxonomic information with site information. (In technical terms, the pseudospecimens serve as a "relating record" for the many-to-many relation between Species and Collections or Localities; see Chapter 2). This technique also allows records for actual specimens of parasites, commensals, herbivores, etc. to be linked with host species records for which no specific specimen exists (pp. 393-396).

Tables and Fields
Core Tables and Core Fields

Biota has 12 *Core tables*. (Appendix A). The Specimen table lies at the bottom of two hierarchies, the place (geographical) hierarchy and the taxonomic hierarchy. In the place hierarchy, the *Specimen* table has a many-to-one (child-parent) relation with *Collection*, which has a many-to-one relation with *Locality*. The Specimen table also has a many-to-one relation with *Species*, Species is linked in a many-to-one relation with *Genus*, and so on for *Family, Order, Class, Phylum,* and *Kingdom*. Finally, the Specimen table also has a many-to-one relation with *Loans*. Several tables have many-to-one links with *Personnel*. The fields of the Core tables are called *Core fields* (Chapter 2).

Peripheral Tables

Biota also has more than a dozen *Peripheral tables* (Chapter 2). These tables keep track of Notes (linked many-to-one with the corresponding Core table) (pp. 179–183), Specimen Determination Histories (linked many-to-one with the Specimen table) (Chapter 19), images (linked many-to-one with the Species table) (Chapter 18), personnel Groups (pp. 174-177), Choice Lists for fields (pp. 315-322), and Auxiliary Field data (Chapter 15).

Field Aliases

For 36 Core fields distributed among the 12 Core tables, you can assign a *Field Alias* (using the Core Field Alias tool from the Special menu; pp. 323-326) to change the name that is displayed for the entry area for the field on input screens, for column headers on output (record listing) screens, and in exported files (Chapter 24) and printed reports (Chapter 14). For example, a marine copepod systematist might rename the Elevation field in the Collection table with the Field Alias *Depth*. As another example, in some groups of arthropods, the rank of Cohort is interposed between Suborder and Superfamily. The Biota field Family Custom 1 could be given the Field Alias *Cohort* to accommodate this requirement (Appendix A).

Alias Help   For some purposes (building an ad hoc search, sorting records, creating custom reports or labels) you must use the use the default, or internal *field names* instead of the Field Aliases you have defined (p. 323). To help overcome this limitation of the 4D engine, Biota automatically displays all the Field Aliases you have defined and their internal field-name equivalents in a floating window whenever you use one of these tools (e.g., p. 194). For the Export Editor (pp. 439–440), Import Editor (pp. 491–492), and the Find and Replace tool (pp. 222–223), you can display either the Aliases or the internal names, as you wish.

Auxiliary Fields   For the Species, Specimen, Collection, and Locality tables, you can define any number of *Auxiliary Fields*, named as you wish, and assign values to them for each record in the related Core Table (Chapter 15). Auxiliary Fields and their values are part of your Data File, not within Biota itself, so you must redefine them for each new database you create. The number of Auxiliary Fields you define and use is limited only by amount of memory you assign to Biota. (If you intend to use Auxiliary Fields, be sure to read about how they are created and be aware of their limitations, pp. 281–282. *Using Core fields is always preferable*, if feasible for your needs.)

Creating and Displaying Auxiliary Fields   You define new Auxiliary Fields, edit or delete existing ones, or re-order Auxiliary Fields using the Field Editor, accessed through the Aux Fields button on input screens for the Species, Specimen, Collection, and Locality tables (pp. 283–295). You can display Auxiliary Fields for the Species, Specimen, Collection, and Locality tables by clicking the Aux Fields button at the bottom of the output screen (pp. 298–304).

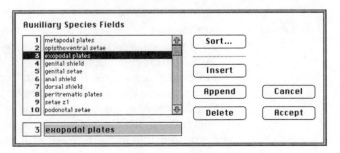

Recursive Specimen Relations (Host Records)

Biota handles recursive relations between specimens—"host-guest" relationships—by including the "host" specimen as one of the attributes of the Collection record of the "guest" specimen (Chapter 21). For example, a DNA sample from the tissue of a wasp that was reared from a caterpillar that was found feeding on the leaves of an epiphyte growing on a tree would produce a five-level host-guest recursive chain. Host data are included as an option on locality labels (p. 245) and Web pages (pp. 479–480).

Information Input

Record Sets

A *Record Set* is a temporary selection of records from among those permanently stored in the database (pp. 12-16). For each Biota Core table (Specimen, Collection, Species, etc.), you can define any number of Record Sets and save them to disk as Record Set Pointer Files for later rapid recall (pp. 16–19). Only one Record Set per table can be active (current) at a given time, however. During data entry, you can elect to add new records to the current Record Set or start a new Record Set (pp. 13–15). Record Sets are also used in printing reports, labels, making or returning loans, and for export options (p. 15). You can display the current Record Set for a table at any time using commands from the Display menu (pp. 15–16).

Images

For each Species record, any number of digitized images can be stored and retrieved simply by clicking the image title in a scrollable list (Chapter 18). Anything that can be digitized can be stored as an image associated with a species—digitized video or microvideo images, scanned photographs or drawings, directly scanned herbarium sheets, original drawings or diagrams produced on the computer. Biota image records can be imported individually from disk files and exported individually or in groups directly to disk files (p. 365) or Web pages (pp. 480–481).

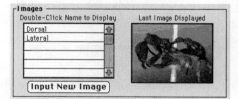

On Macintosh computers, all QuickTime compression options are available automatically during input by depressing the OPTION key while selecting Paste from the Edit menu (p. 349).

Automatic Prompting for Parent Record Entry

Although you can enter data table by table (enter all Genera, then all Species, then all Specimens, etc.), you may also work directly from records lower in a hierarchy—for example, directly from Species entry. With the latter approach, Biota queries you automatically when new entries are required in a related (*parent*) table (e.g., the Genus table, if you enter a new Genus name in a Species record), then returns you automatically to the lower-level (e.g., Species) entry screen (pp. 92–95).

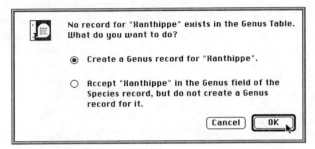

Wildcard Entry for Linking Fields

During data entry, when the appropriate record already exists in a related table (e.g., the Genus record already exists for a Species record currently being entered), the linking field is automatically looked up and entered, based on the first letter or first few letters you enter. (Press TAB to launch the search.) If more than one related record matches the entry, Biota presents you with an alphabetic list to choose from, including an additional, "helper" field from the related table. (For example, if you are entering Genus as a relating field in a new Species record, and you enter the letter P, every Genus that starts with P will appear in a scrollable list along with its Family. You can sort the list by either Family or Genus. You just click the correct record to enter it in the new Species record.) This *wildcard entry* feature minimizes not only input effort, but spelling errors (p. 96).

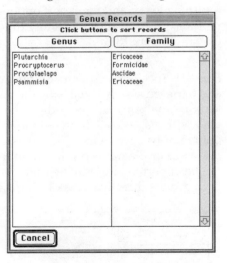

Lookup Tool for Entering Record Codes in Linking Fields

Each record in four of Biota's tables (Specimen, Species, Collection, and Locality) must have a unique Record Code (Chapter 7). Although you can use any name or mnemonic you like for these codes, they can be hard to remember when you need to enter them in relating fields during data entry. Although wildcard entry is the fastest way to enter data for these relating fields, Biota also provides "Look up" buttons that open windows with scrollable lists of all existing records in the related table, sorted at your option by one of several criteria (pp. 147–149). (The Species list is narrowed by genus, pp. 145–146.)

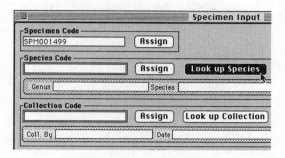

Entry Choice Lists

Some fields in your database will probably have a limited number of repetitive entries—a field for sex or life stage, for example. For any of 36 nonlinking fields in Biota, you can activate an *Entry Choice List* that presents entry options in a floating window whenever the entry area for the field is accessed during input (pp. 315–322). When the window appears, you simply click an item on the list to enter it (or press TAB to cancel the list window and make a manual entry). You construct each list by entering items with the List Editor, which is accessed from the List window itself. Biota saves the lists you have created as part of your Data File, whether activated or not, until you change or erase them. Since Choice lists eliminate misspellings, they are especially useful when someone else enters data for you.

Field Value Defaults

If you repeatedly enter a single value for a field, or one value is used in a large proportion of your records, you may want to assign a *Field Value Default* for that field (Special menu). The default will be automatically entered in the field as soon as a new record is initiated on an input screen. Biota supports Field Value Defaults for 16 Core Fields in the Specimen, Collection, and Locality tables (pp. 313–315).

Using an Existing Record as a Template for a New One

Every input screen includes a *Carry* button, which allows you to make a new record using the current record as a template (pp. 41–42). (If it is a new or edited record, the template record is saved first, automatically, when you click Carry.) Since, in many cases, only one or two fields needs changing between a template record and a new one, this capability saves a great deal of effort. A saved, existing record can also be used as a template (pp. 127–128). Display it in the input screen, click Carry, and a new record will appear, repeating all but the *Key* field (pp. 6–7) of the template record. (Auxiliary Fields, p. 298, and Notes, p. 183, can be carried or not, according to a setting in the Preferences screen, from the Special menu.)

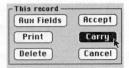

Automated Entry of Specimen Record Series

Extending the Carry concept, two powerful input options (from the Series menu) speed the entry of sets of Specimen records that share either the same Collection data (Input Specimen Series) or both Collection and taxonomic data (Input and Identify Specimen Series) (Chapter 12). These options require entry of only the first and last Specimen Code numbers (with or without prefixes) and one-time data entry for the other fields for each set of specimens. All records in the specified range are then generated automatically, with sequential Specimen Codes.

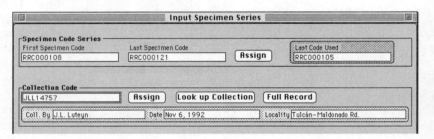

Automatic Record Code Assignment

Record Codes for Specimen, Species, Collection, and Locality records may be any arbitrary, unique (within tables) alphanumeric code you choose. Alternatively, sequential codes can be automatically assigned with either a default alphanumeric prefix (which you define with the Record Code Prefixes utility in the Special menu) for each table, an ad hoc prefix you assign during record input, or no prefix (Chapter 7).

Barcode Entry of Record Codes

All Record Code entry areas (including Specimen Code fields for specimen loans and returns) automatically accommodate barcode entry (pp. 106–108). All record entry and record retrieval areas for Specimen and Species Codes also provide for barcode prefix recognition and optional substitution of an abbreviated or different code prefix in the database (pp. 107–108). Barcode entry can be freely mixed with manual entry in the same Data File.

Record Retrieval and Manipulation: General Tools

Multifunction Output Screens

A uniform interface for displaying records allows you to carry out ad hoc sorts (pp. 131–139), select consecutive or nonconsecutive records to redefine a current selection of records (using the Subselection button, pp. 128–129), delete records (pp. 129–131), modify existing records (double-click and modify the record in the input screen, pp. 121–122), add new records (p. 127), or use an existing record as a template for a new one that shares information with the template (pp. 127–128). In each case, Biota immediately displays the result using the same uniform interface, ready for the next action.

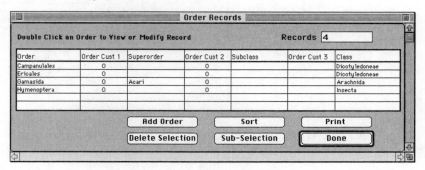

Simultaneous Display of Records from Any Number of Tables

The number of open windows Biota can display simultaneously is limited only by the memory you allocate to Biota (see "Multitasking," p. 63). In fact, you can display different groups of records (or saved Record Sets) in different windows for the same Core table, simultaneously. Displaying records already entered in a related table is sometimes useful during data input.

Displaying Record Sets

The active Record Set (if any) for each Core table can be displayed and manipulated using commands from the Display menu (pp. 15–16).

Finding Records Using the Taxonomic Hierarchy

A Linnean lookup system (in the Tree menu) allows you to navigate rapidly up and down the taxonomic hierarchy by presenting a scrollable list of subtaxa for each taxon. Double-clicking a record brings up a list of the subtaxa of the taxon selected (pp. 186–192; pp. 48–51).

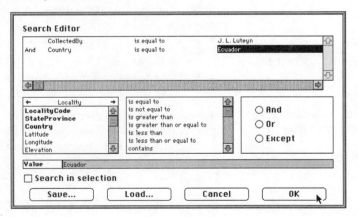

Ad Hoc Queries for All Tables

A general-purpose query construction tool (the Search Editor) allows you to use the content of records, including values in related fields of other tables, to find records. You can choose to search among all records in a table or only among those in a Record Set (pp. 192–200). If your query can be accommodated by using one or more of the special tools described in "Special Tools for Finding Records..." (pp. 73–75, however, the search will invariably be much faster than building the query in the Search Editor.

Comparing and Retrieving Images

The Display Views button on the Species output screen presents thumbnail versions of the first four images for each species in the current selection of Species records, in a scrollable list. Individual images (and those beyond the fourth for a given species) can be displayed in larger format by double-clicking a row of thumbnails, then double-clicking the image title in the species input screen that will appear (pp. 363–365).

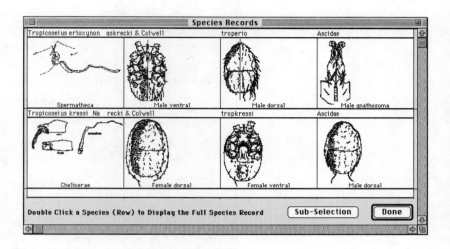

Special Tools for Finding Records in One Table Based on a Set of Records in Another Table

Biota offers four special tools, with a common interface, for the frequent task of finding related records up, down, and across the taxonomic and place hierarchies. As on option, Biota will keep the linking records for all intermediate tables as Record Sets for those tables. These tools support rapid and intuitive query construction and optimized, extremely fast search and display. They can be used sequentially for complex searches. For many users, they are among the most frequently used features of Biota.

Finding Records for Lower Taxa Based on Higher Taxa

The Lower Taxa for Higher Taxa ("top-down") search tool finds all records for a lower taxonomic level that belong to a higher taxon or a selection of higher taxa (pp. 201–205). It does not matter whether the two levels are adjacent in the taxonomic hierarchy or not—for example, you could find all Specimens of a Family or all Genera of three Orders (nonadjacent levels), or all Specimens for a selection of Species or all Classes for a Phylum (adjacent levels). For nonadjacent levels, you can tell Biota to keep all related taxa at intermediate levels as Records Sets.

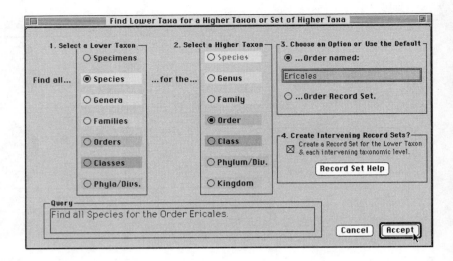

Finding Records for Higher Taxa Based on Lower Taxa

The Higher Taxa for Lower Taxa ("bottom-up") search tool finds all records for a higher taxonomic level that belong to the active Record Set for a lower taxonomic level (pp. 205–208). For example, you could find all Orders for the Specimen Record Set (nonadjacent levels) or all Families for the Genus Record Set (adjacent levels). For nonadjacent levels, you can tell Biota to keep all related taxa at intermediate levels as Records Sets.

Finding Place Records (Collection or Locality) for Specimens or Species

The Places for Specimens or Species search tool finds all Locality or Collection records for the active Specimen or Species Record Set, according to your query (pp. 208–210). For example, your could find all Locality records for a set of Species (a cross-hierarchy query). You can tell Biota to keep the related Specimen and Collection records (intermediate tables in the search) as Record Sets.

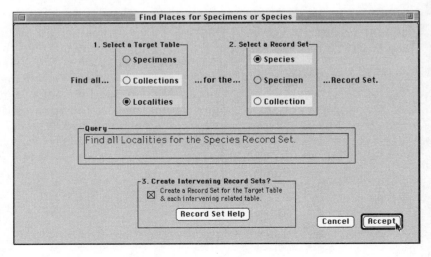

Finding Specimen or Species Records for Places (Collections or Localities)

The Spcms or Species for Places search tool finds all Specimen or Species records for the active Collection or Locality Record Set, as you request (pp. 210–213). For example, your could find all Species records for a Locality (a cross-hierarchy query). You can tell Biota to keep the related Collection and Specimen records (intermediate tables in the search) as Record Sets.

Sequential, Cross-Hierarchy Searches

If you want to find Collections or Localities for taxa above the level of Species, or higher taxa represented in Collections or Localities, you can use the appropriate pair of the above tools sequentially (p. 213). For example, to find all the Localities for an Order, first find all Species for the Order (using the first tool), declare the result the Species Record Set, then find all Localities for the Species Record Set (using the third tool).

Finding Records by Record Code
Single Specimen Records

With the simple Find by Specimen Code tool (Find menu), a single Specimen record can be looked up individually (and instantaneously) by entering the Record Code manually or with a barcode reader (pp. 213–215).

Unordered Specimen Series

The Find Specimen Series tool (Series menu) accepts two kinds of entries (pp. 236–238). In Unordered Series mode (click "In any order"), you simply enter any Specimen Code, press TAB, then toggle on the Auto Accept button. For all subsequent records, just enter the next code, press TAB, and repeat, as Biota retrieves and collects the records. With a barcode reader set to enter carriage returns automatically, this is a completely hands-off operation (neither keyboard nor mouse is needed). When you are done, the records are displayed for examination, modification, or saving as a Specimen Record Set for other operations.

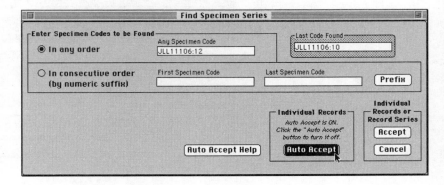

Ordered Specimen Series

In Ordered Series mode (click "In consecutive order"), the Find Specimen Series tool (Series menu; shown above) lets you find, in a single action, any series of records that have numerically sequential Specimen Codes (pp. 238–242). If the codes have an alphabetic prefix, you first enter the prefix using the Prefix button (or assume the default Specimen

Code prefix established in the Set Prefixes screen, Special menu, pp. 102–103). Then enter the first and last integer counters for a consecutive series of Specimen Codes. When the search is complete, the records are displayed for examination, modification, or saving as a Specimen Record Set for other operations.

Finding Individual Records by Species Code, Collection Code, or Locality Code

Species, Collection, and Locality records can be looked up individually (and instantaneously) by entering the Record Code manually or with a barcode reader (pp. 213–215).

Special Tools for Updating Records

Updating Specimen Determinations and Other Specimen Information: Unordered Specimen Series

Two tools for rapid updating of data in existing Specimen records share the Identify or Store Specimen Series screen (Series menu) (pp. 234–242). In Unordered Series (Auto Carry) mode (click "In any order"), you enter a Specimen Code (manually or with a barcode reader), then enter the Species Code for a new determination or other new information (Determined By, Date Determined, Stage/Sex, Storage, Type Status). Then click the Auto Carry button. Press TAB, enter the next code, press TAB, and so on, as Biota retrieves and updates the records with the same new information. With a barcode reader set to enter carriage returns automatically, updating is a completely hands-off operation (neither keyboard nor mouse is needed). When updating is complete, the records are displayed for examination, further modification, or designation as a Specimen Record Set for other operations.

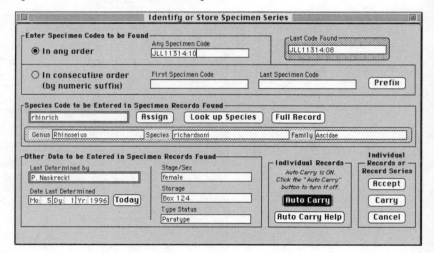

Updating Specimen Determinations and Other Specimen Information: Ordered Specimen Series

In Ordered Series mode (click "In consecutive order"), the Identify or Store Specimen Series screen (Series menu) (pp. 238–242) lets you update, in a single action, any series of records that have numerically sequential Specimen Codes (p. 99). If the codes have an alphabetic prefix, you first enter the prefix using the Prefix button (or assume the default Specimen Code prefix established in the Set Prefixes screen, Special menu, pp. 105–108). Then you enter the first and last integer counters for a consecutive series of Specimen Codes) and the Species Code for a new determination or other new information (Determined By, Date Determined, Stage/Sex, Storage, Type Status). When the updating is complete, the records are displayed for examination, modification, or designation as a Specimen Record Set for other operations.

Updating Records (Any Table) by Importing Information from Text Files

The Import by Tables and Fields tool (Im/Export menu) includes an option for updating existing records in any table (pp. 490–495). Biota matches up the imported data with the existing record based on the Key Field (see pp. 498–499) for each record, which must therefore be one of the fields included in the text file.

The Find and Replace Tool

The Find and Replace tool (Special menu) is a powerful (and therefore dangerous!) utility for correcting errors or updating records in any table in the Biota structure (pp. 220–225). Not only Core tables but Peripheral tables can be accessed. Changes can be made either in records of a current Record Set (for tables that have them) or for all records in a table. Be sure you know what you are doing before using this tool, and *always* be sure you have a backup copy of your Data File first! *Changes made cannot be undone.*). If the user password system has been enabled, only a user with Administration access privileges can use this tool (pp. 409–415).

Find and Replace

1. Choose a Table From the Popup List
   Table Name: Genus

2. Select a Search Option
   ● Search the Genus Record Set
   ○ Search all Genus records
   ○ Select Genus Records using the Search Editor
   [Launch Search Editor]

3. Choose a Field From the Popup List
   Field Name or Alias: Subfamily
   [Show Field Aliases]   *Internal Field Names appear in the lists above. No Field Aliases have been defined.*

4. Enter the Current and New Values for the Field
   Current Value: Myrmicini   New Value: Mymicinae   [Help]

   [Find & Replace]   ☒ Save Find & Replace Setup   [Cancel]

Keeping Track of Synonymies

A species Synonymy system (based on the Valid Species Code field of the Species table) allows you to keep track of historical or new synonymies among Species records (Chapter 20). (In inventory work, this system can be used to document the "synonymies" that often arise in pooling several temporarily determined sets of Specimens.) The Synonymy button in the Species input screen finds and displays either the senior synonym for a Species, or a scrollable list of its junior synonyms, if any. A species "Look up" button speeds the recording of new synonymies.

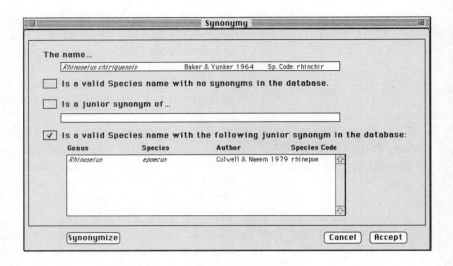

Tools for Maintaining Database Security and Integrity
User Password System

Biota provides an optional user password system that you can configure to your needs (Chapter 23). To activate the system, you give the Administrator a non-blank password. (To deactivate it, assign the Administrator a blank password.)

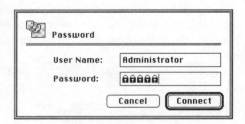

As Administrator, you can use the Password Editor to set up a system of passwords for other users (select Edit Password System from the Special menu). A selection of generic user names (e.g., Collaborator, Browser, etc.) with different access levels (see next paragraph) have been preset, but may be changed as you wish. Any user can change his or her password using the Change Password item in the Special menu.

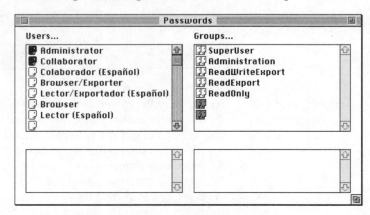

| User Access (Privilege) Levels | The Password Editor allows the Administrator to assign each user to one of four *access levels* (pp. 415–417). (1) Super User (set user passwords, assign access levels, set the Data File password link (see below), plus all privileges of the Administration access level; (2) Administration (create, delete, display, modify, print, export, or import records; change master settings); (3) ReadWriteExport (create, delete, display, modify, print, or export records); (4) ReadExport (display, print, or export records); or (5) ReadOnly (display records only). The Password Editor (a built-in 4D tool, shown above) is a bit tricky; you may need to consult pp. 415–424. |

| Data File Password Link | The user password system protects your copy of the Biota application from unauthorized use but does not in itself protect Biota *Data Files* from unintended or unauthorized access. You can create a secure link between a particular copy of Biota and a particular Data File (or Files) using a *Data File password link*. You need not activate the user password system in order to activate the Data File password link, although it usually makes sense to activate the user password system if the Data File link is used. (The two systems are technically independent.) See pp. 425–431. |

| Spanish Dialog Screens | Although Biota is by no means fully bilingual (menus and input and display screens are strictly English), more than 100 of the most important dialog screens (options, instructions, warnings, etc.) are available in both Spanish and English. English is the default language. Any user can enable Spanish (or switch back to English) by setting an option in the Preferences screen (Special menu). The Administrator can change the default in the Preferences screen or assign individual users to either language by entering "StartupSpanish" or "StartupEnglish" (no spaces, no quotes) as the Startup Procedure in the Password Editor (pp. 418–420). |

| Automatic Updating of Linking Fields in Child Records | If you need to change the name of a higher taxon (Genus, Family, Order, etc.), or the Record Code for a Species, Specimen, Collection, or Locality in the parent (one-table) record, Biota will ask if you want to update the linking field in all linked child (many-table) records automatically (pp. 218–220). For example, suppose you misspelled the genus *Excelsotarsonemus* in the parent (Genus table) record and used the same misspelling in the Genus linking field of the Species records for all species |

of the genus. You can correct the name in the Genus record and in all linked Species records with one entry (or change only the Genus record, if you choose).

Finding Orphan and Childless Records

An *orphan* record is one that is not linked to any record in the next higher table in a hierarchy—for example, a Specimen record that either has no Species Code recorded, or has a Species Code that does not correspond to any Species record. A *childless* record is one that has linked records in the next lower table in a hierarchy—for example, a Genus record that has no linked Species records. Using commands from the Find menu (Find Orphan Records and Find Childless Records) you can find, display, and, if you wish, modify such records easily. See pp. 215–218.

Record Deletion Control

Biota will not allow you to orphan records inadvertently. In other words, if you attempt to delete a record that is linked to records at the next lower table in a hierarchy, Biota warns you that you are about to orphan certain records. You then have the option of doing so in spite of the warning, displaying the child records without making any changes, or simply canceling the action (pp. 122–123, pp. 129–131).

Record Creation Control

Whenever you enter a value in the parent linking field of a child record (e.g., when you enter a family name in the [Genus] Family field while creating a new Genus record), Biota checks automatically to see if the record already exists in the parent table (the Family table, in this example). (See "Automatic Prompting for Parent Record Entry," pp. 68–69.) If no parent record exists, Biota prompts you to create one "on the fly," then returns you to the child record in progress (pp. 93–95). (If you wish, you can go ahead and create an orphan record, as an option. This option is especially useful for the [Collection] CollectedBy field, linked to the Personnel table, for historical specimens when the collector's personal data are either unknown or of no interest.)

Automatic Recording of Specimen Determination Histories

Whenever the identification (determination) of a Specimen is changed, all pertinent information for the old determination (Genus, Species, Author, Determined By, Date Determined) is recorded in a new record in the Determination History table, along with a record of where the determination was changed (in the Specimen record, the Specimen Series tool, the linked Species record, the Synonymy system, or the linked Genus record), when it was changed (current date), and by whom (based on password signon) (Chapter 19). To view the determination history for a Specimen record, click the History button on the Specimen input screen. The History button is enabled only if one or more history records (child records) exist for that Specimen (parent) record. If the user password system has been enabled, only a user with Administration access privileges (pp. 415–417) can enable or disable the Determination History system (Preferences, Special menu) or delete or modify Determination History records (commands in the Special menu; pp. 375–377).

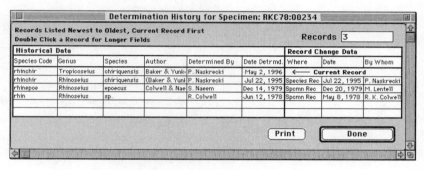

Duplicate Species Checking

Although the Key field (see pp. 6–7 on Key fields) of the Species table is Species Code, not Species Name (specific epithet), Biota checks for uniqueness each time you enter a new Species Name value. (You cannot enter two records with the same Species Code). Of course, more than one record for *sativum* may legitimately exist, in different genera, but you would not want to create two species records for *Pisum sativum* L. Thus, Biota warns you if one or more Species records exist with the same specific epithet (Species Name), and offers to display any such records (pp. 166–167). You must then decide how to proceed. (Exceptions include "sp.," "sp. nov.," and related abbreviations.)

Reports and Labels
Printed Reports

Biota offers a variety of printed reports designed for biodiversity information, as well as user-designed, ad hoc reports (using the on-screen Report Editor). For each level of the taxonomic hierarchy, taxonomically organized, indented reports can be generated for the current selection of records by clicking the Print button on the corresponding display screen, or by selecting Print from the File menu. Collection and Locality reports are also available. Each input screen for Core tables has a Print button for printing individual records. Biota offers several options for sorting records before printing. See Chapter 14.

Specimen Label Printing for Fluid-Preserved, Pinned, Slide-Mounted, or Herbarium Specimens

Based on the current Specimen Record Set, Biota prints locality and determination specimen labels for fluid-preserved (pp. 248–251), pinned entomological (pp. 248–251), or slide-mounted (pp. 251–254) specimens. Herbarium labels are printed in standard format, including (optionally) both locality and determination data (pp. 255–258), as well as an option to include a descriptive field note. Numerous other options allow control over the information included on labels. (All commands are in the Labels menu.)

```
COSTA RICA: Heredia              COSTA RICA: Heredia
La Selva                         La Selva
Elev 150 m 10°26'0"N84°1'0"W     Elev 150 m 10°26'0"N84°1'0"W
coll. J. T. Longino 22 Jan 1991  coll. J. T. Longino 9 Oct 1989
INBIOCRI001B459857               INBIOCRI001459864
```

```
COSTA RICA: Heredia              COSTA RICA: Heredia
La Selva                         La Selva
Elev 150 m 10°26'0"N84°1'0"W     Elev 150 m 10°26'0"N84°1'0"W
coll. J. T. Longino 22 Jan 1991  coll. J. T. Longino 9 Oct 1989
INBIOCRI001B459857               INBIOCRI001459864
```

```
                        COSTA RICA
Ericaceae                                           RKC0061:03
Cavendishia crassifolia (Bentham) Hemsley
Cartago: Cordillera de Talamanca, Cerro de la Muerte, La Georgina, 60 W of
divide
9°41'0"N  83°52'0"W                                      3100 m
Epiphytic shrub on trunk of Quercus costaricensis. Floral bracts red; corolla
bottle-shaped, red with white at base; anthers yellow.


R. Colwell, J. H. Hunt & M. L. Mackey                 22 Jun 1970
                                                         RKC0061
                   The Torrey Herbarium
                of the University of Connecticut
```

```
JLL10953:05   JLL11314:05
Tropicoseius  Tropicoseius
steini         steini
   sp. n.         sp. n.
det. 1993     det. 1993
P. Naskrecki  P. Naskrecki
Protonymph    Adult female
ASCIDAE       ASCIDAE
```

Exporting Text Data for Creating Custom Locality/Collection Labels

Preformatted collection/locality labels do not suit everyone. If you need more control over formatting and content, an option in the Collection Labels command (Labels menu) exports data for the current Collection Record Set (including all standard fields from parent Locality records) to a text file on disk. You can then use a word processor to modify, reorganize, or duplicate label data before printing (pp. 246–248).

Custom Label Printing

Using the Label Editor (a built-in 4D tool), you can design your own labels in any size or format for the current Record Set of any Biota Core table and save the template to disk, if you wish (pp. 261–264). This otherwise clever tool has one unfortunate limitation, however. You can include fields only from parent tables no more than one link away from the focal table.

Species Labels

Species labels (for entomological unit trays or herbarium folders, for example) can be printed from the Species Record Set or from any arbitrary set of Species Codes entered in a special input screen (Labels menu; pp. 259–261).

Specimen Loan System
Lending Specimens

Biota provides a comprehensive bookkeeping system for specimen loans (Chapter 17). Groups of specimens to be loaned can be selected using any of Biota's usual selection tools, or one-by-one by entering each Specimen Code (often the fastest way when barcodes are used) (New Loan, Loan menu; pp. 328–333). The Deposited At field of the Specimen table is used for recording the Specimen Loan Code in each Specimen record.

```
Loans
  New Loan
  Record Returns

  Display All Loans
  Display Loan Record Set

  Print List of All Loans
```

Recording Returns

Specimen returns can be recorded by selecting records on-screen from the list of those in the original loan, by recording the current Specimen Record Set as returned, or by entering Specimen Codes one-by-one (Loan Returns, Loans menu, pp. 339–345). The Unordered option of the Find Specimen Series tool (Series menu) is particularly useful for gathering records from several loans that have been returned together (pp. 339–341).

Loan Forms

For each new loan, Biota generates a standard loan agreement form, with the name of your project, museum, or herbarium if you wish, with complete lender and borrower information (pp. 333–337).

Printed Report of Specimens Loaned

As an option, the loan system produces a full, taxonomically organized, printed listing of all specimens loaned (pp. 333–337).

Text Flatfile Export of Specimens Loaned

As an additional option, Biota will export a text file listing the specimens in any loan, including, for each specimen, all fields from the database that you designate, with blank or partially completed species identification fields (pp. 333–337). The text file can be opened and modified using any spreadsheet program. Identification fields can then be filled in by the borrower as the specimens are determined and the determinations returned to the lender in the updated flatfile, which can be used to update records in the Biota Data File (pp. 490–495).

Importing Data
The Import Editor

New records can be directly imported into Biota tables, one table at a time, from plain-text flatfiles (including plain-text flatfiles exported by other database applications) using Biota's Import Editor (Import by Tables and Fields, Im/Export menu; Chapter 26). You can specify the order of fields to match your flatfile, skip fields in the flatfile, specify field and record delimiters (including either PC or Macintosh standard line terminators), and account for a column header in the flatfile. Biota does extensive error checking (validation for correct field types, field lengths, duplicated Key fields, nonexistent dates, etc.) with informative error dialogs. If an error is found, the guilty record is displayed, and you are offered the option of keeping any records already imported successfully, or deleting them and starting "clean" after fixing the problem in the text file.

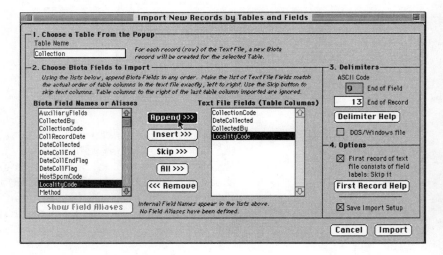

Exporting Data
The Export Editor

Any field of any Biota record or set of records can be exported, one table at a time, to a plain-text flatfile (readable by spreadsheet, word processing, or other database applications) using Biota's Export Editor (Export by Tables and Fields, Im/Export menu; pp. 434–443). For a given table, you can specify any order for the exported fields, specify field and record delimiters (including either PC or Macintosh standard line terminators), and export a column header in the flatfile. Plain-text flatfiles are the lingua franca of data exchange among databases, spreadsheets, and statistical applications. When something you like better than Biota comes along, or you need to send data somewhere in electronic form, you can always get your data out of Biota Data Files.

Exporting Specimen Flatfiles

For any Specimen Record Set, you can export a flatfile with specimens as rows, that includes columns for any set of fields from Biota Core tables (Appendix A). You choose the export fields simply by clicking checkboxes on a selection screen (Im/Export menu; pp. 454–457).

Exporting Taxonomic Flatfiles

A taxonomic flatfile is a list of taxa at a particular level (the *base level*, displaying, for each taxon, its membership in one or more higher taxa (for example, a list of species that includes the genus and family of each, or a list of genera that includes the family, suborder, and order for each genus). Biota can export such files based on any taxonomic base level, with additional higher levels (columns) that you specify by clicking checkboxes (Im/Export menu; pp. 451–453).

Exporting Notes

You can export plain-text files of Notes for Record Sets from the Species, Specimen, Collection, Locality, and Loans tables (pp. 443–448). (You can create any number of Note records per parent record.) This is a good way to launch a manuscript based on details you have recorded in the database as Notes.

Exporting Auxiliary Field Values

If you have defined Auxiliary Fields and entered values for them, you can export matrix with records as rows and Auxiliary Fields as columns directly to a plain-text file, for a Record Set in the Specimen, Species, Collection, or Locality tables (Im/Export menu; pp. 308–310). Exported Auxiliary Field data can then be printed, or joined with Core Field data for the same set of records using a spreadsheet application.

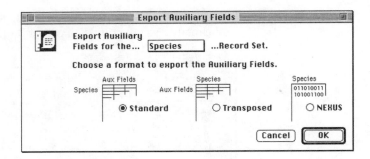

Exporting Character Matrices in the NEXUS Format

If you have defined Auxiliary Species Fields and have entered single-digit or single-letter values in them, you can export a complete, PAUP- or MacClade-ready NEXUS file, using an option in the Export Auxiliary Fields dialog (Im/Export menu; pp. 310–311). Biota also supports this option for Auxiliary Fields of the Specimen, Collection, and Locality tables.

Exporting Collections × Species Matrices

For any Collection, Specimen, or Species Record Set, Biota can export a Collections × Species incidence (presence/absence) or abundance table (Im/Export menu; pp. 469–472). These data are the raw material for graphical and statistical estimation of species richness from samples, and for ordination, sample classification, and other statistical techniques.

Exporting Specimens Examined Lists

For any Species Record Set, Biota can export a virtually journal-ready text file of Specimens Examined (Exsiccatae), organized within species by Locality and Collection, then by Stage/Sex and host (if any) (Im/Export menu; pp. 463–469).

Exporting Images

Biota can export either the first image (alphabetically by Image Name) or all images for a Species Record Set. A separate disc file is automatically created and named for each image (Im/Export menu; pp. 449–451).

Custom Export

The Quick Report Editor (a 4D utility) is a powerful and flexible tool for exporting text files to disk, as well as for printing reports. Quick Report formats are easily created and can, themselves, be stored on disk for repeated use (Im/Export menu; pp. 457–462).

Exporting Web Pages

Biota will guide you through the setup for completely automated export of hyperlinked Web pages, ready to post on a Web server (Chapter 25). (These are "static" pages, not "dynamic," on-line access to your Data File from the Internet.) You can export hierarchically linked pages for any range of taxonomic levels (including intermediate levels such as Subfamily), for Record Sets that you define with either a top-down or bottom-up search (pp. 474–475).

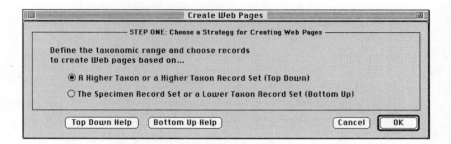

If you include Specimen records, you can select any set of fields in the Specimen, Collection, or Locality tables for inclusion as Specimen data. You can include images for Species records. Other options include custom page footnotes, hyperlinking of host and "guest" Specimen records, and non-ASCII character translation to HTML character codes.

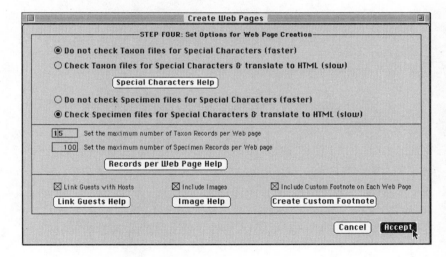

PART 3 Data Entry and Information Retrieval

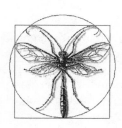

Chapter 6 Entering Data

This chapter outlines conventions and shortcuts common to data entry screens throughout Biota. Details of particular data entry screens appear in Chapter 10.

Screen Colors and Textures

Each of the 12 Core tables (Appendix A) has been assigned it own characteristic screen color, to help you navigate the database. All input and output screens associated with a particular table share the same background color. (On a gray-scale screen these colors appear as different gray levels.)

Information from the related Species record

Information from the related Collection record

On input screens, some fields are used only to display information, rather than for entering data. These fields are consistently displayed inside a stippled panel. For example, information from the Species and

Collection tables are shown in the Specimen input screen, above, in stippled areas. The Record Date field (which is automatically assigned to each Specimen record) in the upper-right corner of the Specimen Input screen is also displayed in a stippled panel. When these information fields display data from a table other than the table for which records are currently being entered, the stippled panel carries the color characteristic of that other table.

Entry Areas and Default Entry Order

In all Biota data input screens, you enter data in a series of empty boxes (entry areas). For each input screen, there is a default entry order. Once you have made an entry, press the TAB key to move on to the next entry area in the default order. Arrows show the default entry order for the Specimen input screen, below.

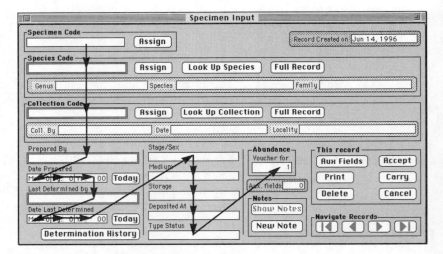

At any time, you can override the default entry order and move directly to any entry area on an input screen, using the mouse. When you click in another area, the entry in the field you just left is registered, whether or not you first pressed the TAB key.

All entries can be edited using standard screen-editing techniques. For example, an entry can be changed by selecting characters with the mouse and cursor or selecting an entire entry area with a double click, then typing in the new information. If you change an entry, but have not yet pressed the TAB key or clicked in another field, you may select Undo from the Edit menu to restore the previous entry.

Keyboard Equivalents for On-Screen Buttons

Many buttons on Biota input and output screens have keyboard equivalents that save time over clicking with the mouse, for many users (see box).

Button Name	Keyboard Equivalent	
	Macintosh	Windows
Accept	COMMAND–j (⌘–j)	CTRL–j
Carry	COMMAND–k (⌘–k)	CTRL–k
Print	COMMAND–p (⌘–p)	CTRL–p
Aux Fields	COMMAND–m (⌘–m)	CTRL–m
Next Record	COMMAND–RIGHT ARROW (⌘–RIGHT ARROW)	CTRL– RIGHT ARROW
Previous Record	COMMAND–LEFT ARROW (⌘–LEFT ARROW)	CTRL– LEFT ARROW
Sub- Selection	COMMAND–n (⌘–n)	CTRL–n

Option Window (Radio) Buttons

There are also keyboard equivalents for most message windows with *radio button* options, such as the Import option window shown here. The first button can be selected by pressing COMMAND–1 (⌘–1; Macintosh) or CTRL-1 (Windows), the second using COMMAND–2 (⌘–2; Macintosh) or CTRL-2 (Windows), and the third (if present) using COMMAND–3 (⌘–3; Macintosh) or CTRL-3 (Windows). Option screens with more than one set of radio buttons, such as the screens for setting label options, do not have keyboard equivalents.

Today Buttons

Several input screens have one or two buttons labeled Today, to input today's date in date fields. For each screen, the keyboard equivalent for the first date field in the default entry order is COMMAND–t (⌘t) (Macintosh) or CTRL–t (Windows).

If there is a second Today button for another field input screen (Specimen and Collection input screens) the keyboard equivalent for the second Today button is OPTION–t (both Macintosh and Windows).

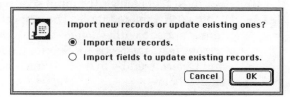

Double-Bordered (Default) Buttons Any button with a double border, such as those shown here, can be activated either by clicking it with the mouse or by pressing the RETURN key.

Using Entry Choice Lists

For fields that have a small number of recurring entries (such as Stage/Sex in Specimen records) Biota can display a list of alternatives in a window when the cursor enters the field during data entry or when you click on the field.

1. **To enter an item from the Choices window**, just click on the item.

2. **To enter an item manually that is not shown in the Choices window**, press the TAB key or click the Cancel button in the Choices window (*not* the Cancel button of the input screen!), then enter the special item manually in the field.

NOTE: For information on how to activate or deactivate the Choice List for a given field and how to set up the options in a list, see pp. 315–322 in Chapter 16.

Entering Data in Linking Fields

For most Biota tables, the input screen includes entry areas that link the table with a related table. These special *linking fields* (pp. 7–8) consistently appear with a *double-bordered data entry area* in all Biota input screens (e.g., the fields for Species Code, Collection Code, Prepared By, and Determined By in the Specimen input screen, a portion of which is shown below).

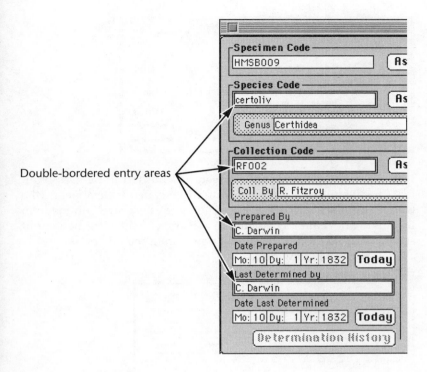

Double-bordered entry areas

"On-the-Fly" Creation of Linked Records

Double-bordered data entry areas for linking fields work in a special way. Once you enter a value in the field and press TAB, Biota checks to see whether a record already exists for that value in the related table. If the record exists, you can move on to the next field. If it does not exist, Biota offers you the option to create it "on the fly" and then return to the table where you started.

1. **Enter a value for a linking field.** Suppose, for example, you enter the genus *Drosophila* in the Genus entry area of the Species input screen.

2. **If your entry already exists in the related table,** your entry will be accepted and the cursor will move on to the next field. In the example, if *Drosophila* were already in the Genus table, Biota would accept the entry and you move on to the Specific Name field.

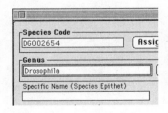

3. **If your entry does not match any entry in the related table,** Biota presents the screen below. (The screen illustrates the *Drosophila* example.)

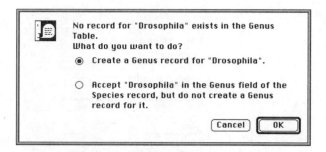

- **If you choose the first option,** "Create a Genus record for 'Drosophila'" (the default), the input screen for the related table appears (the Genus input screen, in this example), with the entry you made in the linking field already entered in the correct field of the new input screen. This method of creating a new parent record "on the fly" is available for all linking fields.

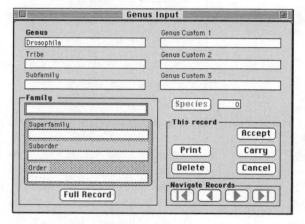

- **If you choose the second option,** "Accept 'Drosophila' in the Genus field of the Species record, but do not create a Genus record for it," no link to the parent table will exist for this record. In other words, it will be an *orphan record* (pp. 7–8). You can add a record in the related table later (a Genus record for *Drosophila*, in this example), but it is generally more efficient to add new, related records when this query appears—or use the table-by-table method outlined in the next section (p. 95).

NOTE: One common exception arises for the Collected By field of the Collection table. With historical collections of specimens, you may not want to create a Personnel record for every collector—some of whom may be long deceased and their particulars unknown.

4. **Repeat at additional levels, if necessary.** If you choose the first option (create a new record "on the fly" in the linked table), you may now need to make an entry in a linking (double-bordered) data entry

area in the second input screen. If you do so, the process repeats itself. In the *Drosophila* example, if you make an entry in the Family field on the Genus input screen that is not in the Family table, Biota will ask if you wish to create a new Family record.

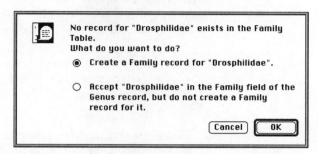

You can continue to create new records in related tables as prompted by Biota. Eventually you will reach a screen for which your entry is already present in the related table, or you will reach a screen that has no linked fields (e.g., Kingdom input at the top of the taxonomic hierarchy or Locality input at the top of the place hierarchy).

5. **Accept and return to previous levels.** When you have created the record or records you wish to in related tables, click the Accept button at the bottom of the current input screen. The input screen for the next lower-level will appear.

6. **Enter data in any remaining fields** that you wish to record, then click the Accept button, at each level.

7. **Continue with this process,** working back down the hierarchy, until you get back to the original screen (the Species input screen, in the *Drosophila* example).

Table-by-Table Creation of Linked Records

In addition to or instead of on-the-fly creation of records in related tables, as described in the previous section, new records can simply be entered directly for any table by selecting the corresponding Input command from the Input menu.

For example, you could begin the entry of a hundred specimens by first entering the names of all genera using Input Genus. Then, when species names are entered, the genera would already be available for wildcard entry (see below).

Which approach is more efficient will depend upon the circumstances. Allowing Biota to request new records for linked fields, when needed, permits each specimen to be handled only once when data are entered directly from specimen labels.

A Powerful Shortcut: Wildcard Data Entry for Linking Fields

For any double-bordered linking field (but not for any other fields), you can save much time and effort by using the *wildcard entry* option.

1. **Enter the first letter, or the first few letters** of a taxonomic name, Personnel Short Name, or Record Code (Specimen Code, Species Code, Collection Code, or Locality Code) that you know or suspect is already recorded in the related table.

   NOTE: To get a full list of *all* records in the related table, enter @ (the wildcard character in Biota) and press the TAB key.

2. **Press TAB.**

3. **If only one record in the related table matches your entry,** the match will be entered in the linking field automatically.

4. **If more than one entry in the related table begins with the sequence of characters you entered,** all matching values will appear in an alphabetized, scrolling list. For example, if you entered the two letters dr in the entry area for Genus in the Species input screen and there were existing records for several genera beginning with Dr in your Data File, Biota would present a list of the options, as shown here.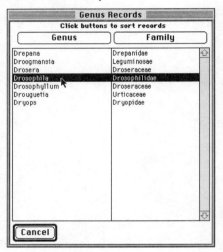

5. **If you see the entry you want**, just click it in the list and it will be entered in the linking field. (*Drosophila* is being clicked in the illustration.)

6. **A second, "helper" field is also listed**, in all cases—and a third, in some cases. In the example above, Family is listed for each genus.

7. **The values in the list can be sorted** by either column by clicking once on the column header button. (For example, to sort the genera in the window above by Family, you would click the Family button in the window.)

8. **If Biota enters a value automatically that you did not intend,** it means your intended value is not present in the related table. Select the incorrect entry with the mouse and enter the correct value manually. For example, suppose you wanted *Drosophila*, so you entered "Dr". If *Drymys* were in fact the *only* genus in the Genus table starting with *Dr*, Biota would complete your entry as *Drymys*. You would then have to change the entry manually to *Drosophila* (spelled out fully), then proceed as in step 3, above.

Chapter 7 Record Codes

As introduced in Chapter 2, the Key fields for the Specimen, Species, Collection, and Locality tables rely on *Record Codes*, rather than actual names of taxa or places, although codes can be names if you wish. (*Record Code* is a general term for Specimen Codes, Species Codes, Collection Codes, and Locality Codes.)

For any Record Code, you may use any combination of up to 20 ASCII characters that you wish. The only restrictions on Record Codes are that each record *within a table* must have a unique Record Code, and a Record Code may not start or end with a space character (Biota checks both restrictions, in case you forget). On the other hand, careful planning of a Record Code system for your particular database can enhance the usefulness of the database you create. Some examples appear later in this chapter

If you want, you can use the same code for records in *different* tables. Although this might seem a useful approach for related records, it makes little sense when many records in one table (e.g., many Specimen records) are linked to a single record in a related table (e.g., a Species record), since the "many" records will each require a different Record Code. In general, it is likely to be less confusing to use Record Codes that are globally unique throughout a Data File.

Why Does Biota Require and Display Record Codes?

In database applications designed for commerce—such as airline reservation systems—or for scholarly applications—such as literature reference programs—unique identifiers for records are normally created automatically and are usually never seen by the user. Each record can be found by other criteria, though often less efficiently. When the airline reservation agent gives you a "record locator" (the automatically generated unique Record Code for your reservation) the airline hopes you will use it when you call to reconfirm or change your plans, because it saves the company time and money over having the agent find your record by date, flight, and name. The Record Code is the fast lane to the record.

Biota gives you direct control over Record Codes so that:

- You can choose a system of codes that is meaningful in your research or institution.
- You can avoid conflicts (identical codes) when combining data from two or more Biota Data Files.
- Museums and herbaria can use their established system of accession numbers or collection codes.
- For data imported from existing flatfiles or field notebooks, you can use whatever unique records codes you have already created. This advantage is especially applicable to Collection Codes and Specimen Codes, for which a collector's own scheme of unique codes may be appropriate.
- Species Codes are necessary for several reasons.
   ◊ Specific epithets (specific names) are obviously unsuitable as unique identifiers of Species records because of the redundancy of published specific epithets among genera.
   ◊ In revisionary systematics and especially in biotic inventory projects, Species records are often needed to unite groups of Specimen records, long before a specific epithet can be applied to those specimens.
   ◊ In a relational database, a Species record is required to connect Specimen records to a Genus record, when the genus is known but the species is uncertain or undescribed.
- Biota allows you to create a meaningful system for assigning Species codes, including the option of using barcodes. Like the airlines' record locator, a Species code is always the fastest way to access a Species record.

Guidelines and Suggestions for Designing Record Code Systems

Although you may use any system you wish, this section provides some suggestions for designing Record Code systems, based on the experience of Biota users. The four tables that require you to create Record Codes for new records each differ according to how Biota users generally assign Record Codes.

Specimen Codes

For some groups of organisms (vertebrates, for example), traditions of unique specimen or accession codes have long existed for individual specimens, in museums and herbaria. Whenever possible, one should probably use such an existing system for assigning Specimen Codes to records in a Biota Data File.

The use of barcodes for individual specimens or specimen lots (pp. 106–107) suggests another obvious recommendation: simply use the specimen barcode (or possibly an abbreviated version of it, pp. 107–108) as the Specimen Code for Biota records (see also Appendix H).

Sequential Specimen Codes

A special consideration in designing a code system applies to Specimen Codes. The Series menu (Chapter 12) provides a set of powerful input, search, and specimen-identification utilities that can be used to greatest advantage only if Specimen Codes for related groups of Specimen records can be assigned Specimens Codes that end with a consecutive *integer counter* (including leading zeros as necessary—see the Warning on p. 101), with or without an *alphanumeric prefix*.

A "group" of specimens, in this sense, may share any characteristic that you consider important or convenient, such as being from the same mass collection or lot, from the same locality, from the same collector, or from the same species.

Detailed instructions appear later in this chapter (pp. 106–108) for setting up Biota to use commercial barcodes with sequential counters, for use with the Series tools.

Unified Record Code Systems: Specimen, Collection, and Locality Codes

In other cases, it may make sense to create a joint system of Collection Codes and Specimen Codes. A group of specimens from the same collecting event share the same Collection Code, which may form a prefix for their Specimen Codes. For example, if the Collection Code is Peru89-01234, the Specimen Codes for three specimens from that collection might be Peru89-01234:001, Peru89-01234:002, and Peru89-01234:003.

A vegetation study that used Biota to keep track of trees, shrubs, and seedlings in a complex layout of sites, transects, and quadrats designed a three-level nested system of Record Codes for the Locality, Collection, and Specimen tables. Each site/transect combination had a Locality record, with the Locality Code made up of a site abbreviation and transect number, for example, LOC2. The quadrats within the LOC2 transect each had a Collection record, with Collection Codes of the form LOC2-T034, LOC2-T035, etc. for the tree (T) quadrats of the transect. Smaller quadrats on the same transect used for shrubs (S) had Collection Codes such as LOC2-S011, LOC2-S012, and so on, with a third series for the even-smaller seedling quadrats. Finally, each plant species that occurred within each quadrat was recorded in a separate Specimen record, with the Abundance field registering the number of individuals of that species in that quadrat. The Specimen Codes for trees were of the form LOC2-T034-003, LOC2-T034-004, or, for shrubs, LOC2-S012-026, LOC2-S012-027, and so on.

Biota could easily find all the Specimen records for a particular locality, transect, or quadrat in this study, using fields and relations, regardless of what values had been assigned for Record Codes. In practice, however, the use of meaningful Record Codes helped the researchers and data input personnel keep track of progress and make sense out of patterns more easily—especially once records were exported to text files for statistical or spatial analysis.

Note: You can ask Biota to enter alphanumeric Record Code prefixes automatically. For manual input of the rest of the code (integer or alphanumeric), after the prefix, you should use Entry Choice Lists to enter the prefix (see pp. 102, 315–322). For automatic generation of unique integer counters appended to a prefix you specify (but with the integers not under your control), see pp. 102–103, later in this chapter.

Locality Codes

For biogeographical databases of all kinds, Locality Codes present special problems. First of all, of course, one must decide at what level in the hierarchy of places a "locality" will be defined, but this is a problem not easily solved in the abstract.

There exist place-name authority files, complete with established locality codes, for some place names within countries and smaller political entities. Obviously, if these are appropriate and available, one would do well to use the established codes as Biota Locality Codes.

Otherwise, if you are on your own to invent Locality Codes. The most meaningful approach is probably to base the Locality Code closely on whatever you enter for the Locality Name field of the Locality record—the Locality Name entry may be up to 50 characters. This entry will have to be abbreviated to 20 characters or fewer (if necessary), to a unique but recognizable form, to create a Locality Code.

Species Codes

When you begin a database with a species list, one obvious way to create Species Codes is to concatenate the first few characters of the generic name with the first few letters of the specific epithet—4 and 4 is a common approach, e.g., *drosmela* for *Drosophila melanogaster*, *brasnigr* for *Brassica nigra*. For some taxa or institutions, there may be established codes for species that should be used instead.

In survey work and systematic revisions, Species Codes are usually needed long before there are secure names to apply to the corresponding Species records. Fortunately, temporary Species Codes are easily accommodated by Biota—there is a special utility to create them automatically, if you wish (see Chapter 22).

Or, you can create your own temporary Species Codes, which are required to link Specimen records (and thus Collection and Locality data as well) to known affinities at higher taxonomic ranks. A "species" that starts out with the Species Code *hysthairy#1* can later be changed to *hysthirs* for *Hystrix hirsuta* when it has been described or determined authoritatively, by making a single change in the Species record (pp. 218–220).

Sorting Records by Record Codes

As the default, Biota displays records for all tables except Species sorted by the Key field for each table in output (listing) screens. This means that Specimen, Collection, and Locality records will be sorted by Record Code, and taxon tables—Genus and above—by taxon name. As the default, Species records are sorted by Genus, then by Species Name (specific epithet), rather than by Species Code.

In the Preferences screen (Special menu), you can switch to either of two other options, or go back to the default.

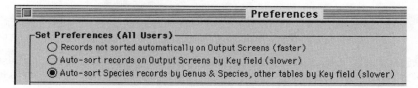

- **Records not sorted automatically on output screens** *(faster)*. Records will appear in the order they were created. This option produces faster displays and is therefore useful for large data files when you intend to sort records on some other criterion anyway, once they are displayed.

- **Auto-sort records on output screens by Key field** *(slower)*. With this option selected, records for *all* tables are sorted by Key field, *including* Species records (which are sorted by Species Code).

- **Auto-sort Species records by Genus and Species, other tables by Key field** *(slower)*. This is the default setting when you create a new Data File or click the Reset to Defaults Now button on the Preferences screen.

NOTE: You can sort records any way you wish by using the Sort button on each output screen, once they are displayed (pp. 131–139).

WARNING: Because Record Code fields are alphanumeric, sorts on record fields treat numerals "alphabetically." Thus, Record Codes A1, A20, and A100 will sort in the order: A1, A100, A20. Likewise, codes 3, 5, 31, and 40 will sort in the order: 3, 31, 40, 5. To avoid this problem, use imbedded or leading zeroes: A001, A020, A100; 03, 05, 31, 40. These series will sort in the order listed.

Using Entry Choice Lists to Enter Record Code Prefixes

Once you decide on a system of Record Codes for your database, Biota offers some help with the input process. If you plan to enter the same prefix for the Record Codes in a series of records, but you need to control the remainder of each Record Code yourself, using an Entry Choice List can save you time and mistakes. (If you want to set the prefix, but let Biota create a unique integer counter to complete each Record Code, see the next section, "Assigning New Record Codes Automatically During Data Entry.")

You can activate and set up an Entry Choice List for any Record Code field ([Species] Species Code, [Specimen] Specimen Code, [Collection] Collection Code, or [Locality] Locality Code), using the procedures described in detail in Chapter 16, (pp. 315–322). If you prepare the list of entry options with the Record Code prefix you are currently using as the first (or only) option, you can enter it quickly simply by pressing the RETURN or ENTER key.

Assigning New Record Codes Automatically during Data Entry

You can create your own complete system of Record Codes (a general term for Specimen Codes, Species Codes, Collection Codes, and Locality Codes), or you can rely on Biota to generate Record Codes automatically, based on settings you specify in the Prefix settings screen.

The Format of Automatically Generated Record Codes

Throughout this book, a Record Code of the form *A1B2CD01234* is said to be made up of the *alphanumeric prefix A1B2CD* and *the integer counter 01234*. In other words, a Record Code prefix ends with the last nonnumeric character in the Record Code and the integer counter comprises all remaining characters (digits), if any.

When Biota generates Record Codes automatically, the *integer counter* is based on the *sequence number* for the corresponding table—the number of records entered in the table since the Data File was created. The sequence number increases by one every time a record is created, no matter how many records are later deleted. Leading zeros are added, if necessary, to make all codes the length you specify in the Prefix settings so that alphabetic sorting will work correctly (see the Warning earlier in this chapter, p. 101).

Changing the Defaults for Record Code Prefixes and Lengths

When you first create a new Biota Data File (an "empty" database; pp. 9–10), Biota assigns the following default Record Code prefixes:

- *LOC* for Location Code
- *COL* for Collection Code
- *SPM* for Specimen Code
- *SPP* for Species Code

For each of these Record Codes, Biota initially assigns six digits as the length of the integer counter, allowing up to 999,999 unique Record Codes for a particular prefix—the maximum Biota allows. (If you need more than that, modify the prefix.)

If you want to change these default prefixes, adjust the length (number of digits) of the corresponding integer counters, or eliminate all default Record Code prefixes, by using the following steps.

1. **Select Record Code Prefixes from the Special menu.** The Prefix setting screen appears.

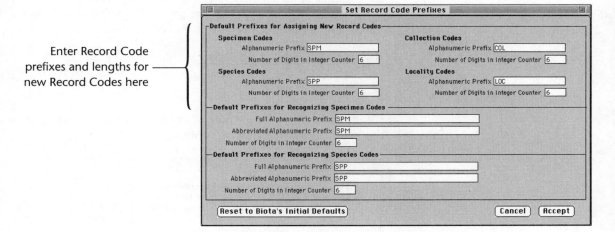

Enter Record Code prefixes and lengths for new Record Codes here

2. **Enter prefixes and lengths (number of digits in the integer counters) in the top panel of the screen.** You can enter any prefix you want for each of the four Record Codes (Locality Code, Collection Code, Specimen Code, and Species Code), up to 14 characters for each prefix. You can enter any value between 3 and 6 for "Number of Digits in Integer Counter." Thus, the maximum number of unique records with a particular prefix can vary from 999(for 3 digits) to 999,999 (for 6 digits).

WARNING: If you set the number of digits for the counters too small and you reach the limit, Record Codes will begin to repeat, and Biota will post a Duplicate Key Errror message (p. 7). At that point, you must change to a different prefix. If you switch back and forth between default prefixes, you may run out of unique counter values for a prefix even sooner, since only the final digits of the sequence number are used.

---

3. **Click the Accept button.** Until you change them again, these prefixes will be used for all subsequent new Record Codes assigned by clicking the Use Default Prefix button in the Assign Record Code window (see the following section of this chapter).

Assigning New Record
Codes Automatically:
Step by Step

NOTE: The prefixes you establish using the Record Code Prefixes screen are recorded in your Biota Data File; they remain just as you set them until you change them, regardless of how many times you start up or shut down Biota. If you start a new Data File, you will have to set them again.

Here are the steps you follow to assign Record Codes automatically while creating or modifying records.

1. **Click the Assign button.** During data entry, you may elect to click the Assign button next to a Record Code entry area, instead of typing in a Record Code. Assign buttons appear in the input screens for the Specimen, Species, Collection, and Locality tables.

2. **Choose an option for assigning the Record Code.** A message window will appear, offering two options.

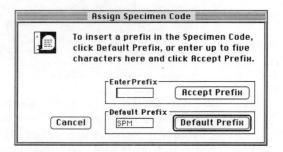

- **Default Prefix.** If you click the Default Prefix button, Biota will create a Record Code that begins with the Default Prefix shown in the window and will automatically append a unique *integer counter* to this prefix. (You can change the default prefix and the number of digits in the integer counter. See the previous section of this chapter.)

- **Enter Prefix, Accept Prefix.** Alternatively, you may enter up to 14 characters in the Enter Prefix area as an ad hoc prefix for the record you are currently creating or editing. Enter the prefix, then click the Accept Prefix button in the message window. In this case, Biota uses your entry as a prefix for the unique integer counter.

Setting Default Prefixes for Recognizing Specimen and Species Record Codes

Biota has the capacity to recognize, automatically, particular Specimen Code and Species Code prefixes that you specify, whether the codes were created by you, generated by a commercial producer of barcodes, or assigned automatically by Biota. The codes you specify are recognized automatically wherever these Record Codes are required in the Biota interface—including input screens, search query screens, and Loans screens.

Here's how it works. When you enter any value in a Specimen Code entry area, either manually or using a barcode reader, Biota checks to see if the value you entered begins with the alphanumeric prefix you specified in the Prefix settings screen. If the entry does not start with the specified barcode prefix, it is treated as an ordinary entry. Thus, Specimen Codes and Species Codes whose attributes you have specified in the Prefix settings screen can be freely mixed with Record Codes with other prefixes, barcodes of any origin, Biota-generated codes, ad hoc codes you make up for each specimen, and any existing Specimen codes (e.g., museum or herbarium accession numbers) or Species codes (from authority files), all in the same database.

There are two reasons you might want to exploit Biota's ability to recognize particular prefixes.

- **To use the Specimen Series tools.** To use the tools of the Series menu (Chapter 12) efficiently, either from the keyboard or using barcodes and a barcode reader, you need to use automatic Specimen Code recognition. See the next section of this chapter.

- **To abbreviate Specimen Codes or Species Codes.** To substitute abbreviated barcode prefixes for long barcode prefixes in a Data File, you will need to set up Biota to recognize Specimen Codes and/or Species Codes automatically. (See pp. 106–108, later in this chapter.)

Using Alphanumeric Specimen Codes with the Series Tools

Much time and effort can often be saved by using commands from the Series menu to enter, update, or find consecutive series of Specimen Codes—although the Series tools can also be used for nonconsecutive Specimen Codes (Chapter 12).

If you are using Specimen Codes that have an alphanumeric prefix followed by an integer counter, you can tell Biota to look for that particular prefix, and when it finds it, to use the integer part of the Specimen Code as a starting or stopping point to create a series of new Specimen records, or to find an existing series.

NOTE: Throughout this book, a Record Code of the form *A1B2CD01234* is said to be made up of the *alphanumeric prefix A1B2CD* and *the integer counter 01234*. In other words, the prefix ends with the last nonnumeric character in the Record Code and the integer comprises all remaining characters (digits), if any.

There are two ways to tell Biota to look for a particular Specimen Code prefix.

- **Manual input of the prefix.** You can enter the alphanumeric prefix manually. Click the Prefix button in any Series screen, then use the Enter Prefix/Accept Prefix option. See p. 104, earlier in this chapter. Unless you need to do this for no more than one or two series of Specimen Codes, this method is likely to be far less efficient than the automatic method (below).

- **Preset Specimen Code Prefix recognition.** With this approach, you use the Prefix setting screen to inform Biota about the alphanumeric prefix for the Specimen Codes you are using. See the next section of this chapter.

Setting Up for Automatic Record Code Prefix Recognition

Follow these steps to set Biota up to recognize automatically, Specimen Code prefixes and/or Species Code prefixes automatically.

1. Select Record Code Prefixes from the Special menu. The Prefix setting screen appears.

Enter Record Code prefixes here for automatic recognition of Codes

2. **Enter the full Record Code prefixes.** In the entry areas labeled Full Alphanumeric Prefix, enter the alphanumeric prefix that appears in the full Specimen Codes and/or Species Codes you will use. (The prefix ends with the *last* nonnumeric character.) *These prefixes may be up to 14 characters in length.*

3. **Enter the Abbreviated (or repeat the Full) Record Code prefixes.**

- **If you want to use full Specimen Code prefixes with the Series tools:** In the entry areas labeled Abbreviated Alphanumeric Prefix, *repeat precisely the full alphanumeric prefix* that you entered in the previous entry area. *These prefixes may be up to 14 characters in length.*

NOTE: Although it does no harm, there is no point in repeating the full Species prefix. Just leave all fields blank in lowest panel of the Prefix settings screen unless you want to abbreviate a Species prefix.

- **If you want to abbreviate barcode prefixes:** In the entry areas labeled Abbreviated Alphanumeric Prefix, enter the abbreviated alphanumeric prefix that you want Biota to substitute for the full Specimen Codes and/or Species Codes you will use. *These prefixes may be up to 14 characters in length.* (See the next section of this chapter for details and warnings about abbreviating barcode prefixes.)

4. **Enter the number of digits for the integer counter.** Finally, in the fields labeled Number of Digits in Integer Counter, enter the number of digits that follow the alphanumeric prefix in your Specimen Codes or Species Codes, so Biota can correctly format the regenerated Record Codes.

Substituting Abbreviated Barcode Prefixes for Long Barcode Prefixes in the Data File

A barcode reader is simply an alternative to the keyboard for entering data. If you read a barcode into a text document or a Biota field, you will simply see whatever is coded on the barcode label. (See Appendix H for general information on barcode readers and barcodes.)

Specimen barcodes usually have a long alphanumeric prefix, identifying country, institution, and series, followed by a long, unique integer counter. (For example, barcodes for specimens collected by or for the Instituto Nacional de Biodiversidad in Costa Rica—INBio—all start with INBIOCRI00, followed by a unique, seven-digit integer.) Some projects (e.g., Project ALAS, see the Preface) also use barcode labels for Species Codes. The barcode for a species can be attached to the species label in specimen storage units.

Although the complete barcode is essential on the specimen itself, it may not be necessary in your database—but see the warning at the end of this section. Biota allows you to use either the full barcode (with any mixture of barcode prefixes from different institutions or collections) or to strip off all or whatever part of the prefix you specify and replace it with an abbreviated or alternative prefix, or with none.

If you want, you can specify an abbreviated or alternative form of a particular alphanumeric barcode prefix—separately and independently for Specimen Codes and Species Codes—using the Prefix setting screen (details in the next section). If you do this, each time a Specimen Code or Species Code is entered, Biota checks to see if it begins with the specified prefix. Is so, the prefix is replaced by the abbreviated or alternative prefix you have specified, followed by the original value of the integer counter.

- For example, if the original barcode reads *IntlMusNatHist001234*, and you have specified that *IntlMusNatHist* is be replaced by *IMNH*, Biota will assign the Specimen Code *IMNH001234* to a new Specimen record when you read in the barcode in the Specimen input screen.

- Later, suppose you search for the record using the By Specimen Code tool from the Find menu or the Find Specimen Series tool from the Series menu. You can find the record by entering either the shortened code, which is actually on the record in the Data File, or the full, original barcode *(IntlMusNatHist001234)*, which Biota will recognize and translate to the abbreviated equivalent to make the search.

If you want to use barcodes for Species Codes, you can shorten or replace Species barcode prefixes in the same way.

WARNING: Abbreviating institutional specimen barcode prefixes is generally not advisable. Prefixes that have been designed to be universally unique are best recorded in full, to avoid ambiguity when data are pooled in aggregated databases. *Even if you do not abbreviate, however, you need to register the prefix, with an identical "abbreviation," in the Prefix setting screen to be able to use the Series commands.* See pp. 106–107.

NOTE: Long barcodes, and long Record Codes in general, will not be visible in full in some Biota output screens, but they are right-justified to show the integer counter and as much of the prefix as possible. To view the complete Record Code, you may have to open the record in the input screen by double-clicking it in the output screen (pp. 121–123).

Chapter 8 Special Data Types

Certain special data types are notoriously troublesome in biodiversity databases. This chapter describes how Biota handles intermediate taxonomic levels (ranks), dates, and geographical coordinates (latitude and longitude).

Intermediate Taxonomic Levels (Subtaxa and Supertaxa)

The taxonomic tables (Species, Genus, etc.; Appendix A) represent only the obligatory levels (ranks) of the Linnean hierarchy, in the sense that every organism must be assigned to a taxon at each of the seven levels. In fact, of course, systematists use many intermediate levels in classifying organisms: Subclasses, Suborders, Superfamilies, Cohorts, Tribes, and so on.

The Dilemma of Intermediate Levels

The problem is that each group of organisms (or each group of taxonomists) uses different intermediate levels, so that even sister taxa are frequently classified to different numbers of levels. For this reason, it is not feasible to create a separate table for each possible intermediate level (e.g., a Subfamily table), since some child records (e.g., Genus records) would have a Subfamily parent but others only a Family parent.

The ideal (and cladistically correct) solution is to implement a recursive taxonomic design (not unlike the recursive system Biota uses for "host-guest" relationships, Chapter 21). A recursive taxonomic rank system allows an indefinite and potentially unlimited number of taxonomic levels within any clade.

There are some practical drawbacks to the recursive design, however. In addition to providing the programmer with a permanent source of migraines, recursive relations can produce a significant computational overhead, slowing operation of many features of a database. For example, in a recursive design, finding all Specimens for an Order requires following each branch and subbranch of a clade from its ordinal root to each specific or subspecific tip, then finding all specimens for each tip, tip by tip. In contrast, a strictly hierarchical model requires a single command to do the same search.

The key consideration in electing to use a traditional design for Biota, however, was to simplify for the user, both in concept and in fact, the processes of importing and exporting text files. It is much easier to conceptualize these processes when the table and field structure of the database is only one step away from the form in which most biologists are used to listing taxa on paper or the computer screen.

How Biota Handles Intermediate Taxonomic Levels

Biota uses fields within the obligatory taxon tables (Species, Genus, Family, Order, etc.) for intermediate taxonomic levels (Appendix A). For example, the Family Formicidae (the ants) is divided into several subfamilies (Myrmicinae, Formicinae, etc.), and each subfamily is further classified into tribes (e.g., within the Subfamily Myrmicinae: Tribes Cephalotini, Attini, Dacetini, etc.).

Biota treats membership in a particular Tribe and subfamily to be attributes of each Genus record. Thus, the Genus table in Biota contains two fields for a taxonomic levels intermediate between the generic and the familial level: [Genus] Tribe and [Genus] Subfamily (Appendix A). Each ant genus gets its own Genus record, with the appropriate values in the [Genus] Tribe and [Genus] Subfamily fields. In other words, there is no single record for a Tribe or Subfamily, as there must be for each taxon in the database at the obligatory levels. Nonetheless, you can search, sort, select, and display, and therefore print or export records based on intermediate taxonomic levels.

For groups not requiring certain intermediate taxonomic levels (or if you choose to ignore them), simply leave the fields for intermediate levels blank.

Default Names for Intermediate Taxon Fields

Biota provides predefined (but renamable) fields for the following intermediate taxonomic ranks (see Appendix A):

[Species]	Subgenus
[Genus]	Tribe
[Genus]	Subfamily
[Family]	Superfamily
[Family]	Suborder
[Order]	Superorder
[Order]	Subclass
[Class]	Subphylum
[Phylum]	Subkingdom
[Kingdom]	Superkingdom

Renaming an Intermediate Taxon Field

If the organisms you work with require a name for an intermediate taxonomic level that does not correspond to the default names for an appropriately placed field in Biota (Appendix A), you can rename any existing intermediate taxon field with a Core Field Alias (pp. 323–326).

Defining Additional Intermediate Levels

If you need more intermediate levels than Biota has provided with default names (Appendix A), use one of the many Custom fields (e.g., [Genus] Genus Custom1) and rename it with an appropriate Core Field Alias (pp. 323–326).

Remember, an intermediate level is always an attribute of the next *lower* obligatory level.

Updating Intermediate Taxa

To add intermediate taxonomic names to existing records, rename an existing intermediate taxon, or split off certain records to be recorded as a new intermediate taxon:

1. **Define a Record Set** (pp. 13–15) for the records you want update.
2. **Use the Find and Replace tool** (pp. 220–225) from the Special menu to add or change the intermediate taxon name for all records in the Record Set.

Dates

Date information appears in several places in Biota tables and records. You can record the date of a collecting event (or the beginning and ending dates for a collecting period), the date each specimen was prepared, the date each was last determined, the date a Note record was added to a parent record, and the date a Specimen Loan was made. Biota automatically registers record creation dates for new Specimen, Species, Collection, and Locality records. If the Determination History option has been activated, the current date is recorded any time the determination of a specimen is updated.

Date Formats in Biota: U.S. Format for Input and Display, International Format for Labels

♦ **Date format for input and display:**

4th Dimension, the engine that drives Biota, records dates in data files in U.S. format only: MM/DD/YYYY (digits for Month/Day/Year). For example, 04/12/1994 means April 12, 1994, not December 4, 1994. For the sake of efficient execution and display, therefore, all Biota input and output screens use this ordering of date elements, wherever dates appear.

♦ **Date format for labels printed or exported to a text file:**

Following accepted practice, Biota uses International Date Format for collection dates on all specimen collection labels: Day Month Year, with the month expressed either by the English three-letter abbreviation (12 Apr 1994) or with Roman numerals (12 IV 1994), as options.

COSTA RICA: Heredia	PERU: Amazonas
La Selva	19 km SW Leimebamba
Elev 150 m 10°26'16"N84°1'56"W	Elev 3020 m 6°45'0"N77°48'0"W
coll. J. T. Longino 22 Jan 1991	coll. J. L. Luteyn 13 Feb 1985
INBIOCRI001B459857	JLL11381

Entering Dates

To enter a date in a record, you enter numerals in three separate, labeled fields, as shown here. (The entries Mo:, Dy:, and Yr: are simply guides to the fields, which are always present. You enter only the numbers.)

Date Collected (or Date Started)
Mo: 5 Dy: 15 Yr: 1984 [Today]

To enter today's date (more precisely, the date set in your computer's operating system), click the Today button next to a date entry area.

You can also use the Field Value Defaults tool (Special menu, pp. 313–315) to enter any date automatically as the default, for any user-enterable date field.

Date Displays on Output Screens

In output screens, Biota displays dates in U.S. format with the standard three-letter English abbreviation for the month, to avoid any ambiguity

Partial dates (see the next section) are displayed as Month (abbreviated) Year, or Year.

Collection Date ranges (see pp. 113–114) are displayed in condensed format, but may nonetheless be too long to be seen in full in an output screen. To view a collection date range in full, you may need to open the record by double-clicking it in the output display.

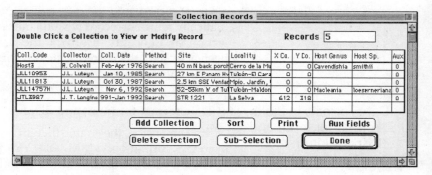

Partial Dates

Partial dates are allowed in several of Biota's date fields. Especially for historic specimens, the exact dates of collection, preparation, and determination are often not known. Instead, collection dates on specimen labels often show only month and year, or just the year. *For collection, preparation, and determination dates,* Biota allows you to enter dates in any of three formats:

♦ **Complete date:** MM/DD/YYY (e.g., enter 05/17/1975 for May 17, 1975)

- **Month-year date:** MM/YYYY (e.g., enter 05/00/1975 for May, 1975)
- **Year-only date:** YYYY (e.g., enter 00/00/1975 for 1975)

How Biota Handles Partial Dates

For most purposes, there is no reason to pay any attention to how Biota handles partial dates internally. *But if you need to import or export partial dates, you will need to read this section and the next one.* In addition, you may want to read this section to understand how records are sorted when some records have complete dates and some have partial dates.

Biota keeps track of complete and partial dates by means of a second, internal Date Flag field paired with each of the date fields that allow partial dates (Appendix A). Thus, Date Determined is paired with a field called Date Det Flag, an integer field of length 1 that has the value 0 if you enter a complete date for Date Determined, the value 1 if you enter a month-year date, and the value 2 if you enter a year-only date.

Partial dates are actually recorded as complete dates (4th Dimension accepts only complete dates in a date field). For screen display, labels, or printed reports, however, Biota displays or prints only month and year from records with month-year dates and only the year for year-only dates (as flagged by the corresponding Date Flag field).

Specifically, Biota uses the following internal conventions to record complete and partial dates:

- **Complete date** (e.g., May 17, 1975): Biota records the complete date plus a 0 in the corresponding Date Flag field.
- **Month-year date** (e.g., May 1975): Biota records the first day of the month (May 1, 1975), plus a 1 in the corresponding Date Flag field.
- **Year-only date** (e.g., 1975): Biota records the last day of the year (December 31, 1975), plus a 2 in the corresponding Date Flag field.

Importing and Exporting Partial Dates

The only time you will ever see or need to be concerned with Date Flag fields is if you need to export partial dates using the Export by Tables and Fields tool (pp. 434–443) or import partial dates using Import by Table and Fields (pp. 488–495). For details, see the pages just referenced.

For importing partial dates, you can use any convention you wish (e.g., middle day of the month for month-year dates, middle day of the year for year-only dates), as long as you import the correct Date Flags.

Collection Date Ranges

Sometimes specimens are collected automatically in traps or nets over a period of time (even seeds in seed traps), so that the exact date of capture for a specimen is not known. Or, specimens are known to have been collected on a certain expedition or cruise whose start and end date are known, even though no specific collection date was recorded for individual specimens.

In such cases, you may enter the first date for the collection period in the Date Collected field and the last date for the collection period in the Date Coll End field (both in the Collection table, see Appendix A). Both fields allow both complete and partial dates.

```
Date Collected (or Date Started)
Mo: 12  Dy: 27  Yr: 1964  [Today]
Date Collection Completed (Optional)
Mo:  1  Dy: 17  Yr: 1965  [Today]
```

Collection Date Ranges on Labels and in Exported Text Files

Collection date ranges are condensed "intelligently" for labels (p. 246) and in some kinds of exported text files (454–457). If both the Date Collected field and the Date Coll End fields have been entered in a Collection record, a Collection or Locality label made using that record will include the date range, condensed if possible. He are the rules with examples:

COSTA RICA: Cartago Cerro de la Muerte, La Georgina Cordillera de Talamanca Elev 3100m, 9°41'0"N83°52'0"W coll. R. Colwell, 9-12 Jan 1976	COSTA RICA: Cartago Cerro de la Muerte, La Georgina Cordillera de Talamanca Elev 3100m, 9°41'0"N83°52'0"W coll. R. Colwell, Jan-Mar 1976
COSTA RICA: Cartago Cerro de la Muerte, La Georgina Cordillera de Talamanca Elev 3100m, 9°41'0"N83°52'0"W coll. R. Colwell, 29 Jan-2 Feb 1976	COSTA RICA: Cartago Cerro de la Muerte, La Georgina Cordillera de Talamanca Elev 3100m, 9°41'0"N83°52'0"W coll. R. Colwell, Jan 1975-Mar 1976

- **Complete dates in the same month:** The label for a collection made between 9 January 1976 and 12 January 1976 would read 9-12 Jan 1976.

- **Complete dates in the same year, but not the same month:** The label for a collection made between 29 January 1976 and 2 February 1976 would read 29 Jan-2 Feb 1976.

- **Month-Year dates in the same year:** A collection made during the period January through March in 1976 would be labeled Jan-Mar 1976.

- **Month-Year dates in the different years:** A collection made during the period December 1975 through March 1976 would be labeled Jan 1975-Mar 1976.

- **One date complete, one partial:** You can even use a complete date for Date Collected and a partial date for Date Coll End, or vice versa, but no attempt is made to condense the label format. A collection made between an unknown day in December 1975 and 20 January 1976 would be written Dec 1975-20 Jan 1976 on a label, to avoid ambiguity.

Geographical Coordinates: Latitude and Longitude

Information on collection of specimens has always included geographical data, with steadily increasing precision as the centuries have passed. At present, two trends in "georeferencing" are evident.

First, museums and herbaria are struggling to add geographical coordinates to historical specimen data based on place names. This is a difficult and tedious task that can automated to only a certain degree by using geographical place name databases and Geographic Information Systems (GIS). How to interpret a collector's notation "25 km NW Guadalajara, Jalisco" requires human judgment: Did the collector mean 25 km along a road or in a direct line? Was the distance measured from the center of the city or from the edge (and if the edge, where was the edge at the time of collection)? How precisely was the distance measured? And, of, course, many older collections use political units as their only locality data: "Bolivia," or "Park County, Colorado" or historical political units like "Veragua." What do you enter in a database for the latitude and longitude of Bolivia?

Second, more recent collections usually carry specific latitude and longitude information (or the equivalent in UTM, Lambert, or other coordinate systems). Collectors have now begun using Geographic Positioning System (GPS) devices that provide latitude and longitude information accurate to within a few meters, and we can expect that GPS units will soon be in the backpack or pocket or on the wrist of every collector of new specimens. Again, however, the problem of precision arises for coordinate data. If a collector used a large-scale map to guess at the coordinates of a collecting site, the precision will less than if a high-resolution topographic map was used, and far less than if a GPS device was used.

For these reasons, specimen databases need to provide a place to record not only geographical coordinates, but also a place to record their estimated level of accuracy.

Options for Recording Geographical Coordinates in Biota

Biota's data structure (Appendix A) provides two places intended for geographical coordinate data:

- The Latitude, Longitude, and LatLongAccuracy fields of the Locality table.
- The XCoordinate, YCoordinate, and XYAccuracy fields of the Collection table.

Here is the rationale for having two places in the structure to record coordinates. Collection records have a many-to-one relation with Locality records (Appendix A). Thus, you can use a single Locality record to register approximate or historical coordinates for a named place (for example, a watershed, region, or County) as the parent record for many Collection records. The set of Collection records for that Locality may represent a mixture of modern and historical records. The modern ones may include GPS readings or other point data in the [Collection] X and [Collection] Y fields, while the historical records leave these fields blank.

NOTE: In both tables, the two coordinate fields are real number fields, whereas Accuracy is an alphanumeric field. Biota validates entries for the Latitude and Longitude fields of the Locality table (Latitude may not be less than –90 nor greater than +90, and Longitude may not be less than –180 nor greater than +180, or the equivalents in DMS; see the next section). Values for the XCoordinate and YCoordinate fields of the Collection table, on the other hand, are not validated; they may take any numeric value, negative or positive. Thus, you can use the XCoordinate and YCoordinate fields for map projection systems, such as Lambert or UTM, or for arbitrary grid systems.

Entry and Display of Coordinate Data in the Locality Input Screen

Biota stores Latitude and Longitude data in units of decimal degrees—the standard system now used in GIS work. Nonetheless, Biota offers three unit systems for the entry and display of latitude and longitude values in the Locality table. You can enter data using any of the three systems and freely use different systems for different records in the same data file. As you make an entry in any of the systems, Biota instantly translates the value into the other two and displays all three versions in the Coordinates panel of the Locality input screen. See pp. 162–164 for details and instructions.

- **System 1: Degrees/Minutes/Seconds.** This is the traditional system found in most gazetteers and atlases—but see the important warning, below.
- **System 2: Integer Degrees, Decimal Minutes.** This is the system used by most Geographical Positioning Units (GPU).
- **System 3: Integer Degrees, Decimal Minutes.** This is the system used in Geographical Information Systems (GPS).

See pp. 000–000 for conversion formulas for the three systems listed above.

WARNING: Some gazetteers list Latitude and Longitude values in DMS, but use a *format* that looks like Integer Degrees with Decimal Minutes. For example, 24° 48.39 actually means 24° 48' 39" in these gazetteers. The easiest way to detect this abomination is to see if you can find any values in the gazetteer to the right of the "decimal point" that exceed 59. If not, those values are Seconds, not the decimal part of true Decimal Minutes.

Setting the Display Format for the Latitude and Longitude in the Locality Output Screen

You set the display format for the [Locality] Latitude and [Locality] Longitude fields in the Locality output screen in the Preferences screen (Special menu).

⦿ Show Latitude and Longitude in decimal degrees (faster)
◯ Show Latitude and Longitude in D/M/S (slower)

Because Latitude and Longitude are stored internally by Biota in decimal degrees, display in the output screen is more rapid if you choose the first option than if Biota must compute the DMS equivalents for each record as it is displayed.

Longitude and Latitude on Printed Labels and in Exported Text Files

In all locality labels that Biota exports, Latitude and Longitude (the Locality table fields) are expressed in the DMS system (see Chapter 13). The format for exported text files depends on the export tool you use (see Chapter 24).

Chapter 9 Working with Records in Output and Input Screens

Every Core table in Biota (p. 5, Appendix A) uses the same standard format for listing records. These *output screens* incorporate many functions that are explained in detail in this chapter. Once you know how to use the standard output screen for one Biota Core table, you know how to use them all.

Biota also uses two other kinds of screens to list records, which differ in some important ways from the standard output screens for Core tables. These are explained in other chapters.

- **Output screens for Auxiliary Fields.** See the chapter on Auxiliary Fields (Chapter 15) if you need help with viewing or manipulating information in Auxiliary Fields.

- **Tree menu output screens.** The screens used to list records in the hierarchical lookup system (the Tree menu) for taxa and specimens have the special attribute of hierarchical linkage. Moreover, the method you use to view and edit individual records in the Tree screens is not the same as for with the standard output screens discussed in this chapter. See pp. 186–192 for help with the Tree menu.

In the rest of this chapter, the term *output screen* will refer the standard output screen.

Displaying Records in a Standard Output Screen

Biota offers many tools for finding records (Chapter 11). For every tool in the Find menu, records are displayed in the standard output screen for the appropriate table.

For example, to display all records for a table in the output screen, select the "All..." command for that table from the Find menu, e.g., All Species.

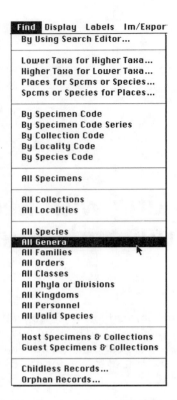

Here are all Genus records for a small data file for Galápagos Finches, shown in the Genus output screen.

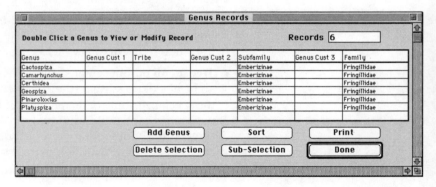

Viewing, Editing, Printing, or Deleting Individual Records from an Input Screen

To view an individual record from the output screen for any Core table, simply double-click the record. (The genus *Geospiza* is being double-clicked in the example, above.) The record will appear in the input (single-record) screen for the table.

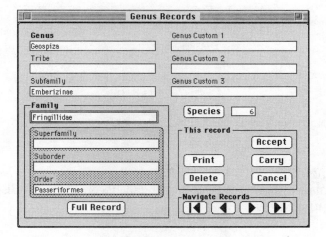

With the record displayed in the input screen, you can view the record, edit any field in the record and save the changes, print the record, or delete the record.

To Return to the Output Screen

To dismiss the record and return to the output screen, without making any changes in the record, click the Cancel button. Biota returns you to the output screen, with the record selected.

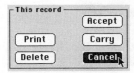

Editing and Saving Changes to the Record

To edit a field, click in the field and use standard editing techniques. Any changes you make can be saved by either of two methods:

- *Either:* **Click the Accept button.** In this case, Biota saves all changes and returns you to the output screen, with the record selected.

- *Or:* **Click any of the Navigation buttons**. To accept any changes and move to another record in the current selection, click any button in the Navigate Records panel (the arrow buttons). This action automatically saves any changes.

NOTE: If the Accept, Carry, and Delete buttons are dimmed on the input screen, it means you do not have Write privileges. The record is in Read-only mode. You must either click Cancel or a Navigation button to dismiss the record.

Printing a Single Record Displayed in an Input Screen

Click the Print button in the input screen to print a replica of the input screen showing the record currently displayed.

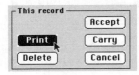

Deleting a Record Displayed in the Input Screen

1. **Click the Delete button.**

2. **Confirm or Cancel the deletion.** Unless the record is a new one that has not yet been saved, Biota presents you with one of two kinds of warnings.

   • **For tables that have no associated Notes, Auxiliary Fields, or Images,** such as the Genus table, Biota will present a warning in the following form.

   • **For tables that have associated Notes, Auxiliary Fields, or Image records** (the Species, Specimen, Collection, Locality, and Loans tables), deleting a record automatically deletes any linked child records in Peripheral tables (Notes, Auxiliary Fields, and/or Images). In such cases, Biota will present a warning of the following form.

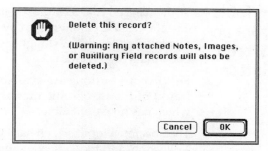

3. **Confirm or Cancel the orphaning of Core table child records.** If the record is linked to one or more child records in other Core tables (but not otherwise), Biota will also post the following warning, with the appropriate table names (Genus and Species, in this example). (For information on displaying linked records, as suggested in the warning, see pp. 125–126, later in this chapter.)

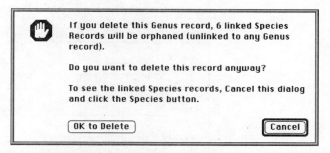

4. **Deleting a Personnel Record.** In the case of Personnel records, which may be linked to child records in several Core and Peripheral tables, the warning screen is more complex.

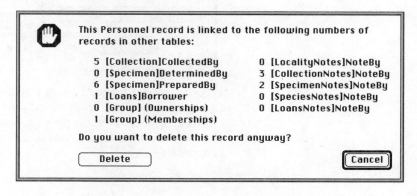

NOTE: See Chapter 10 for detailed information on using input screens to enter and edit records, and using the Carry button.

Moving Up the Table Hierarchies from an Input Screen: Full Record Buttons

With an individual record open in the input screen, you can display and edit the related parent record by clicking the Full Record button in the display area for the next higher table. (The original record will remain open in its own window.)

Full Record buttons are available on the input screen for each level (rank) of the taxonomic hierarchy, with the exception of Kingdom, which has no parent table.

The Specimen input screen has Full Record buttons for both Species and Collection, and the Collection input screen has a Full Record button for Locality. Locality has no parent table, so it has no Full Record button.

Displaying Records with the Full Record Button

For example, with the genus record for *Geospiza* open in the Genus input screen, the principal fields from the related Family record (Fringillidae) are shown in the stippled area at the lower left. To see the full record for Family Fringillidae (the parent record for Genus *Geospiza*), click the Full Record button.

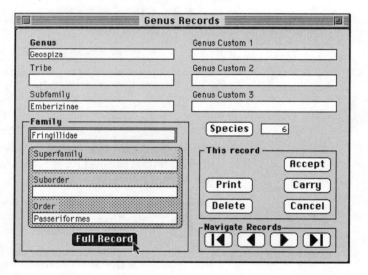

The full record for the Family Fringillidae appears in the Family input screen. Information from the related Order record (for Order Passeriformes) appears in the stippled area at the lower left, with another Full Record button to click if you want to see the full record for the Order Passeriformes.

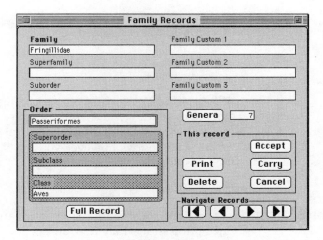

If you want, you can continue on up the taxonomic hierarchy by clicking Full Record buttons. To return to the original record, click the Cancel button—or the Accept button, if you have made changes.

Editing Records Displayed with the Full Record Button

You can open a parent record using the Full Record button, edit it, and save any changes by clicking the Accept button in the input screen for the record you edited.

Moving Down the Table Hierarchies From an Input Screen: Child Records Buttons

With an individual record open in the input screen, you can display all related records in the next lower table in the hierarchy (child records) by clicking the button with the name of the child table, located in the lower-right quadrant of the input screen. (See pp. 7–8 for an explanation of the parent-child terminology.)

These buttons are referred to collectively as *Child Records buttons*, although each button has its own name. For example, the Child Records button in the Family input screen is labeled Genera, whereas the Child Records button in the Locality input screen is labeled Collections. (Consult Appendix A for to see the table hierarchies.)

Child Records buttons are available on the input screen for each level (rank) of the taxonomic hierarchy, as well as for the Locality and Collection input screens. The Specimen table has no child tables (among the Core tables), so the Specimen input screen has no Child Records button.

Biota displays a number next to each Child Records button, showing the number of child records for the parent record currently displayed.

If there are no child records, the number displays zero and the Child Records button is disabled.

When you display child records using a Child Records button, they appear in an the standard output screen for the child table in a separate Biota "process" (task). Meanwhile, the parent record will remain open in its own window; however, you can close it if you want, and the Child Records window will remain open.

Displaying Records with a Child Records Button

For example, with the genus record for *Geospiza* open in the Genus input screen, the Species button is enabled, and the small display area next to the button shows that there are six Species records linked to *Geospiza*.

If you click the Species button, the six species of *Geospiza* finches are displayed in the Species output screen.

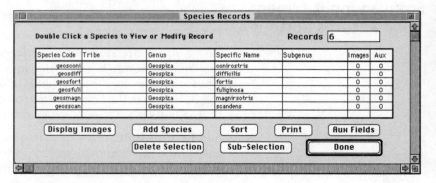

If you wish, you can now open and edit any of the child records you wish to examine or repeat the process to display the child records (Specimen records, in this example) for one of the Species displayed.

To return to the parent record (*Geospiza*, in the example), click the Done button on the child table output screen (above).

Using the Add Record Button

The Add Record button at the bottom left of the output screen for each Core table opens a new, blank record in the input screen for the table. Although these buttons are referred to collectively as *Add Record buttons*, each button has its own name. For example, the Add Record button in the Genus output screen is labeled Add Genera, the Add Record button in the Collections output screen is labeled Add Collection, and so on for each Core table.

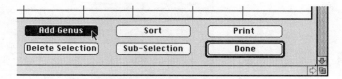

Once the input screen is displayed by clicking an Add Record button, you can create a new record. Click the Accept button to accept the new record and return to the output screen, the Carry button to accept the new record and use it as a template for an additional new record, or Cancel to dismiss the input screen without saving the new record.

In fact, adding a new record in this way is exactly like adding a new record using a command from the Input menu (see Chapter 10, with one exception. When you are done entering data in the record and click the Accept button, you are returned to the output screen, with the new record highlighted. To add another record, click the Add Record button again. (From the Input menu, the Accept button accepts the record and presents a new blank record.)

Using an Existing Record as a Template for a New Record

Sometimes you need to create a new record that duplicates much of the information in an existing record listed in the output screen. You can use the existing record as a template for the new one.

1. **Double-click the existing record in the output screen.** The existing record opens in the input screen. *Make no changes in the existing record.*
2. **Click the Carry button in the input screen.**

A new record appears, with the Key field blank, but all or most other fields (depending on the table) already filled in with the same information as the existing record. (The record below uses the *Geospiza* record as a template.)

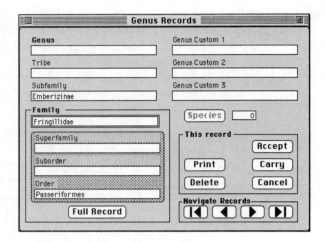

3. **Enter a new value for the Key field.** You can also change any other fields you wish in the new record.

- *Either:* **Click the Accept button** to accept the new record and return to the output screen.

- *Or:* **Click the Carry button again** to accept the new record and use it as a template for an additional new record.

NOTE: As another variation, you can open an existing record, make changes in the existing record, then click Carry to accept the edited record and use it as a template for a new record.

Creating a Sub-Selection of Records

Often, you will want to select one or more of the records displayed in an output screen and dismiss the rest of the records. To do this, you use the Sub-Selection button at the bottom of the output screen.

1. **Select the records** that will form the new Sub-Selection.

- **To select a single record,** click it once to highlight it.

- **To select consecutive records,** click the first record of the group once to highlight it, then press and hold the SHIFT key and click the last record in the group.

- **To select nonconsecutive records,** click the first record once the highlight it. Then hold press and hold the COMMAND (⌘) key (Macintosh) or the CTRL key (Windows) while clicking, once, each additional record you want to select.

2. **Click the Sub-Selection button** when all the records you want to include in the Sub-Selection are highlighted.

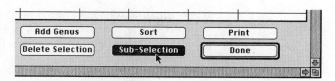

The output screen now displays only the records in the Sub-Selection.

3. **To designate the Sub-Selection as the current Record Set** for the table, click the Done button, then the OK button in the standard Record Set option screen (see pp. 13–15).

Deleting a Group of Records from the Output Screen

In the output screen, you can select records and delete them using the Delete Selection button.

1. **Select the records that to be deleted.**

    - **To select a single record,** click it once to highlight it.

    - **To select consecutive records,** click the first record of the group once to highlight it, then press and hold the SHIFT key and click the last record in the group.

    - **To select non-consecutive records,** click the first record once the highlight it. Then hold press and hold the COMMAND (⌘) key (Macintosh) or the CTRL key (Windows) while clicking, once, each additional record you want to select.

2. **Click the Delete Selection button** when all the records you want to delete are highlighted.

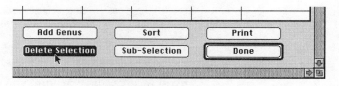

3. **Confirm or Cancel the deletion.** Biota will display one of two kinds of warnings.

    - **For tables that have no associated Notes, Auxiliary Fields, or Images,** such as the Genus table, Biota will present a warning in the following form.

    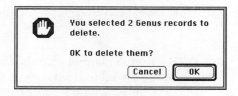

    - **For tables that have associated Notes, Auxiliary Fields, or Image records** (the Species, Specimen, Collection, Locality, and Loans tables), deleting a record automatically deletes any linked child records in Peripheral tables (Notes, Auxiliary Fields, and/or Images). In such cases, Biota will present a warning of the following form.

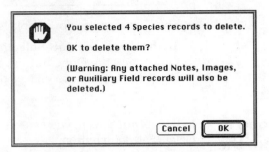

4. **Decide whether to orphan Core table child records.** If the records to be deleted are linked to one or more child records in other Core tables (but not otherwise), Biota will also post the following warning, with the appropriate table names (Genus and Species, in this example). At the same time, the records you selected for deletion will appear by themselves in the output screen behind the message, to help you remember what you selected.

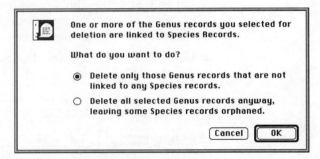

If you receive this warning, you can Cancel the deletion, choose to delete only those selected records that are not linked to any Core table child records, or delete all the selected records anyway, leaving some records orphaned.

NOTE: If you choose the second option (delete all selected records, even though some child records will be orphaned), you can find the orphaned records later using the Find Orphan Records tool from the Find menu.

5. **Deleting Personnel records.** In the case of Personnel records, which may be linked to child records in several Core and Peripheral tables, the warning screen is more complex.

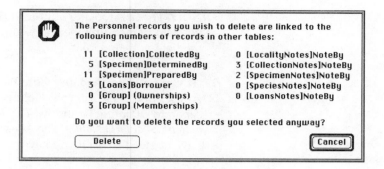

Sorting Records in Output Screens

To sort the records displayed in an output screen, you use the Sort tool, displayed by clicking the Sort button at the bottom of the output screen. As an example, the Genus output screen will be used in this section, but the procedures are the same for all output screens.

1. **Click the Sort button** at the bottom of the output screen.

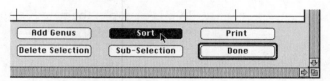

The Sort window appears (labeled Sort Genus in this example).

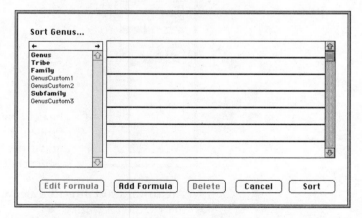

In the accompanying panel, you will see the names of all fields for the table whose records are displayed. Field names in **boldfaced** type are indexed fields, which sort faster than nonindexed fields, but you can sort on any field. (Indexes take up space in data files and require processor time to maintain, so not all fields in Biota are indexed.)

NOTE: The Sort tool is a built-in 4D utility that unfortunately does not list field names alphabetically. You may have to scroll the list to find the field you are looking for.

2. **If you have defined any Core Field Aliases** (pp. 323–326) a list the Aliases and their Internal Field Name equivalents will appear in a separate, "floating" window. *The fields panel in the Sort screen displays only Internal Field Names.* The Alias window does not appear (and you have no need for it) if you have not defined any Aliases.

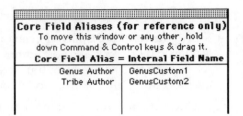

In this example, the field [Genus] Genus Custom 1(Internal Field Name) has been given the Alias *Genus Author*, and the field [Genus] Genus Custom 2 has the Alias *TribeAuthor*.

3. **To sort on a field for the current table** (Genus, in the example), click a field name in the panel at the left of the Sort screen.

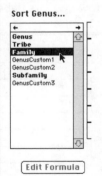

The field name appears in the sort criteria panel of the Sort window, with a downward-pointing, vertical arrow at the far right. This is the *sort direction arrow*.

- **An upward sort direction arrow** means a normal alphanumeric sort (0 to 9, then A to Z).

- **A downward sort direction arrow** means a reverse alphanumeric sort (Z to A, then 9 to 0).

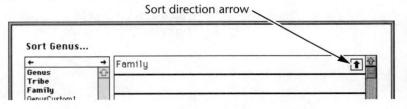

4. **Changing the direction of a sort.** Click the sort direction arrow to toggle between normal and reverse sorts.

5. **Removing a sort level.** If you make a mistake and want to remove a sort level from the list of sort criteria, select the sort level you want to remove (click once on it to highlight it), then click the Delete button in the Sort window, as shown below.

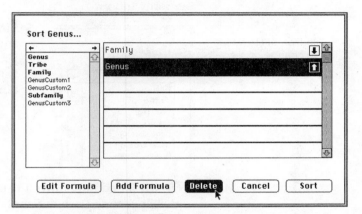

6. **Adding additional sort levels.** To add an additional level to the sort criteria, click another field name in the field name panel, as in step 3. The sort proceeds hierarchically from the top of the list of sort criteria to the bottom. In the example, below, Genus records will be sorted alphabetically by Genus within Tribe, alphabetically by Tribe within Subfamily, alphabetically by Subfamily within Family, and Families will be sorted in reverse alphabetic order (note the sort direction arrow settings at the right).

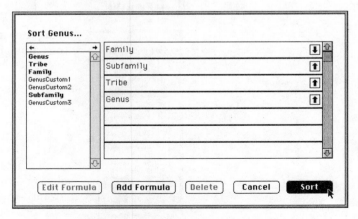

7. **Launching the Sort.** When all sort criteria are ready, click the Sort button in the Sort window to launch the sort.

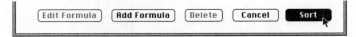

Using Fields from Related Tables for Sorting Records

You can use fields from related tables as sort criteria, as long as a child-parent relation exists between the table whose records you are sorting and the table you use to add a sort criterion. See the next section, "Guidelines for Using Fields From Related Tables as Sort Criteria."

1. **Display the fields for the related table.** There are two ways to access field lists for other tables and display them in the field list panel of the Sort window.

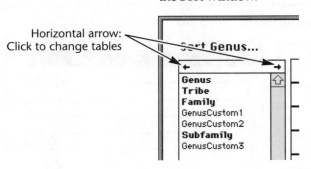

Horizontal arrow: Click to change tables

- *Either:* **Click on the horizontal arrows** at the top of the field list panel at the left of the Sort screen, to move to field lists for other tables, one table at a time.

- *Or:* **Click and hold down the mouse button, in the blank rectangle between the two horizontal arrows,** to display a popup list of all Biota tables.

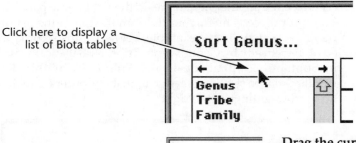

Click here to display a list of Biota tables

Drag the cursor down and release it when the name of the table you want is highlighted.

NOTE: Unfortunately 4D does *not* list table names alphabetically even though they are always shown in the same order. You may have to look for the table you seek until the list becomes familiar.

2. **Select a sort field from the related table.** Once the fields for the related table appear in the field list, you can add fields as sort criteria, using the same techniques described previously (pp. 132–133). Below, the [Family] Order field is being added to the sort criteria for Genus records.

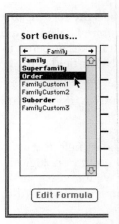

3. **Mixing sort criteria from different tables.** You can mix sort criteria from related tables with sort criteria from the same table as the records to be sorted. In the example below, Genus records will be sorted by the [Family] Order field, then (within Orders) by the [Genus] Family field, and finally (within Families) by the [Genus] Genus field.

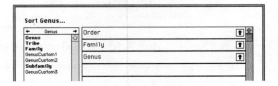

NOTE: When you have a choice between using a relating field and a key field as a sort criterion, the sort will be faster if you use the relating field in the child table (the "foreign key," in database terminology), rather than the Key field in the parent table. In the example above, the [Family] Order field was added to the list instead of the [Order] Order field; the [Genus] Family field was added instead of the [Family] Family field. Functionally, it does not matter which way you do it, however, and the difference in sort speed will not be noticeable unless there are many records.

Guidelines for Using Fields from Related Tables as Sort Criteria

It is up to you to ensure that a sort on a field in a related table makes sense. Here are a few guidelines.

- **You cannot sort the records in a parent table based on a field in a child table.** For example, you cannot sort Species records based on a field in the Specimens table. It makes no sense to sort Species records according to any field in the Specimens table, since there may be many different values in Specimen fields for different specimens of the same Species.

136 Biota: CHAPTER 9

- **You cannot sort the records of a table in one hierarchy (e.g., the taxonomic hierarchy) based on a field in the other hierarchy (the place hierarchy).** It makes no sense to sort Species records, for example, by the Locality Name field of the Locality table, since a species may have specimens from many localities.

Consult Appendix A, if you are unsure how Biota tables are linked.

Using Formulas to Sort Dates by Day, Month, or Year

If you sort on a date field (e.g., [Collection] Date Collected) in the usual way, Biota will sort chronologically according to complete dates. Sometimes you may want to use a sort criterion based on date (either day and month or just month), disregarding year, to study seasonal patterns, for example.

1. **Display the records to be sorted.** Suppose you want to sort Collection records by month of collection, ignoring years.

2. **Click the Sort button at the bottom of the output screen.**

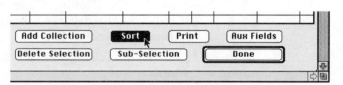

3. **Click the Add Formula button** at the bottom of the Sort window.

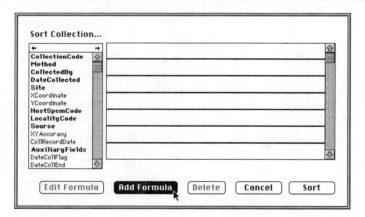

The 4D Formula Editor will appear. It has four panels.

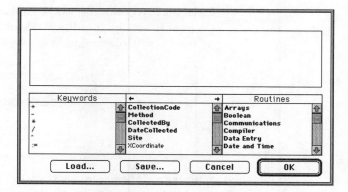

- **The upper panel is where you construct the Sort formula.** You can type all entries in this area manually if you wish. In this example, we will construct the sort criterion:

    Month of ([Collection] Date Collected)

- **The lower-left panel, labeled Keywords, shows a list of operators** that you can enter in a formula by clicking on them. (Frankly, it is faster to enter them manually.)

- **The lower-middle panel offers scrollable lists of tables and fields.** If you click on a field name, it is entered in the formula panel, in the correct format. This panel works exactly like the fields panel in the Sort window itself (see pp. 134–135 for instructions).

- **The lower-right panel, labeled Routines, offers a series of command groups, with boldfaced titles.** If you click and hold down the mouse button on one of these boldfaced titles, a popup list of commands appears.

4. **Find the boldfaced item Date and Time in the Routines panel. Click on the item and hold the mouse button down.** A popup list of date and time functions appears.

5. **Click "Month of," "Day of," or "Year of," as required for the criterion.** The function will be entered in the sort panel. In the accompanying example, "Month of" will be entered.

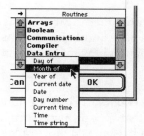

6. **Enter a left parenthesis: (.**

7. **In the lower-middle panel, find and click on the date field that you want to sort by months, days, or years.** In the example, the [Collection] Date Collected field is selected.

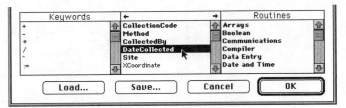

Enter a right parenthesis:) . In the example, the formula will now read: Month of([Collection] Date Collected).

8. **Click the OK button in the Formula Editor.** The Formula Editor goes away and the Sort window reappears, with the formula entered in the sort criterion area. (You cannot type it in directly in this area.)

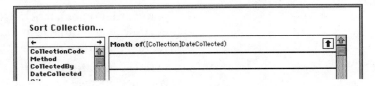

NOTE: If the formula is surrounded by bullets, you have made a syntax error in entering the formula. (In the example below, the parentheses are missing around "DateCollected.")

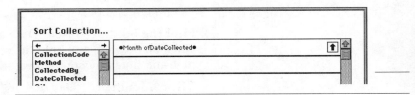

9. **When the formula is entered correctly, click the Sort button on the Sort screen.**

10. **Saving and loading formulas.** Notice that the Formula Editor includes Save and Load buttons. If you use a complicated formula often, you might want to save it to a disk file and reload it when you need it.

CRITICAL WARNING: There are hundreds of other items listed in the Routines list that appears in the Formula Editor. (All 4D commands are listed, as well as all 900 Biota procedures—compliments of 4D.) There are a few other functions that users familiar with 4D might use, but *under no circumstances attempt to use any of the italicized Biota procedures in a formula. You may do irreparable damage to your data file if you do so.*

Using a Formula to Sort Numbers in an Alphanumeric Field

Sometime, you may choose to use an alphanumeric field for numerical data. (See Appendix A for information on Biota field types.) For example, a plant ecologist might rename the [Specimen] Medium field *DBH* and use it to record the diameter at breast height (dbh) of living trees in a quadrat-based study.

Using an alphanumeric field for numerical data causes no problems for data entry or display, but if you *sort* on an alphanumeric field, Biota will sort by ASCII code, left to right. This means that the numbers 1, 2, 3, 9, 10, 20, and 30 will sort in the order 1, 10, 2, 10, 3, 30, 9. To sort then numerically, you must use the *Num* function in a sort formula.

(The Num function ignores any non-numeric characters.)

1. **Display the records to be sorted** in the output screen for the table. Suppose you want to sort Specimen records by numerical values in the [Specimen} Medium field, for example.
2. **Click the Sort button** at the bottom of the output screen.
3. **Click the Add Formula button** at the bottom of the sort window. The Formula Editor appears.
4. **Review the introduction to the Formula Editor** in the previous section ("Using Formulas to Sort Dates by Day, Month, or Year"), if necessary.
5. **Enter the formula *Num ([Field Name])*.** The easiest way to do this is to type in "Num(" and click the appropriate field name in the lower middle panel, then add the final parenthesis ")" to the formula. In the illustration below, the sort criterion Num ([Specimen]Medium) was entered in this way.

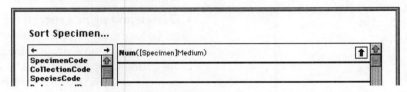

Printing Reports or Creating Text Files Based on Records in an Output Screen

For each Core table and for the Determination History table, Biota offers a preformatted report that you can print, listing all records in the current Selection, including fields from related tables where appropriate (pp. 266–268). Or, you can print the records using a custom report layout that you design yourself using the Quick Report Editor (pp. 269–278). As a third option, you can use the Quick Report Editor to export information from the records and related tables to a text flatfile (457–462).

Printing a Report Based on Records in an Output Screen

1. **Display the records to be printed.**
2. **Click the Print button at the bottom of the output screen.**

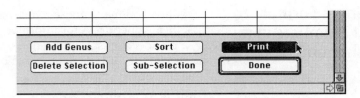

Alternatively, you can select Print from the File menu while the records are displayed in the output screen.

3. **Choose a print option.** A print option window appears offering two or three choices, depending on the table for which you are printing records. (The Determination History table has no options.) Three different option windows are illustrated below.

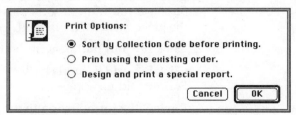

- **Sort by Record Code before printing.** Use this option to print a report with the records sorted by Record Code (Species Code, Specimen Code, Collection Code, or Locality Code). The Loans report can be sorted by Loan Code.

- **Print using the existing order.** If you have used the Sort tool to create a special order for the records that you want maintained in the report, choose this option.

- **Design and print a special report.** If you choose this option, Biota will present the 4th Dimension Quick Report Editor, which you can use to design, save, and/or load custom report formats that you create yourself (pp. 269–278). The report will be based on the selection of records in the output screen when you clicked the Print button or selected Print from the File menu.

- **Sort taxonomically before printing.** For records from taxonomic tables (Species, Genus, Family, etc., and Specimen as well), this option prints the records for the table in alphabetical order within taxonomic levels (ranks) in a hierarchical report organized by rank (as in the table on p. 2).

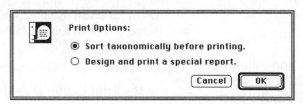

- **Include Host data in the report.** For the Specimen table, an additional option is offered. If the "Include Host data" checkbox is checked, the report will include, for each Specimen record, the Specimen Code of the Host for that specimen (see Chapter 21 on the Host-Guest system).

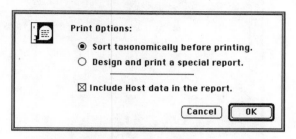

NOTE: This option is automatically checked if any Specimen records in the current selection have host data. Otherwise, it appears unchecked. In either case, you can override the setting manually. The host data checkbox is disabled when the "Design and print a special report" option is clicked.

4. **Click the OK button in the print option screen.**
   - **If you choose the standard Biota report,** see "Printing Procedures," in Chapter 13, pp. 243–245.
   - **If you choose Design and Print a Special Report,** see "Designing and Printing Reports with the Quick Report Editor" in Chapter 14, pp. 269–278.

Exporting a Text Flatfile Based on Records in an Output Screen

1. **Display the records to be exported to a text file.**
2. **Click the Print button at the bottom of the output screen.** Alternatively, you can select Print from the File menu.

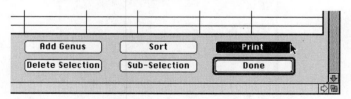

3. **When the print option window appears, choose "Design and print a special report."** The option windows are illustrated in the previous section. Biota will present the Quick Report Editor.
4. **Follow the instructions in Chapter 24, in the section "Exporting Custom Flatfiles,"** pp. 457–462.

Chapter 10 Input—Table by Table

As much as possible, all of Biota's input screens share the same features. These common elements are discussed in Chapter 6, "Entering Data," Chapter 7, "Record Codes," and Chapter 9, "Working With Records in Output and Input Screens." This chapter focuses on features restricted to one or a few input screens.

Specimen Input

You can enter a new record in the Specimen input screen by selecting Specimen from the Input menu or by clicking the New Specimen button on the Specimen output (record listing) screen.

The illustration below shows the default order (p. 90) for entry areas in the Specimen input screen. Press the TAB key to move between entry areas in default order, or click in any entry area to move to it directly.

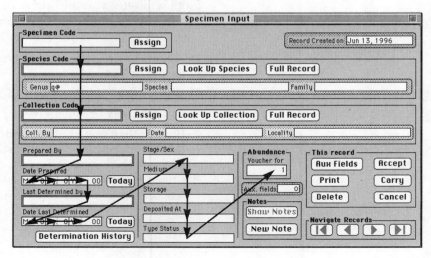

Entering Specimen Data: Step by Step

1. **Enter a Specimen Code.** You can enter the Code by typing it in the Specimen Code entry area, by using the adjacent Assign button (pp. 102–103), or by reading in a barcode (pp. 105–108). Specimen Code, the Key field for the Specimen table, is an obligatory entry, which must be made first. All remaining entry areas are optional. The Specimen Code you enter must not duplicate the Specimen Code of any existing Specimen Record, or you will receive an error message (p. 7).

2. **Enter a Species Code.** The Species Code links this Specimen record with a record in the Species table (pp. 164–168). There are four *alternative* ways to enter a Species Code:

   - *Either:* **Enter a complete Species Code manually or with a barcode reader.**
     ◊ **If a Species record exists with the Code you entered,** the stippled Species display area will show the Genus, Species (specific name), and Family, and the cursor will move on to the next entry area. (Be sure to confirm that the correct Species record was found. If your entry uniquely matches the first part of a different Code, the wrong link will be made.)
     ◊ **If no Species record exists with the Code you entered,** Biota will offer you the option of creating a new Species record "on the fly," with the Code you entered, or accepting an orphan Specimen record. See pp. 92–95 for details.
   - *Or:* **Use the adjacent Assign button to create a new Species Code.** Biota will offer you the option of creating a new Species record "on the fly," with the Code you entered, or accepting an orphan Specimen record. See p. 94 for details.
   - *Or:* **Enter the first letter or first few letters of a Species Code, then press TAB** (wildcard lookup).
     ◊ **If only one Species Code matches the entry,** the entry will automatically be completed.
     ◊ **If more than one existing Code matches the entry,** a scrollable list of matching Codes for existing Species records will appear, from which you can choose the correct code by clicking it (p. 96). (If you enter the @ character, and nothing else, *all* Species codes will appear in the scrollable list.)

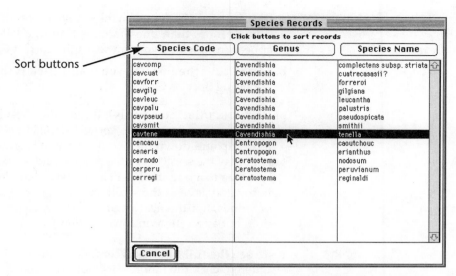

Sort buttons

To sort the entries in the list by Genus or Species Name (specific name) or to resort by Species Code, click the sort buttons at the top of the screen. When you sort by Genus, records are sorted within Genus by Species Name.

◊ **If no existing Species Code matches your entry,** Biota will offer you the option of creating a new Species record "on the fly," with the Code you entered, or accepting an orphan Specimen record. See p. 94 for details.

♦ *Or:* **Click the Look Up Species button.** The Look Up Species window appears.

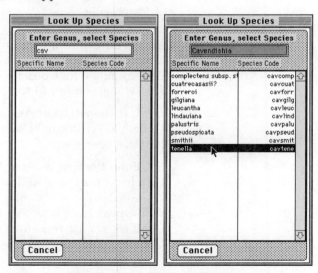

a. **Enter a Genus, or the first letter or letters of a Genus and press TAB.** If a matching Genus record is found, all Species records for the Genus will be appear in the scrollable area of the Lookup window. If more than one match is found, a list of matches will appear from which you can select the correct Genus.

b. **If you find the Species you want, click it** to enter the corresponding Species Code.

> WARNING **(Macintosh only):** If the list of Species is longer than will fit in the window, the vertical scroll bar will be enabled. To move up and down the list, use only the arrows at the top and bottom of the scroll bar or the "thumb-slider" box inside the scroll bar. Due to a bug in 4D, clicking above or below the slider box in the gray area of the scroll bar displays the first or last screen of values, not the next or previous screen of values. (This bug does not affect the Windows version.)

c. **To see the full Species record,** once the Species Code has been entered, click the Full Record button in the Species input panel of the Specimen input screen.

3. **Enter a Collection Code.** You can enter a Collection Code using any of the same four *alternative* techniques described above for Species Codes. (See step 2, above for details.)

- *Either:* **Enter a complete Collection Code manually.** (You can use a barcode reader, but Biota does not offer automatic barcode prefix recognition for Collection Codes. See pp. 105–108.) You can create a Collection record "on the fly" for a new Collection Code.

- *Or:* **Use the adjacent Assign button to create a new Collection Code.** You can create a Collection record "on the fly" for a new Collection Code.

- *Or:* **Enter the first letter or first few letters of a Collection Code, then press** TAB to use wildcard lookup for existing Collection Codes. A scrollable list of matching Codes for existing Collection records will appear, from which you can choose the correct Code by clicking it (p. 96). (If you enter the @ character, and nothing else, *all* Collection records will appear in the scrollable list.)

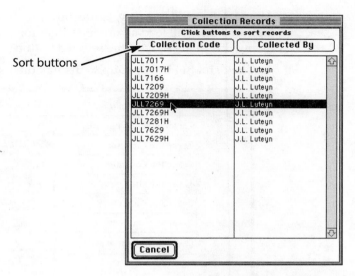

Sort buttons

To sort the entries in the list by Collector (Collected By) or resort by Collection Code, click the sort buttons at the top of the screen.

- *Or:* **Click the Look Up Collection button.** The Look Up Collection options window appears.

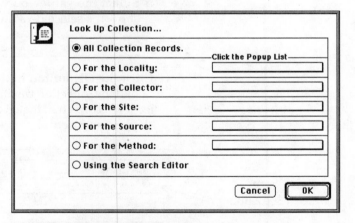

a. **Select a display option by clicking a button in the Look Up Collection options window.** You can display All Collection Records (a slow process if there are hundreds or thousands of records) or restrict the choices displayed to a certain value for Locality Code, Collector (the Collected By field), Site, Source, or Method. Or, you can launch the Search Editor (pp. 192–200) to find a record using ad hoc criteria.

> **NOTE:** If you have defined an Alias for the [Collection] Site, [Collection] Source, or [Collection] Method fields (pp. 323–326), the Alias will appear in the corresponding button text in the Look Up Collection window in place of the Internal Field Name.

b. **If you select the Locality, Collector, Site, Source, or Method option,** Biota will load all unique values for selected field in the popup list to the right of the selected button.

  i. **Click and hold the popup to display the list of values.**

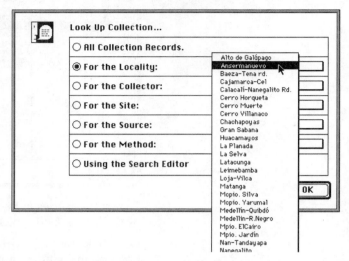

  ii. **Select the value you want from the popup list.** Biota will show the selected value in the popup area.

c. **Click the OK button in the Look Up Collection window.** The Look Up Collection display screen will appear, listing all Collection records that match the option and criteria you selected in steps a and b above.

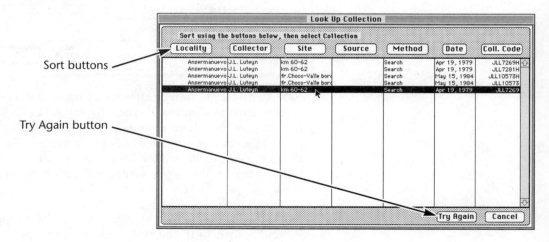

◊ **If there are many records in the display screen,** you may first want to use the sort buttons at the top of each column to sort the records displayed by the values for that column.

◊ **If you want to try a different selection option or field value** to display records, click the Try Again button at the bottom of the Look Up Collection selection screen.

> **WARNING (Macintosh only):** If the list of Collections is longer than will fit in the window, the vertical scroll bar will be enabled. To move up and down the list, use only the arrows at the top and bottom of the scroll bar or the "thumb-slider" box inside the scroll bar. Due to a bug in 4D, clicking above or below the slider box in the gray area of the scroll bar displays the first or last screen of values, not the next or previous screen of values. (This bug does not affect the Windows version.)

d. **When you find the record you want in the Lookup Collection display screen,** click it to enter its Collection Code in the Specimen input screen.

e. **To see the full Collection record,** once the Collection Code has been entered, click the Full Record button in the Collection input panel.

4. **Record who prepared the specimen, who last determined it, and when.**

   The techniques for entering data in the entry areas for specimen preparation and specimen determination (and their dates) are identical. Because information on specimen determination is more frequently entered, determination will be used as an example.

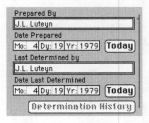

> **NOTE:** You can set Field Value Default entries for the entry areas Prepared By, Date Prepared, Last Determined by, and Date Last Determined. See pp. 313–315.

a. **Enter the Determiner's Short Name in the Last Determined By entry area.** Because this field is linked to the Short Name field of the Personnel table, the entry area has a double border. You can use all the usual techniques for entering data in linking fields described in Chapter 6 (pp. 92–95).

◊ **If you know that the Determiner has no existing record in the Personnel table,** enter a Short Name for the Determiner (e.g., Albert B. Smith or A. B. Smith) and press the TAB key. The name may have up to 25 characters.

NOTE: On Pin, Slide, Vial, and Herbarium labels, Biota uses the Short Name field for the Collector's name (p. 173). Since Determiners are very often also Collectors and Preparators, be sure to choose a version of the Determiner's name that is appropriate for this purpose—e.g., initials and last name, or first name and last name. You can also use a Group Name in the Determined By field, e.g., *A. B. Smith & C. D. Jones* (pp. 174–177).

◊ **If you think the Determiner may already have a Personnel record, but you are not sure what Short Name the record carries,** enter the @ character and press the TAB key. A scrollable list of all existing Personnel records will appear (below), from which you can choose the correct record by clicking it. You can sort the records by Short Name, Last Name, or Institution by clicking the buttons above the columns.

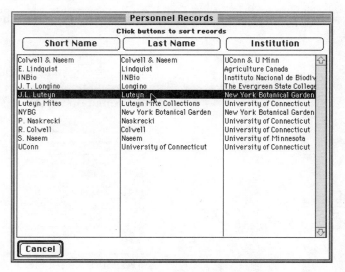

◊ **If you know the Short Name for the Determiner's existing Personnel record,** enter the first letter, the first few letters of the Short Name, or the full Short Name, and press the TAB key. If a single match is found, the name will be completed and the cursor will move into the next field. If more than match is found, a choice list like the one above will appear, with all matches shown.

If the Determined By entry you make does not match the Short Name field of any existing Personnel record, Biota will offer you a choice between creating a new Personnel record for that name, "on the fly," or accepting the Specimen record without creating a link to the Personnel table (pp. 92–95) (a "Personnel-orphan" Specimen record).

Note: If you are working with historical specimens, you may not want to create a Personnel record for every Determiner, Preparator, or Collector—some of whom may be long deceased and their particulars unknown.

b. **Enter the Date Last Determined.** You can enter either a full date (entered as MM/DD/YYYY), a Month-Year date (entered as MM/00/YYYY), or a Year-only date (entered as 00/00/YYYY). See pp. 112–113 for details on how Biota handles partial dates. (Date Prepared also accepts partial dates.)

  i. **Enter the Month number** in the "Mo" entry area and press TAB—or just press TAB without entering a number for a Year-only date.

  ii. **Enter the Day number** in the "Dy" entry area and press TAB—or just press TAB without entering a number for a Month-Year or Year-only date.

  iii. **Enter the Year number** in the "Yr" entry area and press TAB. If you enter a two-digit number "YY," Biota will complete the Year entry as "19XX." To enter a date for another century, enter all four digits.

  iv. **To enter Today's date, click the Today button next to the date entry area.** The full date from your computer's internal clock is entered. (See pp. 91–92 for keyboard equivalents for the Today buttons.)

NOTE: The Determination History button is not enabled in a new Specimen record. The button becomes active when an existing determination is changed. See Chapter 19.

5. **Enter data for Stage/Sex, Medium, Storage, Deposited At, and Type Status.** Enter each value and press TAB key to advance to the next entry area.

You can set any of the following options for each of these five Specimen fields:

- **Aliases.** Each field can be renamed using the Core Field Alias tool from the Special menu (pp. 323–326).
- **Field Value Defaults.** You can ask Biota to enter a particular value in each new record automatically. (Special menu; pp. 313–315.)
- **Entry Choice Lists.** Each field can have an Entry Choice List enabled (Special menu; pp. 315–322).

6. **Enter a value for Abundance.** Biota enters a "1" as the default for the Abundance field.

- **If this record is for a single physical specimen,** leave the value at 1. (The entry area then reads "Voucher for 1"—the record is a "voucher" for one specimen). Some Biota tools sum the values in this field for sets of specimen records, so a record representing a single physical specimen should read "1" for Abundance.

- **If this record represents more than one physical specimen,** enter the total number it represents. If the record for one physical specimen represents n additional, unmounted specimens (a physical voucher), enter the value $n + 1$.

7. **Enter any Specimen Notes or Specimen Auxiliary Values.** See pp. 179–183 for details on entering Notes, and Chapter 15 for information on Auxiliary Fields.

8. **Accept or Carry the Specimen record.** See pp. 41–42 for information on how the Carry button works.

Collection Input

You can enter a new record in the Collection input screen by selecting Collection from the Input menu, by clicking the New Collection button on the Collection output (record listing) screen, or "on the fly" from the Specimen input screen.

The illustration below shows the default order (p. 90) for entry areas in the Collection input screen. Press the TAB key to move between entry areas in default order, or click in any entry area to move to it directly.

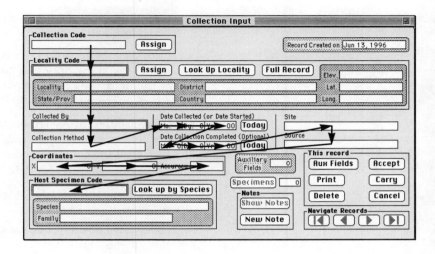

Entering Collection Data: Step by Step

1. **Enter a Collection Code.** You can enter the Code either by typing it in the Collection Code entry area or by using the adjacent Assign button (pp. 102–103). Collection Code, the Key field for the Collection table, is an obligatory entry, which must be made first. All remaining entry areas are optional. The Collection Code you enter must not duplicate the Collection Code of any existing Collection Record, or you will receive an error message (p. 7).

NOTE: If you are creating a Collection record "on the fly" while working on a Specimen record (pp. 92–95), the Collection Code you entered in the Specimen input screen will already be displayed and the cursor will be in the next field (Locality Code).

2. **Enter a Locality Code.** The Locality Code links this Collection record with a record in the Locality table (pp. 160–164). There are four *alternative* ways to enter a Locality Code:

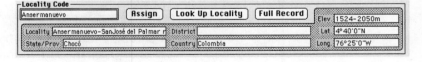

- *Either:* **Enter a complete Locality Code manually.**
   ◊ **If a Locality record exists with the Code you entered,** the stippled Locality display area will show the Locality (the [Locality] Locality Name field) and five other fields from the related Locality record, and the cursor will move on to the next entry area. (Be sure to confirm that correct Locality record was found. If your

entry uniquely matches the first part of a different Code, the wrong link will be made.)

◊ **If no Locality record exists with the Code you entered,** Biota will offer you the option of creating a new Locality record "on the fly," with the Code you entered, or accepting an orphan Collection record. See pp. 92–95 for details.

♦ *Or:* **Use the adjacent Assign button to create a new Locality Code.** Biota will offer you the option of creating a new Locality record "on the fly," with the Code you entered, or accepting an orphan Collection record. See p. 94 for details.

♦ *Or:* **Enter the first letter or first few letters of a Locality Code, then press** TAB (wildcard lookup).

◊ **If only one Locality Code matches the entry,** the entry will automatically be completed.

◊ **If more than one existing Code matches the entry,** a scrollable list of matching Codes for existing Locality records will appear, from which you can choose the correct code by clicking it (p. 96). (If you enter the @ character, and nothing else, all Locality codes will appear in the scrollable list.)

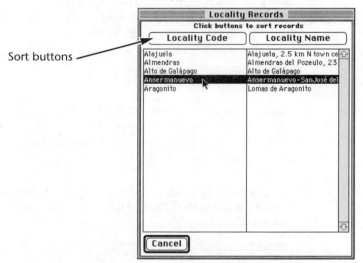

Sort buttons

To sort the entries in the list by Locality Name or resort by Locality Code, click the sort buttons at the top of the screen.

◊ **If no existing Locality Code matches your entry,** Biota will offer you the option of creating a new Locality record "on the fly," with the Code you entered, or accepting an orphan Collection record. See p. 94 for details.

♦ *Or:* **Click the Look Up Locality button.** The Look Up Locality options window appears.

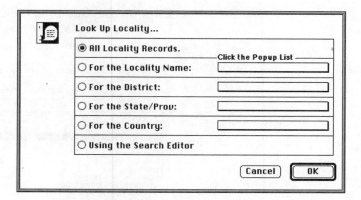

a. **Select a display option by clicking a button in the Look Up Locality options window.** You can display All Locality Records (a slow process, if there are many Locality records in the database) or restrict the choices displayed to a certain value for Locality Code, State/Province, or Country. Or, you can launch the Search Editor to find a record using ad hoc criteria.

NOTE: If you have defined an Alias for the [Locality] State/Province, or [Locality] Country fields (pp. 323–326), the Alias will appear in the corresponding button text in the Look Up Locality window in place of the Internal Field Name.

b. **If you select the Locality, State/Province, or Country option,** Biota will load all unique values for the selected field in the popup list to the right of the selected button.

   i. **Click and hold the popup to display the list of values.**

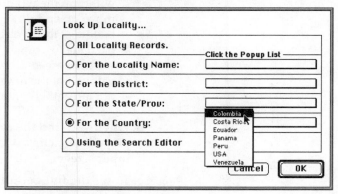

   ii. **Select the value you want from the popup list.** Biota will show the selected value in the popup area.

c. **Click the OK button in the Look Up Locality window.** The Look Up Locality display screen will appear, listing all Locality records that match the option and criteria you selected in steps a and b above.

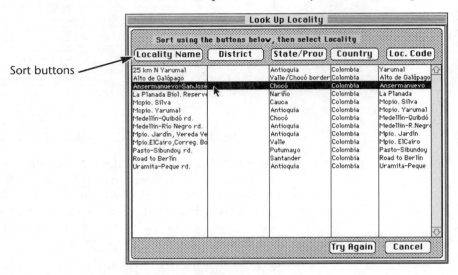

◊ **If there many records in the display screen,** you may first want to use the sort buttons at the top of each column to sort the records displayed by the values for that column.

◊ **If you want to try a different selection option or field value** to display records, click the Try Again button at the bottom of the Look Up Locality selection screen.

WARNING (Macintosh only): If the list of Localities is longer than will fit in the window, the vertical scroll bar will be enabled. To move up and down the list, use only the arrows at the top and bottom of the scroll bar or the "thumb-slider" box inside the scroll bar. Due to a bug in 4D, clicking above or below the slider box in the gray area of the scroll bar displays the first or last screen of values, not the next or previous screen of values. (This bug does not affect the Windows version.)

d. **When you find the record you want in the Lookup Locality display screen, click it to enter its Locality Code in the Collection input screen.**

e. **To see the full Locality record,** click the Full Record button in the Locality input panel.

3. **Enter data on who made the Collection, by what Method, and when.**

NOTE: You can set Field Value Default entries for the entry areas Collected By, Collection Method, Date Collected, and Date Collection Complete. See pp. 313–315. You can activate a Entry Choice List for Collection Method. See pp. 315–322.

a. Enter the Collector's Short Name in the Collected By entry area.
Because this field is linked to the Short Name field of the Personnel table, the entry area has a double border. You can use all the usual techniques for entering data in linking fields described in Chapter 6.

◊ **If you know that the Collector has no existing record in the Personnel table,** enter a Short Name for the Collector (e.g., Albert B. Smith or A. B. Smith) and press the TAB key. The name may have up to 25 characters.

NOTE: On Pin, Slide, Vial, and Herbarium labels, Biota uses the Short Name field for the Collector's name (p. 173). Be sure to choose a version of the Collector's name that is appropriate for this purpose—e.g., initials and last name, or first name and last name. You can also use a Group Name in the Collected By field, e.g., "A. B. Smith & C. D. Jones." Group names link records for individuals in the Personnel table. See pp. 174–177 for information on how to use Group Names.

◊ **If you think the Collector may already have a Personnel record, but you are not sure what Short Name the record carries,** enter the @ character and press the TAB key. A scrollable list of all existing Personnel records will appear, from which you can choose the correct record by clicking it. You can sort the records by Short Name, Last Name, or Institution by clicking the buttons above the columns.

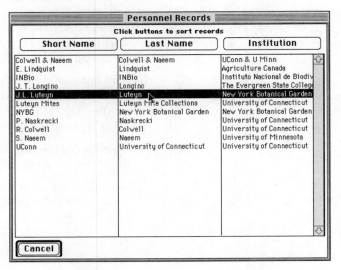

◊ **If you know the Short Name for the Collector's existing Personnel record,** enter the first letter, the first few letters of the Short Name, or the full Short Name, and press the TAB key. If a single match is found, the name will be completed and the cursor will move into the next entry area. If more than match is found, a choice list like the one above will appear, with all matches shown.

If the Collected By entry you make does not match the Short Name field of any existing Personnel record, Biota will offer you a choice between creating a new Personnel record "on-the-fly," for that name, or accepting the Collection record without creating a link to the Personnel table (pp. 92–95) (a "Personnel-orphan" Collection record).

NOTE: If you are working with historical Collections, you may not want to create a Personnel record for every Collector—some of whom may be long deceased and their particulars unknown.

b. **Enter the Collection Method.** You can set any of the following options for the [Collection] Method, [Collection] Site, and [Collection] Source fields:

◊ **Aliases.** Each field can be renamed using the Core Field Alias tool from the Special menu (pp. 323–326).

◊ **Field Value Defaults.** You can ask Biota to enter a particular value in each new record automatically. (Special menu; pp. 313–315.)

◊ **Entry Choice Lists.** Each field can have an Entry Choice List enabled (Special menu; pp. 315–322).

c. **Enter the Date Collected and Date Collection Completed.** To enter a single collection date, use the Date Collected entry area and leave the Date Collection Completed entry area blank ("TAB though" it). To enter a Date Range, use both areas. See Chapter 8 (pp. 113–114) for details on how Biota handles Date Ranges.

You can enter either a full date (entered as MM/DD/YYYY), a Month-Year date (entered as MM/00/YYYY), or a Year-only date (entered at 00/00/YYYY) in either date entry area. See Chapter 8 (pp. 112–113) for details on how Biota handles partial dates.

i. **Enter the Month number** in the "Mo" entry area and press TAB—or just press TAB without entering a number for a Year-only date.

ii. **Enter the Day number** in the "Dy" entry area and press TAB—or just press TAB without entering a number for a Month-Year or Year-only date.

iii. **Enter the Year number** in the "Yr" entry area and press TAB. If you enter a two-digit number "YY," Biota will complete the Year entry as "19XX." To enter a date for another century, enter all four digits.

iv. **To enter Today's date, click the Today button next to the date entry area.** The full date from your computer's internal clock is entered. (See pp. 91–92 for keyboard equivalents for the Today buttons.)

4. **Enter data for Site and Source.** You can change the name of these fields using Aliases (pp. 323–326), set Field Value Defaults (pp. 313–315), or establish Entry Choice Lists (pp. 315–322).

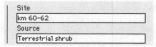

5. **Enter Coordinate data and a code or description for the Accuracy of the Coordinates.**

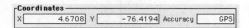

The X Coordinate and Y Coordinate fields (which are numeric) can be used for a variety of purposes. If you use the Latitude, Longitude, and Lat Long Accuracy fields of the Locality table for low-resolution coordinates from physical maps or gazetteers, you can use the X Coordinate, Y Coordinate, and XY Accuracy fields of the Collection table for GPS readings pinpointing precise collection sites—of which there may be many for each Locality record. The X Coordinate and Y Coordinate fields can also be used for grid positions at local sites or for quadrat-based studies. See Chapter 8, pp. 115–116, for further discussion.

6. **If the Collection site includes Host data,** enter the Specimen Code for the Host. See Chapter 21 for details on Biota's Host-Guest system before using this field.

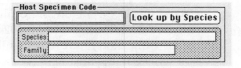

7. **Enter any Collection Notes or Collection Auxiliary Values.** See pp. 179–183 for details on entering Notes, and Chapter 15 for information on Auxiliary Fields.

8. **Accept or Carry the Collection record.** See pp. 41–42 for information on how the Carry button works.

160   Biota: CHAPTER 10

Locality Input

You can enter a new record in the Locality input screen by selecting Locality from the Input menu, by clicking the New Locality button on the Locality output (record listing) screen, or "on the fly" from the Collection input screen.

The illustration below shows the default order (p. 90) for entry areas in the Locality input screen. Press the TAB key to move between entry areas in default order, or click in any entry area to move to it directly.

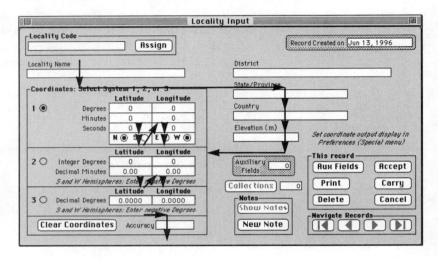

Entering Locality Data: Step by Step

1. **Enter a Locality Code.** You can enter the Code either by typing it in the Locality Code entry area or by using the adjacent Assign button (pp. 102–103). Locality Code, the Key field for the Locality table, is an obligatory entry, which must be made first. All remaining entry areas are optional. The Locality Code you enter must not duplicate the Locality Code of any existing Locality Record, or you will receive an error message (p. 7).

NOTE: If you are creating a Locality record "on the fly" while working on a Collection record (pp. 92–95), the Locality Code you entered in the Collection input screen will already be displayed and the cursor will be in the next field (Locality Name).

2. **Enter Locality Name.** Often, this entry will be an official geographical place name, from a geographical authority list or gazetteer—up to

50 characters in length. The Locality Code (step 1), on the other hand, can be any unique alphanumeric value (p. 100), but in the absence of official locality codes (which do exist in some countries), you may find it convenient to use a short form (up to 20 characters) of the Locality Name, as in the illustration here. (Compare the illustration above with the one below.)

```
Locality Name
Ansermanuevo-San José del Palmar rd.
```

3. **Enter District, State/Province, Country, and Elevation.**

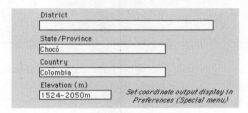

You can set any of the following options for each of these four Locality fields—as well as for the Locality Name field (step 2 above):

- **Aliases.** Each field can be renamed using the Core Field Alias tool from the Special menu (pp. 323–326). For example, you may want to rename the District field *County* or the Elevation field *Depth*.

- **Field Value Defaults.** You can ask Biota to enter a particular value in each new record automatically. (Special menu; pp. 313–315.)

- **Entry Choice Lists.** Each field can have an Entry Choice List enabled (Special menu; pp. 315–322).

NOTE: The [Locality] Elevation field is an alphanumeric, not a numeric, field. This design allows elevation ranges (as illustrated above) and nonmetric units to be included in the entry. If you enter a numeral, with no units, Biota will assume that the elevation is in meters in printing specimen locality labels and will append "m" to the elevation (Chapter 13). If you use feet instead of meters, include the abbreviation "ft" (or anything with the letter "f" in it) and Biota will print the value on locality labels just as you entered it, without adding "m." You may want to change the default Alias for the [Locality] Elevation field from "Elevation (m)" to "Elevation (ft)" if all your elevation data are in feet. See pp. 323–326.

4. **Enter the Latitude and Longitude (and their Accuracy) for the Locality.**

Biota offers three unit systems for the entry and display of latitude and longitude values. You can enter data using any of the three systems and freely use different systems for different records in the same data file. As you make an entry in any of the systems, Biota instantly translates the value into the other two and displays all three versions in the Coordinates panel of the Locality input screen.

NOTE: Biota validates entries for the Latitude and Longitude fields of the Locality table. Latitude may not be less than –90 nor greater than +90, and Longitude may not be less than –180 nor greater than +180, or the equivalents in degree/minutes/seconds. Values for the XCoordinate and YCoordinate fields of the Collection table (p. 159), on the other hand, are not validated; they may take any numeric value, negative or positive. Thus, you can use the XCoordinate and YCoordinate fields for map projection systems, such as Lambert or UTM, or for arbitrary grid systems.

a. **System 1: Degree/Minutes/Seconds.** This is the traditional system found in most gazetteers and atlases. Be sure to see the important warning (under System 2) before using values from a gazetteer.

   i. Click the "1" button in the Coordinates panel to select System 1.
   ii. **Enter Degrees, Minutes, and Seconds of Latitude,** pressing TAB after each entry.
   iii. **Continue with Degrees, Minutes, and Seconds of Longitude.**
   iv. **Click the Hemisphere buttons (N or S, E or W) if necessary.** (The default Hemispheres are N and W, reflecting Biota's heritage.)

b. **System 2: Integer Degrees, Decimal Minutes.** This is the system used by most Geographical Positioning Units (GPUs). See pp. 116–117 for details.

  i. **Click the "2" button in the Coordinates panel to select System 2.**

  ii. **Enter Integer Degrees and Decimal Minutes of Latitude,** pressing TAB after each entry. The Integer Degree value must be negative for Latitudes in the Southern Hemisphere.

  iii. **Continue with Integer Degrees and Decimal Minutes of Longitude.** The Integer Degree value must be negative for Longitudes in the Western Hemisphere.

WARNING: Beware! Some gazetteers list Latitude and Longitude values in DMS, but use a *format* that looks like Integer Degrees with Decimal Minutes, e.g., 24° 48.39 actually means 24° 48' 39" in these gazetteers. The easiest way to detect this abomination is to see if you can find any values in the gazetteer to the right of the "decimal point" that exceed 59. If not, those values are Seconds, not the decimal part of true Decimal Minutes.

c. **System 3: Integer Degrees, Decimal Minutes.** This is the system used in Geographical Information Systems (GPS's). See pp. 116–117 for details.

  i. **Click the "3" button in the Coordinates panel to select System 3.**

  ii. **Enter Decimal Degrees of Latitude** and press TAB. The value must be negative for Latitudes in the Southern Hemisphere.

  iii. **Enter Decimal Degrees of Longitude.** The value must be negative for Longitudes in the Western Hemisphere.

d. **Once the Coordinates have been entered, enter a code or description for the Accuracy of the Coordinates.**

NOTE: If you use the Latitude, Longitude, and Lat Loc Accuracy fields of the Locality table for low-resolution coordinates from physical maps or gazetteers, you can use the X Coordinate, Y Coordinate, and XY Accuracy fields in Collection records for GPS readings that pinpoint precise collection sites—of which there may be many for each Locality record. See pp. 115–116 for further discussion.

e. **To clear all Coordinate entry areas,** click the Clear Coordinates button at the bottom of the Coordinates panel.

5. **Enter any Locality Notes or Locality Auxiliary Values.** See pp. 179–183 for details on entering Notes, and Chapter 15 for information on Auxiliary Fields.

6. **Accept or Carry the Locality record.** See pp. 41–42 for information on how the Carry button works.

Species Input

You can enter a new record in the Species input screen by selecting Species from the Input menu, by clicking the New Species button on the Species output (record listing) screen, or "on the fly" from the Specimen input screen.

The illustration below shows the default order (p. 90) for entry areas in the Species input screen. Press the TAB key to move between entry areas in default order, or click in any entry area to move to it directly.

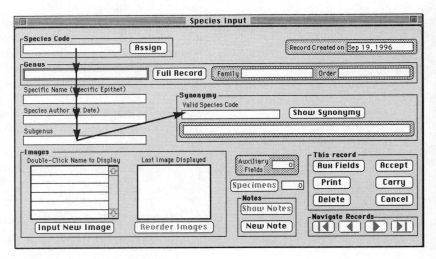

Entering Species Data: Step by Step

1. **Enter a Species Code.** You can enter the Code either by typing it in the Species Code entry area, by using the adjacent Assign button (pp. 102–103), or by reading in a barcode (pp. 105–108). Species Code, the Key field for the Collection table, is an obligatory entry, which must be made first. All remaining entry areas are optional. The Species Code you enter must not duplicate the Species Code of any existing Species Record, or you will receive an error message (p. 7).

NOTE: If you are creating a Species record "on the fly" while working on a Specimen record (pp. 92–95), the Species Code you entered in the Specimen input screen will already be displayed and the cursor will be in the next field (Genus).

2. **Enter the Genus.**

Because the [Species] Genus field is linked to the Genus field of the Genus table (Appendix A), the entry area has a double border. You can use all the techniques for entering data in linking fields described in Chapter 6 .

- **If you know that the Genus name you intend to enter has no existing record in the Genus table,** enter a the new Genus name and press TAB key. The name may have up to 40 characters.

- **If you think the Genus may already have a Genus record,** enter the first letter, the first few letters of the Genus, or the full Genus, and press the TAB key. If a single match is found, the name will be completed and the cursor will move into the next entry area (Specific Name). If more than one match is found, a scrollable list with matching values for existing Genus records will appear, from which you can choose the correct record by clicking it (p. 96). (If you enter the @ character, and nothing else, *all* Genus records will appear in the scrollable list.)

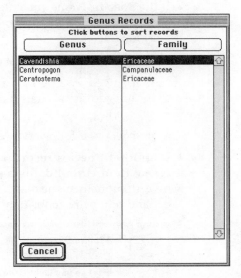

If the Genus entry you make does not match the Genus field of any existing Genus record, Biota will offer you a choice between creating a new Genus record for that name, "on-the-fly," or accepting the Species record without creating a link to the Genus table (pp. 92–95) (an orphan Species record).

NOTE: In two special circumstances, you might need to create a new Genus record that legitimately duplicates an existing Genus name: (1) when genera in two kingdoms have the same name and (2) when you want to record a generic synonym. See "A Special Problem for Genera: Legitimate Duplicate Generic Names" (pp. 169–170) if either case arises.

3. **Enter Specific Name (Specific Epithet).**

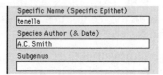

When you enter a Specific Name (the Species Name field of the Species table), Biota checks for identical Species Name entries in existing records, *disregarding Genus*. The purpose of this check is to prevent the accidental creation of two or more Species records for the same species (albeit with distinct Species Codes). The reason for disregarding Genus is to take into account changed or alternative generic concepts.

- **If the Specific Name you entered is unique among Species records in the database,** Biota will accept your entry and move on to the next entry area (Species Author).

- **If the Specific Name you entered is any variation on *sp.*,** Biota will accept your entry whether or not it is unique. Entries in this category include:
  ◊ *sp* without any additional characters
  ◊ *sp* followed by a period or a space character, alone or followed by any additional characters, including numbers (e.g., *sp 1, sp. 3, sp. nov.*)
  ◊ *n. sp* alone or followed by any additional characters

- **If another Species record with the same value in the Species Name field is found,** Biota presents the option window shown here. (If more than one match is found, the message gives the number and omits the genus name.)

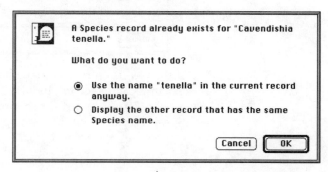

◊ **If you select the first option (the default),** Biota will accept your entry and move on to the next entry area (Species Author).

◊ **If you select the second option,** Biota will accept your entry and move on to the next entry area, but a new window will open showing the existing Species records that share the same Species Name value. Based on this information, it is then up to you whether to modify the Species record you are working on, Delete it, or Accept it as is.

◊ **If you click Cancel** in the option window, the entry you made in the Specific Name entry area will be cleared.

You can enable an Entry Choice List for the Specific Name entry area (Special menu; pp. 315–322), although it is difficult to think of a use for it, since most entries will not be repetitive.

4. **Enter Species Author (and Date).** The Species Author field is an ordinary alphanumeric field (up to 50 characters), not linked to the Personnel table. The date (usually only the year), if you include one, is part of the entry in the Species Author field and can therefore be in any format, without reference to Biota date formats (pp. 111–114).

   You can enable an Entry Choice List for the Species Author entry area (Special menu; pp. 315–322). A Choice List can speed up data entry when certain authors (and dates) apply to a many Species records. However, if you are entering a list of species all described in the same publication (same entry in the Species Author field), using the Carry button (pp. 41–42) is more efficient than using an Entry Choice List.

5. **Enter Subgenus.** In Biota's structure, Subgenus is an attribute (field) of the Species table. See pp. 109–111 on how Biota handles Intermediate Taxonomic levels (ranks).

   You can enable an Entry Choice List for the Subgenus entry area (Special menu; pp. 315–322).

   You can rename the Subgenus field (and entry area) using a Field Name Alias (Special menu, pp. 323–326).

6. **If the Species record is for a junior synonym,** click the Show Synonymy button in the Species input screen and follow the instructions in Chapter 20 to set up the synonymy.

7. **If you want to include images for this Species record,** click the Input New Image button in the Species input screen and follow the instructions on in Chapter 18, where image input is explained in detail.

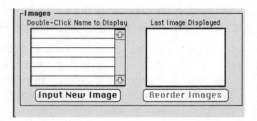

8. **Enter any Species Notes or Species Auxiliary Values.** See pp. 179–183 for details on entering Notes, and Chapter 15 for information on Auxiliary Fields.

9. **Accept or Carry the Species record.** See pp. 41–42 for information on how the Carry button works.

Genus, Family, Order, Class, Phylum, and Kingdom Input

Data entry in the higher taxon (Genus, Family, Order, Class, Phylum, and Kingdom) input screens is virtually identical for all levels. You can enter a new record in one of these screens by selecting the level from the Input menu, by clicking the New Record (New Genus, New Family, etc.) button on the output (record listing) screen for that level, or "on the fly" from the input screen of a child table.

Differences among the Higher Taxon Input Screens

The input screens for Genus, Family, and Order are identical in layout. The illustration below shows the default order (p. 90) for entry areas in the Genus input screen, as an example. Press the TAB key to move between entry areas in default order, or click in any entry area to move to it directly.

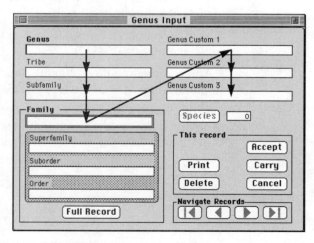

In the Genus input screen, there are input areas for two intermediate taxon fields—Tribe and Subfamily—and for three Custom fields—Genus Custom 1 through 3. Likewise, the Family and Order input screens also have entry areas for two intermediate taxon levels (Superfamily and Suborder in the Family input screen; Superorder and Subclass in the Order input screen) and three Custom fields (labeled Family Custom 1–3 and Order Custom 1–3).

The input screens for Class and Phylum differ from this design by offering only a single intermediate taxon field (Subphylum in the Class input screen and Subkingdom in the Phylum input screen) and two instead of three Custom fields. The Class input screen is shown below as an example, with the default order (p. 90) for entry areas indicated.

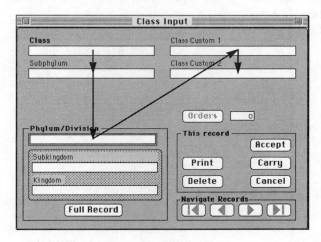

The input screen for the Kingdom table, which has no parent table (Appendix A) is simplified by the lack of the parent table panel at the lower left, but is otherwise like the Class and Phylum input screens. (The single intermediate taxon level is labeled Superkingdom.)

A Special Problem for Genera: Legitimate Duplicate Generic Names

The Genus field of each Genus record must contain a unique value within a Biota Data File. There are two cases in which Biota's use of generic names as the Key field (pp. 6–7) for the Genus table may cause problems.

First, although the rules of nomenclature require unique names for Genera within Kingdoms, a number of cases exist in which the same name is legitimately used in two Kingdoms. For example, *Ammophila* and *Dryas* are each valid genera of both plants and animals. The uniqueness problem arises only if you need to use both names in the same Biota Data File. Aside from general cross-Kingdom databases, host-parasite or herbivore-plant databases provide potential examples.

Second, if you create Genus records to record generic junior synonyms, you will need to handle duplicate names in the Genus field.

In these cases, you will need to create two (or more) Genus records, with distinct values in the Genus field. For synonyms within a Kingdom, appending the genus author makes sense. For valid genera in different Kingdoms, you can add a an asterisk (*) or a pound sign (#) to distinguish the two records (e.g., *Dryas* and *Dryas** or *Dryas#* and *Dryas**). Or, you could add an explanatory suffix: *Dryas-Plant* and *Dryas-Butterfly*.

With this approach in view, Biota looks for the characters * and # in Genus names when printing determination labels and deletes the characters from the printed version (Chapter 13).

Entering Data in a Higher Taxon Input Screen: Step by Step

In this section, the Genus table will be used as an illustration. Input for the other higher taxon tables is precisely analogous.

1. **Enter the taxon name** (a Genus name, in this example). Genus, the Key field for the Genus table, is an obligatory entry, which must be made first. All remaining entry areas are optional. The Genus name you enter must not duplicate the Genus entry of any existing Genus Record, or you will receive an error message (p. 7).

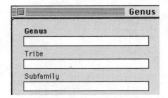

NOTE **(Genus input only):** In two special circumstances, the requirement that each Genus record have a unique value for the Genus field may cause problems: (1) when genera in two kingdoms have the same name and (2) when you want to record generic junior synonyms. Please read the previous section (pp. 169–170) if either of these situations arises.

2. **Enter intermediate taxon name or names** (Tribe and Subfamily, in the Genus input screen).

   You can enable an Entry Choice List for each intermediate taxon entry area (Special menu; pp. 315–322).

   You can rename each intermediate taxon field (and entry area) using a Field Name Alias (Special menu, pp. 323–326).

3. **Enter the parent taxon name** (the Family name, in the Genus input screen). Because the [Genus] Family field is linked to the Family field of the Family table (Appendix A), the entry area has a double border. You can use all the techniques for entering data in linking fields described in Chapter 6.

NOTE: This step does not apply to the Kingdom input screen, since the Kingdom table has no parent.

- **If you know that the Family name you intend to enter has no existing record in the Family table,** enter the new Family name and press TAB key. The name may have up to 40 characters.

- **If you think the Family may already have a Family record,** enter the first letter, the first few letters of the Family, or the full Family, and press the TAB key. If a single match is found, the name will be completed and the cursor will move into the next entry area (the first Custom field). If more than one match is found, a scrollable list with matching values for existing Family records will appear, from which you can choose the correct record by clicking it (p. 96). (If you enter the @ character, and nothing else, *all* Family records will appear in the scrollable list.)

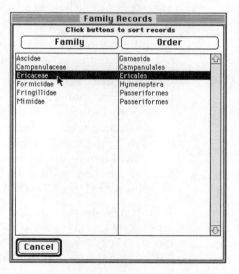

If the Family entry you make does not match the Family field of any existing Family record, Biota will offer you a choice between creating a new Family record for that name, "on the fly," or accepting the Genus record without creating a link to the Family table (pp. 92–95) (an orphan Genus record).

4. **Enter data for the Custom fields.**

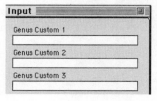

You can enable an Entry Choice List for each Custom field. (Special menu; pp. 323–326).

You can rename each Custom (and entry area) using a Field Name Alias (Special menu, pp. 313–315).

5. **Accept or Carry the record.** See pp. 315–322 for information on how the Carry button works.

Personnel and Project Name Input

Through relational links, Biota uses records in the Personnel table to register the names of specimen collectors, preparers, determiners, and loan borrowers, as well as authors of Specimen, Species, Collection, Locality, and Loan Notes (Appendix A). But Personnel records need not be linked to other records. You can also use the Personnel table as an address book for colleagues and institutions you contact frequently, if you wish.

The Personnel table holds three kinds of records:

- **Individual records.** An Individual Personnel record generally registers information for one person, but you can also create an Individual record in the Personnel table for an institution or company.

- **Group records.** A Group record in the Personnel table links a set of Individual personnel records. A collecting team needs a Group record linking the Individual Personnel records of its members, so that all names in the group are linked to the appropriate Collection records (pp. 156–158). Or, you might link all the individual records for your contacts at a museum or herbarium to a Group record for the institution. (You can transform an Individual record for an institution into a Group record, pp. 174–177.)

- **A Project Name record.** One record in the Personnel table is reserved for a name you give your database, with associated address and contact data (pp. 177–179). The Notes field in the Project Name record has a special function in printing herbarium specimen labels and Specimen Loan Invoices (p. 179).

You can enter a new Individual or Group record in the Personnel input screen by selecting Personnel from the Input menu; by clicking the New Personnel button on the Personnel output (record listing) screen; or "on the fly" from the Specimen, Collection, or Notes input screens. (See pp. 177–179 for instructions on entering the Project Name record.)

The illustration below shows the default order (p. 90) for entry areas in the Personnel input screen. Press the TAB key to move between entry areas in default order, or click in any entry area to move to it directly.

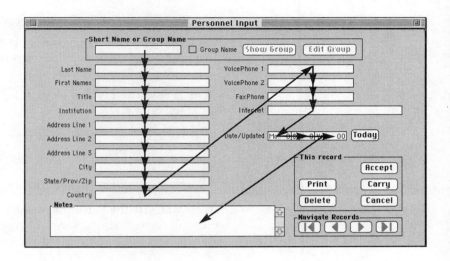

Entering Individual Personnel Records

1. **Enter a Short Name for the record.** Short Name, the Key field for the Personnel table, is an obligatory entry, which must be made first. All remaining entry areas are optional. The Short Name you enter must not duplicate the Short Name of any existing Personnel Record or you will receive an error message (p. 7).

   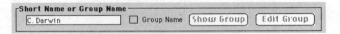

   What should you assign for a Short Name? The Short Name you assign appears in Specimen records in the Prepared By and Determined By fields, in Collection records in the Collected By field, and in Notes records in the Note By field. On pin, slide, and vial labels (pp. 246–254), and on herbarium labels with a single collector (pp. 255–256), Biota uses the Short Name field for the Collector's name. Be sure to choose a version of the each person's name that is appropriate for this purpose—e.g., initials and last name, or first name and last name. A Group Name is often appropriate in the Collected By field, e.g., "A. B. Smith & C. D. Jones." Group names can link records for individuals in the Personnel table. See pp. 174–177 for information on how to use Group Names.

   NOTE: If you are creating a Personnel record "on the fly" while entering data in the [Specimen] Prepared By (pp. 149–151), [Specimen] Determined By (pp. 149–151), [Collection] Collected By (pp. 156–158), or a Note By field (p. 180), the Short Name Code you entered in the Specimen input screen will already be displayed and the cursor will be in the next field (Last Name).

2. **Enter the surname of the individual in the Last Name entry area.** The Last Name field is not obligatory.

3. **Complete the rest of the Personnel record.**

    ♦ **The Updated/Date entry area** accepts only full dates (pp. 112–113). You can use the Today button to enter the current date (from the computer's internal clock).

    ♦ **The [Personnel] Notes field** accepts a single note (up to 32,000 characters).

4. **Accept or Carry the Personnel record.** The Carry button is particularly useful for entering a series of records for people who share an address. See pp. 41–42 for information on how the Carry button works.

Entering a New Group Personnel Record

A Group record in the Personnel table links sets of Individual Personnel records. You use the same input screen for creating both Group and Individual records (see the previous section). Once you have created a Group record, you use the Group Editor to establish the membership of the Group.

1. **Enter a Short Name for the Group record.** Short Name, the Key field for the Personnel table, is an obligatory entry, which must be made first. All remaining entry areas are optional. The Short Name you enter must not duplicate the Short Name of any existing Personnel Record, or you will receive an error message (p. 7).

    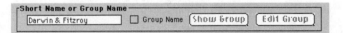

    What should you assign for a Group Short Name? If you use a Group record to link members of an institution, project team, or other organization, choose a Short Name for the Group that best identifies the organization. If you use a Group record to identify a group of individuals who were jointly responsible for collecting, preparing, or determining certain specimens or for writing a Note, form the Short Name just as you want it to appear on pin or vial collection labels; in the Collected By field of Collection records; in the Prepared By and Determined By fields of Specimen records; or in the Note By field of Notes records. On herbarium labels, a Group Record is used to create a list of collectors from the linked Individual records (pp. 225–226).

    NOTE: If you are creating a Group Personnel record "on the fly" while entering data in the [Specimen] Prepared By (pp. 149–151), [Specimen] Determined By (pp. 149–151), [Collection] Collected By (pp. 156–158), or a Note By field (p. 180), the Short Name Code you entered in the Specimen input screen will already be displayed and the cursor will be in the next field (Last Name).

2. **Click the Group Name checkbox** to declare this a Group record.

The Edit Group button is enabled when you check the Group Name box.

3. **Click the Edit Group button.** A query window appears.

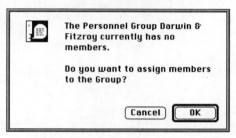

4. **Click the OK button** in the query window to proceed. The Group Editor appears.

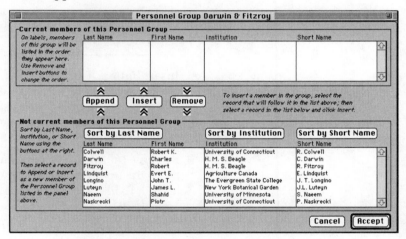

The Group Editor screen has two scrollable panels. For a new Group record, the lower panel (labeled "Not current members of this Personnel Group") initially displays all Individual Personnel records (all Personnel records for which the Group Name box is not checked).

You can sort the records in the lower panel by Last Name, Institution or Short Name using the buttons just above it.

The upper panel (labeled "Current members of this Personnel Group"), initially blank in a new Group record, will show the members of the new Group once you form it.

5. **Use the Append, Insert, and Remove buttons** in the Group Editor window to establish and order the members of the Group in the upper panel.

- **Append button.** To enter the first Individual record or add an Individual to the end of the "Current members" list in the upper panel, click the record you want in the lower panel to select it, then click the Append button. Notice that the record is "moved" to the upper panel—it is no longer listed in the lower panel.

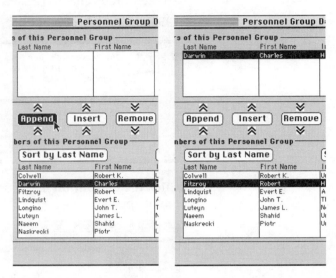

- **Insert button.** To place an Individual record between two records already listed in the upper panel, click the record in the upper panel that you want to follow the record to be inserted. Click the record in the lower panel that you want to insert, then click the Insert button.

- **Remove button.** To remove an Individual record from the Group list in the upper panel, click the record in the upper panel to highlight it, then click the Remove button. Notice that the record now appears in the lower panel.

6. **Click the Accept button in the Group Editor** to record the Group's composition and the order of its members. The Personnel input screen reappears with the Group record displayed. Notice that both the Show Group and Edit Group buttons are now enabled. The Group's Short Name has been entered automatically in the Last Name field (you can change it if you want) and the code "[GROUP NAME]" has been automatically entered in the First Name field for this record.

7. **Complete any additional fields (or none)** in the Personnel record for this Group. For an organizational Group record, you might record the address and contact information for the organization. For a Group record that links specimen co-collectors, preparators, or determiners, you would probably leave the rest of the record blank or possibly make an entry in the Notes field with details of the collaboration.

> NOTE: The entry "[GROUP NAME]," automatically placed in the First Name field, is not necessary for Biota to keep track of the Group . (The fact that this is a Group record is registered in the [Personnel] Group field.) The reason for the "[GROUP NAME]" entry is to allow you to identify Group records easily in the Personnel *output* screen. If this is not a concern, you can use the First Name field of Group records for other information if you wish.

8. **Accept the Group Personnel record.**

Displaying, Changing, or Reordering Group Membership for an Existing Group Personnel Record

1. **Display the Group record in the Personnel input screen.** (See pp. 121–123 if necessary.)

- **To display the members of the Group,** click the Show Group button in the Personnel input screen. The records will be displayed in a new window (a new process) in the Personnel output screen. Any number of Groups can be displayed simultaneously, each in its own window. The original Personnel output screen from which you "clicked open" the Group record(s) remains open also.

- **To change or reorder Group membership,** click the Edit Group button in the Personnel input screen. The Group Editor will appear, with the current Group members listed in the upper panel. Use the tools of the Group Editor, as described on pp. 174–176, to add, remove, or reorder members of the Group. When you are done, click the Accept button in the Group Editor window to record the changes and return to the Group Personnel record.

2. **If you made any changes, Accept the Group Personnel record.**

Entering the Project Name Record

The Project Name record is a special record in the Personnel table, which Biota identifies by the presence of the entry "[PROJECT NAME]" in the First Name field. There can be only one Project Name record in a particular Biota Data File.

1. **From the Input menu, select Project Name.** A query screen appears.

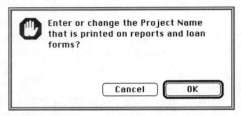

2. **Click OK in the query screen.** A special version of the Personnel input screen appears.

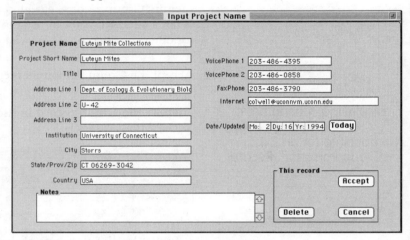

3. **Enter the name for your project, collection, organization, or institution in the Project Name entry area.** The name can have up to 30 characters. (This entry is recorded in the Last Name field of the Project Name record.)

    ♦ **The Project name appears as a title** on all standard printed Biota reports (Chapter 14).

    ♦ **If you use the Specimen Loan system** (Chapter 17), the Project Name and the address and contact information from the Project Name record will appear on all printed loan reports, in the "Lender" section (p. 334).

4. **Enter a Short Name for your project.** This entry cannot exceed 20 characters but it can be identical to the Project Name itself (above). Because Short Name is the Key field for the Personnel table, this an obligatory entry. All remaining entry areas are optional. The Short Name you enter must not duplicate the Short Name of any existing Personnel Record or you will receive an error message (p. 7).

5. **Complete the rest of the Project Name record.**

    ♦ **The Updated/Date entry area** accepts only full dates (pp. 112–113). You can use the Today button to enter the current date (from the computer's internal clock).

- **The [Personnel] Notes field** accepts a single note (up to 32,000 characters).

  Approximately the first 100 characters of the Notes field of the Project Name record are printed in full at the bottom of herbarium labels. See pp. 255–256 for details.

  If not blank, approximately the first 100 characters of the Notes field are used as the institutional/project heading on Specimen Loan Invoices. See p. 334.

6. **Accept the Project Name record.**

Notes Input

You can attach an unlimited number of Notes to each record in the Species, Specimen, Collection, Locality, and Loans tables.

For each of these tables, a special Notes table (linked as a child table to the Species, Specimen, Collection, Locality, or Loans table) accommodates these Notes records (Appendix A).

This structure means you can import and export Notes records in the same way you would any other records, using the Import (Chapter 26) and Export Editors (pp. 434–443). In addition, the special Export Notes tool exports certain fields from each parent record, along with full records for each Note (pp. 443–448).

When you create a new record in the Species, Specimen, Collection, Locality, or Loans table, it has no Notes. When you delete a record from one of these five tables (or Cancel a new record not yet Accepted), any associated Notes you have created for it are deleted automatically (after Biota's warning and your confirmation), since they would otherwise be "orphaned."

The Personnel table accommodates notes in a simpler way. Each Personnel record has a Notes *field*. This means you can enter or update just one note for each Personnel record. The note can be up to 32,000 characters long. The Personnel Notes field is treated elsewhere (p. 174).

Entering a New Note

This section will use Collection Notes screens as an example, but all procedures are identical and all screens are analogous for Locality, Specimen, Species, and Loans Notes.

1. **In a new or existing Collection, Locality, Specimen, Species, or Loans record, click the New Note button.**

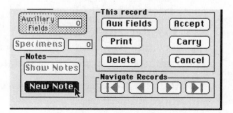

The Notes screen appears, with the insertion point (blinking cursor) in the Note By entry area and today's date (the date set in your computer's internal clock) appearing in the Note Date entry area.

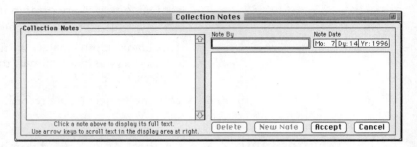

This single screen is used not only for creating new Notes and for displaying and editing existing individual Notes, but also for listing Notes, in the same way an output screen lists records for Core tables.

Note: If you click in the Core table input screen (the Collection input screen, in the illustrations) while the Notes window is open, the cursor will change to the "hidden window" cursor.

To return to the Notes window, press any key on the keyboard. (The keystroke is not recorded.)

2. **Enter the Note author's Short Name in the Note By entry area.**

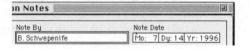

Because this field is linked to the Short Name field of the Personnel table, the entry area has a double border. You can use all the same techniques for entering data in this field that are described earlier in this chapter for the Determined By field of the Specimen table (pp. 149–151).

The Note By field is optional. Just "TAB through" the field if you do not want to use it.

3. **Enter the Note Date.** This field is obligatory, since Notes are sorted by Note Date.

   ♦ **To accept today's date,** just "TAB through" the date entry area, or click in the Note Text area below.

   ♦ **To enter a different date,** enter Month (two digits), Day (two digits), and Year (four digits). You must use a full date in this field; you cannot enter a month-year date or a year-only date (pp. 112–113).

4. **Enter the Note Text in the lower-right panel of the Notes screen.** (Not required.)

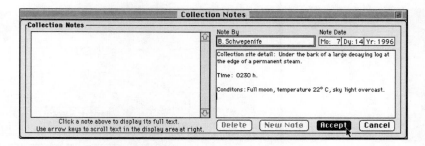

If the length of the Note Text exceeds the size of the display panel, use the UP and DOWN arrow keys to scroll in the panel.

In the example above, notice that you can use line breaks (RETURN key in the Macintosh operating system or ENTER key in Windows) within the Note Text field to organize the entry.

NOTE: When exporting Notes using the Export Editor (pp. 443–443), if you indicate carriage return (Macintosh) or carriage return plus line feed (Windows) as the End of Record delimiter (p. 441), line breaks *within* Note Text fields are replaced by blanks to keep the Export Editor from interpreting them as End of Record characters. To preserve internal line breaks in exported Notes records, specify a different End of Record delimiter, ideally one not used in Note Text fields, then change this character globally to line breaks with a text editor in the exported records. (The Export Editor automatically replaces your specified End of Record delimiter with blanks *within* text and alphanumeric fields, regardless of which delimiter you specify.)

You can leave the Note Text field blank if you can think of a use for Notes records with no Note Text.

5. **Click the Accept button in the Notes screen (illustrated above).** The Notes window closes. The Show Notes button in the Core table (Collection, Locality, Specimen, Species, or Loans) input screen is enabled.

NOTE: You can also use the New Note button in the *Notes screen* to enter a new note. This button is enabled only when you display the Notes screen using the Show Notes button (see below).

Viewing, Editing, or Deleting a Note

This section will use Collection Notes screens as an example, but all procedures are identical and all screens are analogous for Locality, Specimen, Species, and Loans Notes.

1. **In an existing Collection, Locality, Specimen, Species, or Loans record, click the Show Notes button.** The Show Notes button is enabled only if at least one Note has previously been recorded for this

record. To enter the first Note for a record, use the New Note button instead (pp. 179–181).

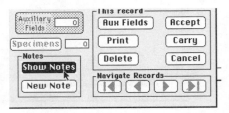

The Notes screen appears, with all Notes for this Core table record (Collection record, in the illustration) displayed in the left panel, sorted by Note Date.

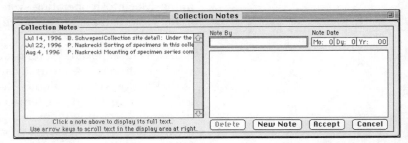

Notice that long note author (Note By) names (e.g., "B. Schwepenife" in the first note above) are truncated to make room for the first part of the Note Text in the left panel.

NOTE: If you click in the Core table input screen (the Collection input screen, in the illustrations) while the Notes window is open, the cursor will change to the "hidden window" cursor.

To return to the Notes window, press any key on the keyboard. (The keystroke is not recorded.)

2. **To display an individual Note in full,** click it in the left panel. The Note By, Note Date, and Note Text fields of the note you clicked appear in the right panels.

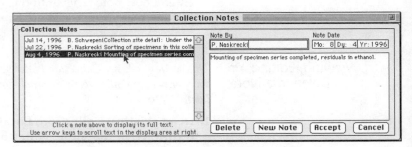

If the length of the Note Text exceeds the size of the display panel, use the UP and DOWN ARROW keys to scroll in the panel.

3. **To edit a Note,** display it, make changes, then click the Accept button in the Notes screen. The Notes screen closes.

4. **To delete a Note,** display it then click the Delete button in the Notes screen.

5. **To enter a new Note,** click the New Note button in the Notes screen, then proceed as in the previous section of this chapter ("Entering a New Note").

6. **To dismiss the Notes screen,** click the Cancel button in the Notes screen.

Carrying Notes

The Carry button in Biota input screens allows you to use an existing record (or a record you have just entered) as a template for a new record (pp. 41–42, 127–128). As an option, you can choose to Carry (copy) all Notes from the existing record to the new record (for the tables that support Notes).

For example, a series of Collection records might share the same note describing meteorological conditions or special equipment used at the time the collections were made.

To enable or disable this option:

1. **Choose Preferences from the Special menu .** The Preferences screen appears.

2. **Click the "Carry Notes from the template records" checkbox** to enable (check) or disable (uncheck) the Carry Note option. The default setting is disabled (unchecked).

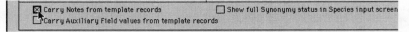

3. **Click Accept** in the Preferences screen.

Chapter 11 Finding and Updating Records

This chapter explains how to use tools from the Find, Tree, and Special menus to find records already in a Biota Data File—and update them if you wish. Before using these tools, you need to understand the ideas introduced in Chapter 2, "Key Concepts," including the distinction between Selections and Record Sets, and the basic procedures outlined in Chapter 9, "Working with Records in Output and Input Screens."

Biota offers several different kinds of tools for finding records, ranging from the general to the specialized; this chapter follows the same organization, from general tools to the more specific. Like tools from the hardware store, a specialized software tool applied to the task for which it was designed is almost invariably more efficient and effective than a general-purpose tool applied to the same task, even though the latter might work. Investing a few minutes in learning what the tools described in this chapter can do for you may save you much time in the long run.

Finding and Displaying All Records for a Table

You can find and display all existing records for one of Biota's Core tables (p. 5) using the "All…" commands from the Find menu.

For example, for Specimen records, choose All Specimens from the Find menu. For Locality records, choose All Localities from the Find menu, and so on for each Core table

- **If there are no records in the table,** Biota displays a message like the one below for the Kingdom table.

- **If there are records in the table,** Biota displays them in the standard output (record listing) screen for the table, where you can work with them using all the techniques outlined in Chapter 9.

  When you click the Done button in the output screen, Biota will display the standard Record Set options screen for the table (for the Order table, in the illustration below). See pp. 13–15 for explanation of the options offered.

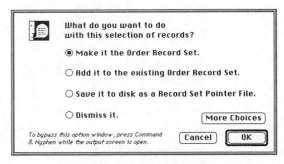

If you choose to create or modify the active Record Set for a table, you can display the records in the new or changed Record Set using the command from the Display menu for that table.

NOTE: The All Loans command is in the Loans menu, discussed in Chapter 17. The command All Valid Species, in the Find menu, has a special function; this command is explained in Appendix I.

Finding and Updating Records Using Tools from the Tree Menu

The Tree menu offers two sets of tools for finding records based on the taxonomic hierarchy, which, at its best, reflects the structure of the "tree of life." The Specimen table is included as the lowest level of this hierarchy (Appendix A).

Finding Lower Taxa for Higher Taxa or Higher Taxa for Lower Taxa

In the upper section of the Tree menu, the Lower Taxa for Higher Taxa ("top-down") tool finds all records for a lower taxonomic level that belong to a higher taxon or to taxa in the active Record Set for a higher taxonomic level. The Higher Taxa for Lower Taxa ("bottom-up") search tool finds all records for a higher taxonomic level that belong to the active Record Set for a lower taxonomic level taxa. It does not matter whether the lower and higher levels are adjacent in the taxonomic hierarchy or not. Because these two tools also appear in the Find menu together with an analogous pair of tools for the place hierarchy (the Locality, Collection, and Specimen tables), all four are described together later in this chapter (pp. 200–213).

Finding and Updating Records by Navigating the Taxonomic Hierarchy

The lower section of the Tree menu offers a lookup system that lets you quickly navigate the taxonomic hierarchy to find and update specific records or establish Record Sets for other uses.

You can enter the hierarchy at any of the principal levels (Kingdom Down, Phylum Down, Class Down, Order Down, Family Down, Genus Down, or Species Down), then navigate all lower levels, and back up again, along any branch of the hierarchy.

The Family Down entry point will be used in the illustration below, but the principles and techniques are identical for all entry points.

1. **From the Tree menu, select Family Down.** A scrollable list of all Family records in your Data File appears in a special Family output screen entitled "Family-Down Tree Lookup."

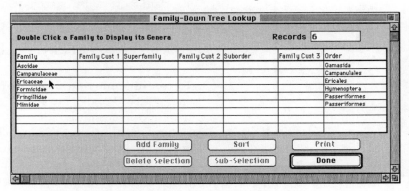

- **All records for the entry level (Family) are shown.** When you choose an entry level from the lower section of the Tree menu, the *first* Lookup screen shows all taxa for the entry level, regardless of their classification at higher levels. In the illustration above, all Family records appear, regardless of Order, Class, etc. (Family records in the illustration include plants, mites, insects, in birds.)

- **Most buttons are disabled.** Notice that all buttons except the Done button are disabled in the entry-level lookup screen above. To use the other buttons, you need to display records using commands from the Find menu, (pp. 185–186, 192–218.)

2. **Double-click a Family record to display all Genus records linked to it.** In the illustration above, the record for the plant family Ericaceae is being double-clicked in the Family-Down Tree Lookup screen. The Genus records for the this Family, exclusively, then appear in the Genus Lookup screen, below.

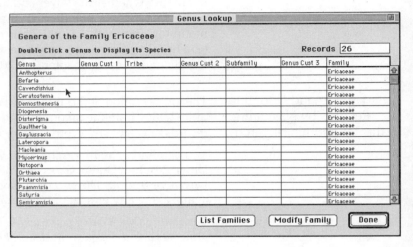

NOTE: When you clicked on a record in the Family Lookup screen, all Genus records appeared in the Genus Lookup screen. In contrast, in standard Biota output screens (Chapter 9), if you double-click a record, that record appears in the input screen for the table. The instruction line just above the display area in both the standard output and Lookup (Tree menu) screens specifies this distinction.

3. **To continue moving down through the hierarchy,** double-click a Genus record to display all its linked Species records in the Species Lookup screen. In the illustration above, the record for the genus "*Cavendishius*" is being double-clicked in the Genus Lookup screen. The Species records for this genus then appear in the Species Lookup screen, below, with the title "Species of the Genus *Cavendishius*." (Botanists, do not jeer, the genus is misspelled on purpose; see step 5, below.)

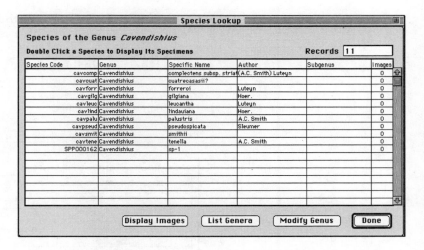

- **To see the Specimen records for a particular Species,** double-click the line for that Species in the Species Lookup screen. The Specimen records will appear in the Specimen Lookup screen.

- **Then, to see the record for an individual Specimen record,** click it in the Specimen Lookup screen. The Specimen record will appear in the Specimen *input* screen.

  When you click on a record in a Lookup (tree menu) screen—at any level in the hierarchy—all linked records at the next lower will be displayed.

4. **To move back up the hierarchy to the Genus Lookup screen,** click the List Genera button at the base of the Species Lookup screen.

In the example, all genera of the family Ericaceae will again be displayed, exactly as shown in step 2, above.

- **To continue moving back up the hierarchy,** click the List Families button in the Genus Lookup screen (see the illustration in step 2). In the example, all Family records in the database will again be displayed (exactly as shown in step 1, above), because the entry level was Family.

- *In general, when you click the "List..." button at the base of a Lookup (tree menu) screen for a taxon—at any level in the hierarchy—all records at the next higher level that share the same parent record as that taxon will be displayed.*

There is one exception. If you click the "List..." button at the level immediately below the entry level, *all* records for the entry level are displayed, just as they were when you entered the hierarchy.

5. **To display and/or update an individual Genus record,** click the Modify Genus button at the base of the Species Lookup screen, while the name of the Genus record to be shown or modified appears in the title of the screen. In the example below, the misspelled Genus record for *"Cavendishius"* will be displayed so the name can be corrected.

Genus name to be modified   Modify Genus button

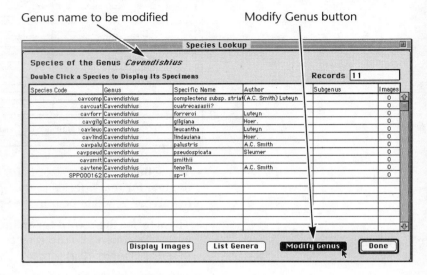

The record appears in the standard Genus input screen (with the special title "Modify Genus").

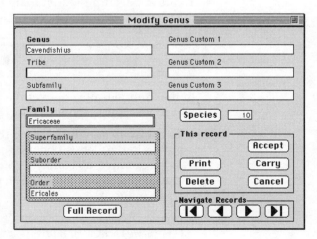

All buttons in the Modify Genus screen are active:

♦ **To make changes in the record** (e.g., to correct *"Cavendishius"* to *Cavendishia*), edit the record, then click the Accept button to register the changes. If you change the Key field (Genus, in the illustration) and the record is linked to child records, Biota offers the option explained in the screen below.

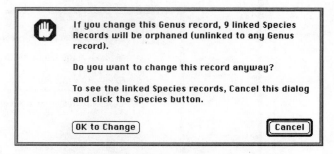

See pp. 218–220, later in this chapter for a full explanation of this options.

- **To use the record as a template for a new record,** click the Carry button (pp. 41–42).
- **To print the individual record,** click the Print button (p. 266).
- **To delete the record,** click the Delete button. If the record is linked to any child records, Biota will warn you and ask for a confirmation (pp. 122–123).
- **To display (and modify) other records in the current selection for the table**—in the illustration, other genera in the family Ericaceae—use the navigation buttons. (If you do this, then return to the Species Lookup screen, the Species records listed will be for the last Genus record you displayed in the Modify Genus screen.)
- **To display linked child records** in a new window, click the Child Records button (the Species button, in the screen illustration above) (pp. 125–126).
- **To display a linked parent record,** click the Full Record button (pp. 124–125).
- **To return to the Lookup screens** without making any changes, click the Cancel button.

6. **To leave the taxonomic hierarchy (Tree menu) screens,** click the Done button in a Lookup screen. (The Species Lookup screen is illustrated below.)

The standard Record Set options screen appears. (See pp. 13–15 for explanation of the options offered.)

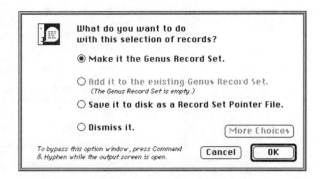

The Search Editor, a General-Purpose Tool for Finding Records Based on Content

When none of the specialized tools from the Find or Tree menus fits the record-finding task you have in mind, the Search Editor may be the answer. The Search Editor is often the tool of choice when you need to find records based on their content, although displaying all records for a table (pp. 185–186), then using the Sort Editor (pp. 131–139) is often just as efficient. The Search Editor is restricted to finding Core table records, but the query can be based on fields and values in virtually any Biota table.

Using the Search Editor: Step by Step

1. **From the Find menu, choose By Using Search Editor.** A table selection window appears.

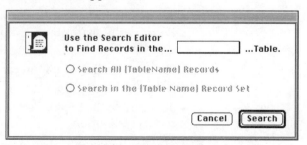

2. **Click and hold the popup list** to select the table containing the records you seek—called the *target table*. In the illustration below, the Genus table is being selected as the target table for the search.

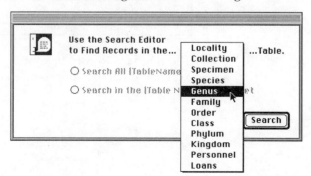

When you have chosen a table, the search option buttons are activated.

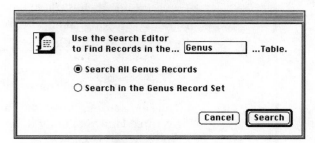

3. **Choose a search option in the window (above).**
   - *Either:* **Search all records in the selected table.**
   - *Or:* **Search only in the active Record Set for the table**.

   If there are no records in the active Record Set for the table, the lower button will not be activated. If there are no records in the table, Biota informs with a message; neither button is activated and the search is canceled.

4. **Click the Search button** to launch the Search Editor. The Search Editor window appears.

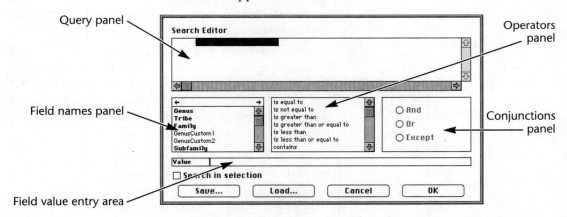

- The *query panel* of the Search Editor (initially blank) will display the search query. This is the upper panel in the window.
- The *field names panel* (the scrollable panel at the lower left) is used to display and select fields.

  Field names in **boldfaced** type are indexed fields. Searches for values in indexed fields proceed faster than searches for values in nonindexed fields—but you can search on any field. (Indexes take up space in data files and processor time to maintain, so not all fields in Biota are indexed.)

NOTE: As explained in the Idiosyncrasies section (pp. 60–61), the Search Editor is a built-in 4D utility, which, unfortunately, does not list field names alphabetically. You may have to scroll the list to find the field you are looking for.

- **The *operators panel*** (the scrollable panel in the lower center) is used to display and select comparison operators for building queries.
- **The *conjunctions panel*** (the panel at the lower right) offers buttons for three logical operators for building queries: the conjunctions And, Or, and Except.
- **The *field value entry area*** (running across the base of the window above the buttons) is the entry area for field values used in search queries.
- **The Save and Load buttons** allow you to save queries (which are sometimes complex to set up) to disk file, then reload them at a later date.
- **The Cancel and OK buttons** are used to dismiss the Search Editor and launch a search, respectively.
- **The "Search in selection" checkbox** has no function in Biota. If you want to search in a particular selection of records, declare them the current Record Set for the table (pp. 13–15) and choose the second option in step 3, above.

5. **If you have defined any Core Field Aliases** (pp. 323–326), a list of the Aliases and their Internal Field Name equivalents will appear in a separate, "floating" window. The fields panel in the Search Editor displays only Internal Field Names. The Alias window does not appear (and you have no need for it) if you have not defined any Aliases.

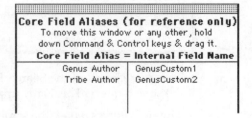

In the example above, the field [Genus] Genus Custom1 (Internal Field Name) has been given the Alias *Genus Author*, and the field [Genus] Genus Custom2 has the Alias *Tribe Author*.

6. **In the Search Editor, click a field name from the field name panel to select it.** When the Search Editor first opens, this scrollable panel is enclosed in a blinking marquee and shows the names of all fields for the table you selected in step 2. (You may have to scroll to see them all and they are not listed alphabetically; see the note in step 4.)

In the illustration below, the [Genus] Family field is being selected for the query.

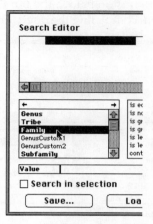

The field name you selected appears in the query panel (see the illustration below) and the operators panel now has the blinking marquee.

7. **Click an operator from the operators panel** to add it to the query. In the illustration below, the operator "is not equal to" is being selected.

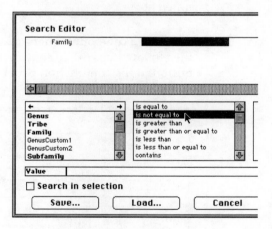

The operator you selected appears in the query panel (see the illustration below) and the field value entry now has the blinking marquee and a blinking cursor.

NOTE: The quantitative operators (*is greater than*, *is greater than or equal to*, and *is less than*) can be used not only on numeric fields, but also on values in alphabetic fields. When used on alphanumeric values, these operators evaluate the ASCII code of the characters, left to right. Thus *01* is less than *10*, but *1* is evaluated as "greater than" *10*. The digits 0–9 are "less than" the letters A–Z. In short, the order of alphanumeric values is the same as their alphanumeric sort order.

8. **Enter a field value** in the field value entry area to complete the query line. (The field value appears simultaneously in the query panel.) In the illustration below, the family name Ericaceae is the field value entered, completing the query line "Family is not equal to Ericaceae."

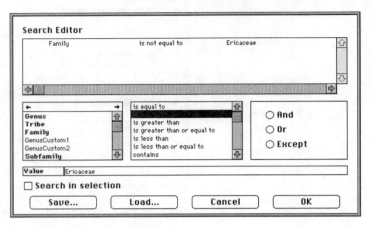

NOTE: Field values are not case-sensitive: Ericaceae, ericaceae, and ERICACEAE would all have the same meaning in the query illustrated above.

9. **To create a compound query,** click a button in the conjunction panel, then follow steps 6–8 to construct an additional query line. The And button in the conjunction panel is being clicked in the illustration below.

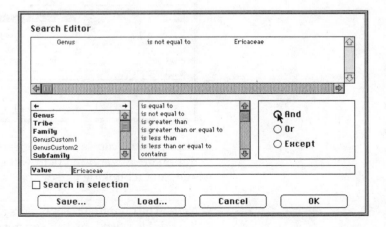

In the completed example below, the compound query will find all Genus records for which the [Genus] Family field is *neither* Ericaceae *nor* Campanulaceae.

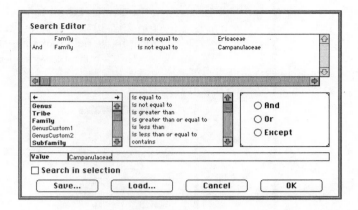

NOTE: Compound searches are performed in the order the lines appear in the search query, top to bottom, regardless of the conjunction used (i.e., there is no fixed precedence among the three conjunctions).

10. **To launch the search,** click the OK button in the Search Editor.

- ◆ **If records are found that match the criteria in the search query,** they will be displayed in the standard output screen for the target table, where you can work with them using all the techniques outlined in Chapter 9. When you click the Done button in the output screen, Biota will display the standard Record Set options screen. (See Chapter 2 for explanation of the options offered).

- ◆ **If no records are found that match the criteria in the search query,** Biota displays the message screen below. If you click the Try Again button, the Search Editor reappears, set to search for records in the same target table as before.

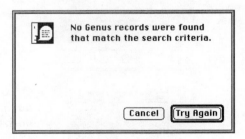

Using Fields from Related Tables in the Search Editor

Before consulting this section, please read the previous section, "Using the Search Editor: Step by Step" (pp. 192–197).

When you open the Search Editor, you specify the table containing the records you seek—the *target table* for the search (step 2 in the previous section). In the previous section, the search queries illustrated all relied solely on fields of the target table itself. You can also use fields from related tables in search queries you build with the Search Editor.

1. **Launch the Search Editor** following steps 1 through 4 in the previous section (pp. 192–194.) The fields for the target table you selected will initially appear in the field name panel of the Search Editor. In the illustration below, the Genus table was selected as the target table for the search.

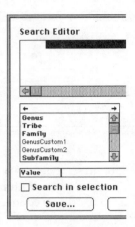

2. **Display the fields for the related table.** There are two ways to access field lists for other tables and display them in the field name panel of the Search Editor.

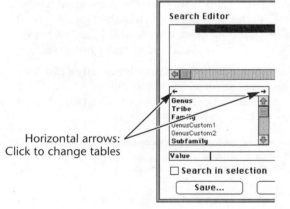

- *Either:* **Click on the horizontal arrows** at the top of the field list panel at the left of the Sort screen to move to field lists for other tables, one table at a time.

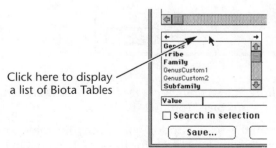

- *Or:* **Click and hold down the mouse button in the blank rectangle between the two horizontal arrows** to display a popup list of all Biota tables.

Drag the cursor down and release it when the name of the table you want is highlighted. The illustration at the left shows the Family table being selected.

NOTE: Unfortunately 4D does *not* list table names alphabetically—although the order is always the same. You may have to look for the table you seek until the order becomes familiar.

3. **Select a field from the related table to use in the search query.** Once the fields for the related table appear in the field list, you can include these fields in search criteria, using the same techniques described in the previous section (pp. 194–195). Below, the [Family] Suborder field is being used in constructing a search query for Genus records.

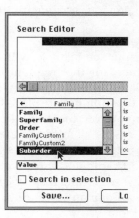

Fields from child tables as well as parent tables can be used in search criteria.

- **If the search criterion refers to a field in a parent table,** any record in the target table with linkage to a parent record that meets the search criterion will be selected. In the illustration above, the target table is Genus. The related table selected is the Family table—the parent table to the Genus table. If the search query were "Suborder is equal to Nematocera," all Genus records linked to Families in the Suborder Nematocera (long-horned flies) would be found. (Suborder is a field of the Family table, pp. 109–111.)

  In the case of parent (many-to-one) links, you can reference fields from a related table *at any level up a hierarchy*. For example, you could select all Genus records linked to a specified Kingdom or any other field from a higher taxon table. You could select all Specimen records linked to Locality records from a certain Region (see Appendix A).

- **If the search criterion refers to a field in a child table,** each record in the target table that is linked to *at least one* child record that meets the search criterion will be selected. For example, with the Genus as the target table, you could use the Number Images field from the Species table (which specifies the number of Image Archive records linked to each Species record) to set up the search query "Number Images is greater than 0." This search would find all Genus records that are linked to at least one Species record that has at least one linked Image Archive record.

  In the case of child (one-to-many) links, you can use fields *only from a directly linked child table*. For example, you could not use a field of the Specimen table in a search criterion for Genus records, because the Species table is interposed.

4. **When the query is ready, complete the search** as in described in step 10 of the previous section. (p. 197).

Special Tools for Finding Records in One Table Based on a Set of Records in Another Table

Biota offers four special tools, with a common interface, for the frequent task of finding related records up, down, and across the taxonomic and place hierarchies. As on option, Biota will keep the linking records for all intermediate tables as Record Sets for those tables. These tools support rapid and intuitive query construction and optimized, extremely fast search and display. They can be used sequentially for complex searches. For many users, they are among the most frequently used features of Biota.

Finding Records for Lower Taxa Based on Higher Taxa

The Lower Taxa for Higher Taxa ("top-down") search tool finds all records for a lower taxonomic level (the *target table*) that belong to a higher taxon or a selection of higher taxa (in the *criterion table*). It does not matter whether the two levels are adjacent in the taxonomic hierarchy or not—for example, you could find all Specimens of a Family or all Genera of three Orders (nonadjacent levels), or all Specimens for a selection of Species or all Classes for a Phylum (adjacent levels). For nonadjacent levels, you can tell Biota to keep all related taxa at intermediate levels as Record Sets.

This tool does exactly the opposite task from the Higher Taxa for Lower Taxa ("bottom-up") search tool described in the next section (pp. 205–208).

1. **From the Find menu, choose "Lower Taxa for Higher Taxa."** The query setup screen below appears.

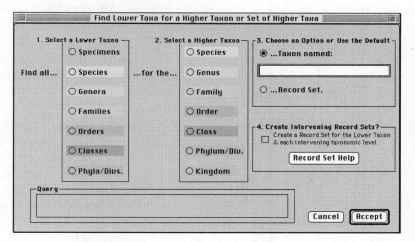

2. **From the list of tables in the left panel (the panel labeled "1. Select a Lower Taxon"), click the button for the target table**—the table containing the records you seek. In the illustration below, the Specimen table is being selected.

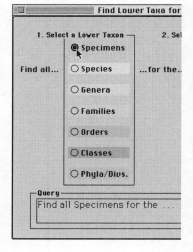

The query panel displays the beginning of the query.

3. From the list of tables in the right panel (the panel labeled "2. Select a Higher Taxon"), click the button for the table to be used in the selection criterion (the criterion table).

◆ **If you choose Species and the Species Record Set is not empty**, the option window below appears.

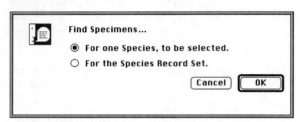

◊ **If you choose the first (default) option** in this window ("For one Species, to be selected"), Biota presents the Look Up Species tool so you can look up a Species record according to its Genus (illustrated below).

◊ **If you choose the second option** ("For the Species Record Set"), Biota automatically sets the Species Record Set option (see step 4, below). The Species Record Set may contain many records or only one.

◆ **If you choose Species and the Species Record Set is empty,** the Look Up Species tool appears. Enter a Genus name or the first few letters of a Genus name in the "Enter Genus, select Species" entry area, then press TAB (wildcard entry method) to display the Species records for that Genus. Click the Species you seek, to complete the search query. See pp. 145–146 if you need help using the Look Up Species tool.

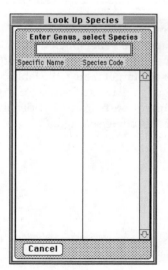 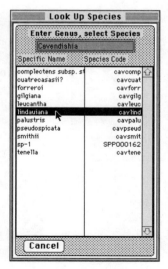

Here is a completed search query, using the illustration above.

- **If you choose a criterion table other than Species,** the option panel (labeled "3. Choose an Option or Use the Default") will be set up for the next step.

4. **Choose to use a single record or the current Record Set from the criterion table, in constructing the search query.** You make this choice in the option panel at the upper right, labeled "3. Choose an Option or Use the Default."

    - **If you choose Species as the criterion table,** the option you chose in step 3 will already be set (one of the two options below).

    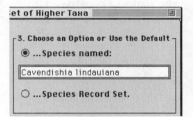 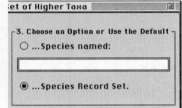

    - **If you choose a table other than Species as the criterion table,** select the individual record option (below, left) or the Record Set option (below, right). If you choose the Record Set option, the Record Set for the criterion table may contain many records or only one. The Order table was selected as the criterion table in the illustrations below.

    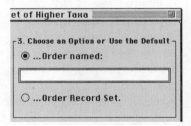 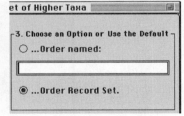

    ◊ **If you choose the individual record option** (above, left), you can use the wildcard entry method to enter the name of the criterion taxon. (Enter the first letter or first few letters of the taxon name and press TAB. See p.96 if you need help.)

    ◊ **If you choose the Record Set option** and there are no records in the current Record Set for the criterion table, Biota will post a message like the one below. If you like, you can leave the Lower Taxon for Higher Taxa screen open while you create the appropriate Record Set in another window.

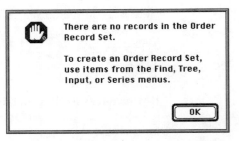

The completed query will appear in the search query panel.

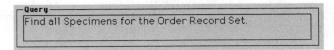

5. **To create Record Sets automatically for the target records and for the linking records in any intervening tables, click the checkbox.** You make this choice in the option panel at the lower right, labeled "4. Create Intervening Record Sets?"

For example, if you set up the query "Find all Specimens for the Order Hymenoptera" and check this option, Record Sets for the Family, Genus, and Species tables will automatically be created, based on the records linking the Hymenoptera record in the Order table with the Specimen records Biota finds. A Record Set will also be automatically created for the Specimen records found, if the box is checked.

It is not necessary to check this option in order to find the target table (Lower Taxon) records you are searching for, nor to create a Record Set for them. If you want to create a Record Set for the target records without checking this option, you may do so by means of the standard Record Set option screen (pp. 13–15) when leaving the output screen for the target table (see step 6, below).

WARNING: Creating Record Sets for intervening tables during the search will replace any existing current Record Sets for those tables.

6. **To launch the search, click the Accept button.**
   - **If target table records matching the search query are found,** they will be displayed in the standard output screen for the target table, where you can work with them using all the techniques outlined in Chapter 9. When you click the Done button in the output screen, Biota will display the standard Record Set options screen. (See Chapter 2 for explanation of the options offered.)

- **If no target table records matching the search query are found,** Biota displays the message screen below. If you click the Try Again button, the search setup screen reappears, with settings as you left them.

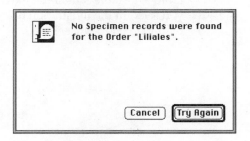

Finding Records for Higher Taxa Based on Lower Taxa

The Higher Taxa for Lower Taxa ("bottom-up") search tool finds all records for a higher taxonomic level (the target table) that belong to the active Record Set for a lower taxonomic level (the criterion table). For example, you could find all Orders for the Specimen Record Set (nonadjacent levels) or all Families for the Genus Record Set (adjacent levels). For nonadjacent levels, you can tell Biota to keep all related taxa at intermediate levels as Record Sets.

This tool does exactly the opposite task from the Lower Taxa for Higher Taxa ("top-down") search tool described in the previous section (pp. 201–205). In addition to working in the reverse direction along the same set of linked tables, this tool also differs in requiring you to designate a Record Set for the criterion table before setting up the search—an optional method in the Higher Taxa for Lower Taxa tool. The Record Set for the criterion table may contain many records or only one.

1. **From the Find menu, choose Higher Taxa for Lower Taxa.** The query setup screen below appears.

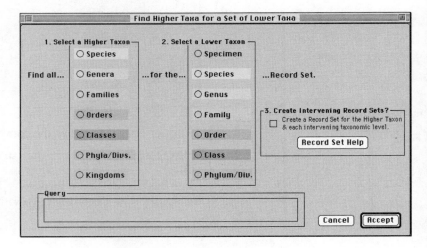

2. **From the list of tables in the left panel (the panel labeled "1. Select a Higher Taxon"), click the button for the target table**—the table containing the records you seek. In the illustration below, the Order table is being selected.

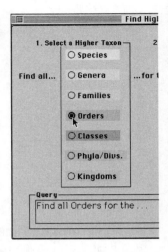

The query panel displays the beginning of the query.

3. **From the list of tables in the right panel (the panel labeled "2. Select a Lower Taxon"), click the button for the table to be used in the selection criterion (the criterion table).** In the illustration below, the Species table is being selected. The Record Set for the criterion table may contain many records or only one.

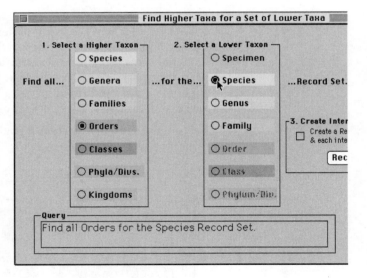

The query panel displays the completed query.

4. **If there are no records in the current Record Set for the criterion table,** Biota will post a message like the one below. If you like, you can leave the Higher Taxon for Lower Taxa screen open while you create the appropriate Record Set in another window.

5. **To create Record Sets automatically for the target records and for the linking records in any intervening tables, click the checkbox.** You make this choice in the option panel at the right, labeled "3. Create Intervening Record Sets?"

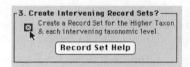

For example, if you set up the query "Find all Orders for the Species Record Set" and check this option, Record Sets for the Genus and Family tables will automatically be created based on the records in these tables linking the Species records in the criterion Record Set with the Order records that Biota finds. A Record Set will also be automatically created for the Order records found, if the box is checked.

It is not necessary to check this option in order to find the target table (Higher Taxon) records you are searching for, nor to create a Record Set for them. If you want to create a Record Set for the target records without checking this option, you may do so by means of the standard Record Set option screen (pp. 13–15) when leaving the output screen for the target table (see step 6, below).

WARNING: Creating Record Sets for intervening tables during the search will replace any existing current Record Sets for those tables.

6. **To launch the search, click the Accept button.**
   - **If target table records matching the search query are found,** they will be displayed in the standard output screen for the target table, where you can work with them using all the techniques outlined in Chapter 9. When you click the Done button in the output screen, Biota will display the standard Record Set options screen. (See Chapter 2 for explanation of the options offered.)
   - **If no target table records matching the search query are found,** Biota displays the message screen below. If you click the Try Again button, the search setup screen reappears, with settings as you left them.

Biota: CHAPTER 11

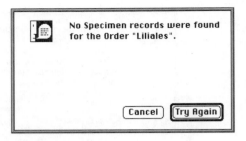

Finding Place Records (Collection or Locality) for Specimens or Species

The Places for Specimens or Species search tool finds all Locality or Collection records for the active Specimen or Species Record Set or all Locality records for the Collection Record Set, according to your query. For example, you could find all Locality records for a group of Species (a cross-hierarchy query), to study their geographic range in the database. You can tell Biota to keep the related Specimen and Collection records (intermediate tables in the search) as Record Sets.

This tool does exactly the opposite task from the Specimens or Species for Places search tool described in the next section (pp. 210–213).

1. **From the Find menu, choose Places for Specimens or Species.** The query setup screen below appears.

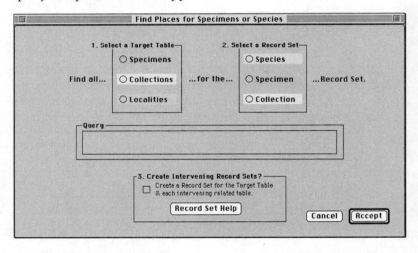

2. **From the list of tables in the left panel (the panel labeled "1. Select a Target Table"), click the button for the target table**—the table containing the records you seek. In the illustration below, the Localities table is being selected.

The query panel displays the beginning of the query.

3. **From the list of tables in the right panel (the panel labeled "2. Select a Record Set"), click the button for the table to be used in the selection criterion (the criterion table).** The Record Set for the criterion table may contain many records or only one. In the illustration below, the Species table is being selected.

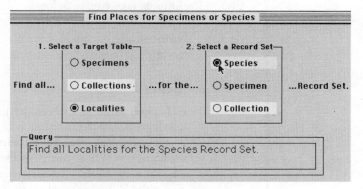

The query panel displays the completed query.

4. **If there are no records in the current Record Set for the criterion table,** Biota will post a message like the one below. If you like, you can leave the Places for Specimens or Species screen open while you create the appropriate Record Set in another window.

5. **To create Record Sets automatically for the target records and for the linking records in any intervening tables, click the checkbox.** You make this choice in the option panel at the right, labeled "3. Create Intervening Record Sets?"

For example, if you set up the query "Find all Localities for the Species Record Set" and check this option, Record Sets for the Specimen and Collection tables will automatically be created,  based on the records in these tables linking the Species records in the criterion Record Set with the Locality records that Biota finds. A Record Set will also be automatically created for the Locality records found, if the box is checked.

It is not necessary to check this option in order to find the target table records you are searching for, nor to create a Record Set for them. If you want to create a Record Set for the target records without checking this option, you may do so by means of the standard Record Set option screen (pp. 13–15) when leaving the output screen for the target table (see step 6, below).

WARNING: Creating Record Sets for intervening tables during the search will replace any existing current Record Sets for those tables.

6. **To launch the search, click the Accept button.**

   - **If target table records matching the search query are found,** they will be displayed in the standard output screen for the target table, where you can work with them using all the techniques outlined in Chapter 9. When you click the Done button in the output screen, Biota will display the standard Record Set options screen. (See Chapter 2 for explanation of the options offered.)

   - **If no target table records matching the search query are found,** Biota displays the message screen below. If you click the Try Again button, the search setup screen reappears, with settings as you left them.

Finding Specimen or Species Records for Places (Collections or Localities)

The Specimens or Species for Places search tool finds all Specimen or Species records for the active Collection or Locality Record Set, or all Collection records for the Locality Record Set, according to your query. For example, you could find all Species records for a group of Localities (a cross-hierarchy query) to create a regional fauna or flora list. You can tell Biota to keep the related Collection and Specimen records (intermediate tables in the search) as Record Sets.

This tool does exactly the opposite task from the Places for Specimens or Species search tool described in the previous section (pp. 208–210).

1. **From the Find menu, choose Specimens or Species for Places.** The query setup screen below appears.

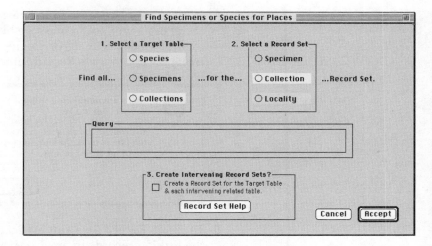

2. **From the list of tables in the left panel (the panel labeled "1. Select a Target Table"), click the button for the target table**—the table containing the records you seek. In the illustration below, the Species table is being selected.

The query panel displays the beginning of the query.

3. **From the list of tables in the right panel (the panel labeled "2. Select a Record Set"), click the button for the table to be used in the selection criterion (the criterion table).** The Record Set for the criterion table may contain many records or only one. In the illustration below, the Locality table is being selected.

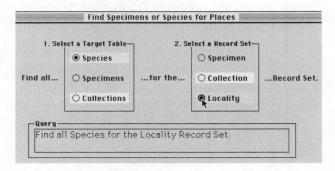

The query panel displays the completed query.

4. **If there are no records in the current Record Set for the criterion table,** Biota will post a message like the one below. If you like, you can leave the Specimens or Species for Places screen open while you create the appropriate Record Set in another window.

5. **To create Record Sets automatically for the target records and for the linking records in any intervening tables, click the checkbox.** You make this choice in the option panel at the right, labeled "3. Create Intervening Record Sets?"

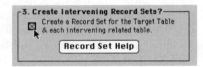

For example, if you set up the query "Find all Species for the Locality Record Set" and check this option, Record Sets for the Collection and Specimen tables will automatically be created based on the records in these tables linking the Locality records in the criterion Record Set with the Species records Biota finds. A Record Set will also be automatically created for the Species records found, if the box is checked.

It is not necessary to check this option in order to find the target table (Higher Taxon) records you are searching for, nor to create a Record Set for them. If you want to create a Record Set for the target records without checking this option, you may do so by means of the standard Record Set option screen (pp. 13–15) when leaving the output screen for the target table (see step 6, below).

WARNING: Creating Record Sets for intervening tables during the search will replace any existing current Record Sets for those tables.

6. **To launch the search, click the Accept button.**
   - **If target table records matching the search query are found,** they will be displayed in the standard output screen for the target table, where you can work with them using all the techniques outlined in Chapter 9. When you click the Done button in the output screen,

Biota will display the standard Record Set options screen. (See pp. 13–15 for explanation of the options offered).

- **If no target table records matching the search query are found,** Biota displays the message screen below. If you click the Try Again button, the search setup screen reappears, with settings as you left them.

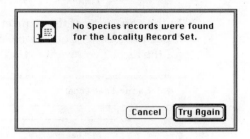

Sequential, Cross-Hierarchy Searches

If you want to find Collections or Localities for taxa above the level of Species, or discover which higher taxa are represented in Collections or Localities, you can use the appropriate pair of search tools sequentially, choosing among the tools introduced in the preceding four sections of this chapter.

For example, to find all the Localities for an Order:

1. **Find all Species for the Order using the Lower Taxa for Higher Taxa tool** (pp. 201–205).

2. **Declare the group of Species records found to be the current Species Record Set** (pp. 13–15).

3. **Find all Localities for the Species Record Set using the Places for Specimens or Species tool** (pp. 208–210). The Locality records found will encompass all Localities in which any Specimen of the original criterion Order has been found (as recorded in the database).

Finding Records by Record Codes

Because Record Codes uniquely identify each record in the Specimen, Collection, Locality, and Species tables (Chapter 7), finding a record in one of these tables by its Record Code is sometimes the fastest way to retrieve, display, and—if necessary—edit an existing record. If you use barcodes (pp. 105–108, Appendix H) for Specimen Codes or other Record Codes, these tools are especially fast to use.

Four commands from the Find menu (By Specimen Code, By Collection Code, By Locality Code, and By Species Code), all with the same interface, take care of this straightforward task. A fifth command with a different interface, By Specimen Code Series, is covered in Chapter 12.

1. **From the Find menu, choose By Specimen Code, By Collection Code, By Locality Code, or By Species Code.** The Record Code request window appears (illustrated below for the Specimen table).

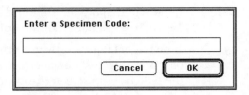

2. **In the entry area of the request window, enter the full Record Code** (Specimen Code, Collection Code, Locality Code, or Species Code) for the record you want to find, manually or with a barcode reader.

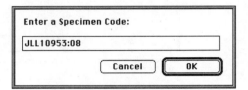

NOTE: You cannot use the wildcard entry method with these tools.

3. **Click the OK button in the request window** to launch the search.
   - **If the record is found,** it will be displayed in the standard input screen for target table, where you can work with it using the techniques outlined in Chapter 9.

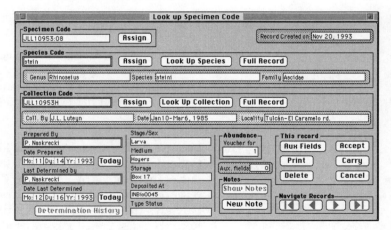

When you click the Cancel button in the input screen, above, Biota will display the standard Record Set options screen for the target table (the Specimen screen is illustrated below). See pp. 13–15 for explanation of the options offered.

   - **If the record is not found,** Biota displays a message like the one below (illustrated for the Specimen table). If you click the Try Again button, the Record Code request screen for the same target table reappears.

Finding Host and Guest Specimens and Collections

Four commands in the Find menu—Host Specimens & Collections, Guest Specimens & Collections, Host Specimens for Guest Specimens, and Guest Specimens for Host Specimens—find and display Specimen and Collection records linked through the [Collection] Host Spcm Code field to the [Specimen] Specimen Code field.

These commands are discussed in Chapter 21.

Finding Childless and Orphan Records

Chapter 2 introduces the concepts of *child* and *parent* records. For example, as a consequence of the many-to-one relation between the Species and Genus tables (Appendix A), a Species record (if linked) is a child record of a particular Genus record; the Genus record is the parent record of the Species record (and perhaps the parent of other Species records as well).

The metaphor is carried further in the concepts of *orphan* records—those that are not linked to any record in the parent table, and *childless* records—those not linked to any record in the child table, for a particular database relation. For example, a Genus record not linked to any Family record is an orphan record. A Genus record not linked to any Species record is a childless record. Orphan records, especially, and childless records, to a lesser extent, tend to be "invisible" in relational searches. This section explains Biota's optimized search tools for finding and displaying orphan and childless records.

Finding Childless Records

1. **From the Find menu, select Find Childless Records.** A table selection window appears.

2. **Using the mouse, select a parent table from the popup list.** The Species table is being selected in the illustration below.

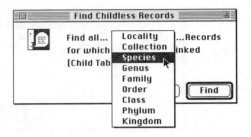

3. **When the table has been selected, click the Find button** in the table selection window to launch the search. The progress indicator appears during the search, which may be lengthy if there are many records in the tables.

You can click the Cancel button in the progress indicator at any time to halt the search.

Note: You can carry on other Biota tasks during the search, as long as you do not edit records, delete records, or add new records to either of the two tables involved in the search.

- **If no childless records are found,** Biota will display a message like the one below.

- **If one or more childless records are found,** Biota will display the records in the standard output screen for the parent table, where you can work with them using all the techniques outlined in Chapter 9.

Finding Orphan Records

1. **From the Find menu, select Find Orphan Records**. A table selection window appears.

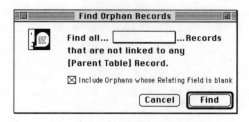

2. **Using the mouse, select a child table from the popup list.** The Specimen table, with the Species table as parent, is being selected in the illustration below.

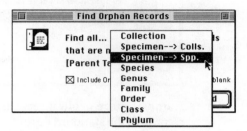

Notice that the popup includes two options for the Specimen table, since it is a child of both the Species and the Collection tables. A Specimen record can be an orphan of either table or of both.

3. **If you want to limit the search to child records that have an entry in the relating field for the parent table,** *uncheck* the checkbox in the table selection window. (It is checked, by default.) If you uncheck the checkbox, the search will gather only those child records for which the (nonblank) entry in the relating field matches no Key value (pp. 6–7) in the parent table.

4. **Click the Find button** in the table selection window to launch the search. The progress indicator appears during the search, which may be lengthy if there are many records in the tables.

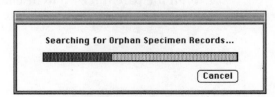

You can click the Cancel button in the progress indicator at any time to halt the search.

Note: You can carry on other Biota tasks during the search, as long as you do not edit records, delete records, or add new records to either of the two tables involved in the search.

- **If no orphan records are found,** Biota will display a message like the one below.

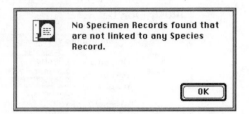

- **If one or more orphan records are found,** Biota will display the records in the standard output screen for the child table, where you can work with them using all the techniques outlined in Chapter 9.

Automatically Updating Child Records by Changing a Parent Record

Sometimes, you need to change the Key field in a record (Key fields are introduced in Chapter 2). For example, you may discover that a Genus name is misspelled in your database, or perhaps the name of a Genus is changed for nomenclatural reasons and you want to update the record. If the Genus record to be changed has any linked Species records, however, the new version of the generic name must also replace the old version in the [Species] Genus relating field. Otherwise, the Species records will be orphaned (they will not longer be linked to any Genus record).

Biota takes care of updating child records for you. If you confirm the change, Biota updates the relating field in all child records when you change the value in the Key field of the parent record. This feature applies to all relations between Core tables. Updates of relating values in child tables are made automatically, without confirmation, when the child table is a Peripheral table (Image Archive, Notes, Determination History, and Auxiliary Field Value tables). (See Chapter 2 and Appendix A.)

To illustrate, suppose you need to correct the misspelled genus name "*Cavendishius*" to *Cavendishia*.

1. **Find the parent record to be changed and display it in the input screen for the parent table.**

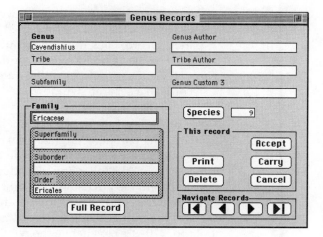

2. **Edit the Key field, then click the Accept button in the input screen** to register the changes. If the record is linked to child records, Biota offers two options, explained in the screen below.

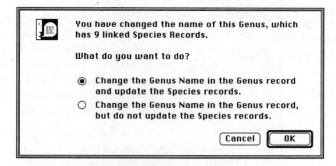

- **If you choose to update the child records** (the default option), all relating fields in linked child records will be updated to match the change in the parent record Key field.

- **If you choose to change the Key field in the parent record without updating the relating field in the child records** (the second option), Biota offers you another chance to change your mind.

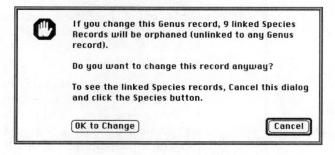

As suggested in the warning above, if you want to take a look at the child records that will be orphaned before severing their link with the parent record, click the Cancel button in the warning screen above. Then use the "Child Records" button (125–126)—the Species button in the Genus input screen illustrated in step 1—to display the child records in a separate window.

Note: If you need to "split" a taxon, leaving some child records linked to an existing parent record (an old name) while reassigning other child records to a new parent record, you can speed the process and avoid errors by using the following procedure: (1) Display all child records for the old parent record, using the Child Records button. (2) Make a Subselection of the child records to be assigned to the new parent record. (3) Declare the Subselection a Record Set. (4) Use the Find and Replace tool (pp. 220–225) to change the relating field in the child table Record Set (only!) to the new parent record name. (5) Create the new parent record. (6) Click the Child Record button in the new parent record to check the new links.

Updating Records Using the Find and Replace Tool

The Find and Replace tool (in the Special menu) is a powerful—and therefore dangerous—utility for correcting errors or updating records in any table in the Biota structure. Not only Core tables but Peripheral tables can be accessed with this tool. The search be restricted to the records of a current Record Set (for tables that have them), limited to a selection of records you find using the Search Editor (pp. 192–200), or can encompass all records in a table.

Because of the potential for massive changes in your Data File, this tool can be used only by a user with Administration access privileges if you have enabled Biota's user password system (Chapter 23). If you have not enabled the user password system, anyone can use this tool. In any case, be sure you understand how it works and be sure to make a backup copy of your Data File before you use the tool, just in case (see Appendix F).

WARNING: Changes made with this tool cannot be undone except by using the tool in reverse to change records back to their previous values.

The Find and Replace tool searches for and replaces field values in one field at a time. Like the Replace command in a text processing application, the Biota tool looks for a "current value" that you specify. When Biota finds the current value in a record, it replaces it with the "new value" that you specify.

In the example below, suppose that a botanist has replaced the fluid in which 23 plant specimens are stored, and he or she needs to update the Specimen records to reflect the change. The entry "FAA" (for formalin and acetic acid) will be replaced by the entry "90% Ethanol" in the [Specimen] Medium field, in all records in which "FAA" currently appears in the active Specimen Record Set.

1. **Create a Record Set, if necessary, for the records to be searched.**
   - **If you want to restrict the search to certain records** in a Core table or the Determination History table, create a Record Set for the records to be changed (pp. 13–15 for Core tables; pp. 375–377 for the Determination History table).
   - **If you want the search to encompass all records in a table, or the table does not support Record Sets** (among Peripheral tables, only the Determination History table supports Record Sets), proceed to step 2.
   - **If you want to use the Search Editor** to find the records to be searched by the Search and Replace tool, proceed to step 2.
2. **From the Special menu, choose Find & Replace.** The screen below appears.

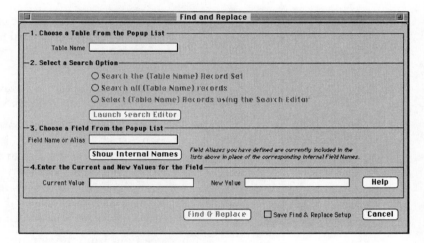

3. **Choose a table from the popup list in the top panel.** Table names are listed alphabetically.

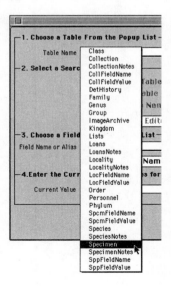

The text of the search option button in the second panel is updated to reflect the choice of table. The field list in the third panel is created for the table you chose.

4. **Select a search option from the second panel.**

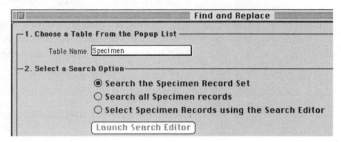

- **The Record Set option will be enabled only if the table you chose is a table that supports Record Sets and the current Record Set for the table is not empty.** Otherwise, Biota posts a message highlighting the other two options in this panel.

- **The "all records" option is always enabled,** and is the default when the Record Set option is disabled.

- **The Search Editor option is always enabled**. If you select it, the Launch Search Editor button is enabled. If you click the button, the Search Editor will be displayed for selecting records for the Find and Replace search. (The Find and Replace tool will nonetheless replace values only in records in which the Current Value matches, not in *all* records found by the Search Editor.) See pp. 192–200 if you need help using the Search Editor.

5. **Choose a field from the popup list in the third panel.** Field names are listed alphabetically.

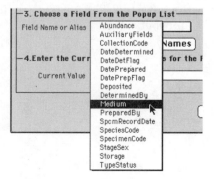

6. **Using the Field Name/Alias Button.** The button below the field name popup list controls whether Biota lists strictly Internal Field Names (as given in Appendix A) or substitutes a Field Alias for the corresponding Internal Field Name for each field you have given an Alias. (See pp. 323–326 to learn about Field Aliases.)

- **If no Aliases have been defined,** the button is disabled and reads "Show Field Aliases." The caption to the right of the button reads "Internal Field Names appear in the lists above. No Field Aliases have been defined."

- **If any Aliases have been defined,** when you first open the Find and Replace screen, the button is enabled and reads "Show Internal Names." The caption to the right of the button reads "Field Aliases you have defined are currently included in the lists above in place of the corresponding Internal Field Names."

- **If you click the Show Internal Names button,** the button text changes to "Show Field Aliases" and the caption reads "Internal Field Names appear in the lists above. Click the button to the left to include the Field Aliases you have defined."

7. **Enter the Current Value to be replaced and press TAB.** In the illustration below, "FAA" has been entered as the Current Value to be sought.

Biota now searches, within the selection of records indicated in step 4, for records with the Current Value entry in the field you selected in step 5. In the example, Biota looks for "FAA" in the Medium field of the Specimen records in the current Specimen Record Set.

- **If no records are found that meet these criteria,** a message like the one below appears, and the Find & Replace button remains disabled.

If you click the Try Again button in the message window, the Current Value entry area is cleared for a new value. If you click the Cancel button, the Find and Replace screen is dismissed.

- **If records are found the meet the criteria,** the Find & Replace button is enabled (see below) and the cursor moves to the New Value entry area.

NOTE: You can use the wildcard character, @, in the Current Value entry area. For example, entering *B@* will find all records (among those you have designated) that have a value beginning with the letter *b* (either lower or uppercase) for the field indicated in step 5. Entering *Bal@* will find values that begin with *Bal*, and so on. If you enter only the wildcard character @, all values for the field will be found.

8. **Enter the New Value to be entered and press TAB.** In the illustration below, "90% Ethanol" has been entered as the New Value to be substituted for "FAA."

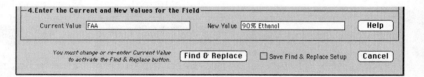

NOTE: If you leave the New Value field blank, the Current Value will be replaced with a blank entry in each record found.

9. **If you want to save the Find and Replace setup, click the checkbox.**

10. **Click the Find & Replace button to launch the search and replacement process.** Biota asks you to confirm the proposed changes.

Notice the default button is Cancel, for safety's sake.

NOTE: The records changed are not displayed after the changes are made. If you changed values in a Record Set, you can examine the changes in the Record Set easily using the Display menu command for the appropriate table.

Updating Records Using the Import Editor

You can use Biota's Import Editor (Chapter 26) to update fields in existing records in a Biota data file.

The Import Editor can update data in any Biota table using input from free-format, delimited text (ASCII) files, such as the column-by-row text files you can create with a spreadsheet application or export as text flat-files from database management applications.

There is no search for "current values" when you use the Import Editor to update records. When you use the Import Editor to update a record in a Core table, Biota matches the Key field in the text file with the Key field in the existing Biota record, then replaces field values in the existing record with the corresponding values from the text file record, regardless of field values.

The complete process is described in detail in Chapter 26, which you should consult before attempting to update records with the Import Editor.

Chapter 12 Creating, Finding, and Updating Specimen Series

Biota's origin as a data management system for a biodiversity inventory involving large numbers of new specimens (see the Preface) is nowhere more apparent than in the tools of the Series menu. The "industrial strength" Series tools are designed to carry out several kinds of repetitive operations on groups of Specimen records, with an absolute minimum of human effort.

The Series menu offers four special tools for efficient entry, updating, and retrieval of groups of Specimen records. All four are optimized for use of Specimen barcodes, but can be used efficiently with manual Specimen Code entry as well.

- **Input Specimen Series** creates groups of *new* Specimen records—with consecutive Specimen Codes—that share the same Collection data.
- **Input and Identify Specimen Series** creates groups of *new* Specimen records—with consecutive Specimen Codes—that share the same Collection data, the same identification data (Species Code and determination information), or both collection and identification data.
- **Find Specimen Series** (also in the Find menu as "By Specimen Code Series") finds groups of *existing* Specimen records. The Specimen Codes for these records can be consecutive or in random order.
- **Find and Identify Specimen Series** finds groups of *existing* Specimen records and adds identification data (Species Code and determina-

tion information) and/or updates other Specimen fields (Stage/Sex, Storage, Type Status) in these records. The Specimen Codes for these records can be consecutive or in random order.

Using the Input Specimen Series and Input and Identify Specimen Series Tools

In the field collection of specimens, it is common for a group of specimens to be collected in the same "collecting event"—from the same site, quadrat, trap, trawl, extraction, host, or modular organism—on the same day by the same collector(s). In the Biota data model (Appendix A), the Specimen records for such a group of specimens share the same Collection Code, representing the single Collection record that shows the details of the collecting event.

In some situations, a group of new specimens not only share the same Collection Code, but also have been identified as representatives of the same species—and thus require the same Species Code in their Specimen records.

It is certainly possible to create individual Specimen records for each specimen in a group in either of these situations using the ordinary Specimen input screen (pp. 143–152), speeded up by using the Carry button (pp. 41–42), but Biota offers a much more efficient way to do the job.

With the Input Specimen Series tool, you set up the input screen, shown below, by entering the first and last consecutive Specimen Codes for a group of records, plus the Collection, preparation, and related data that the Specimen records share. Biota then creates all the records automatically when you are ready.

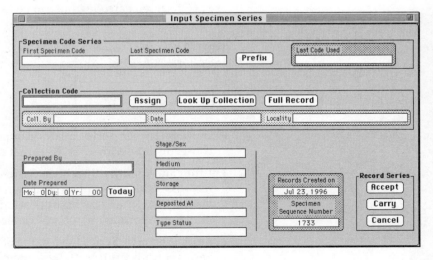

The Input and Identify Specimen Series tool, below, is identical to the Input Specimen Series tool, except for the addition of the Species Code and Determination input areas, familiar from the standard Specimen input screen.

Creating, Finding, and Updating Specimen Series

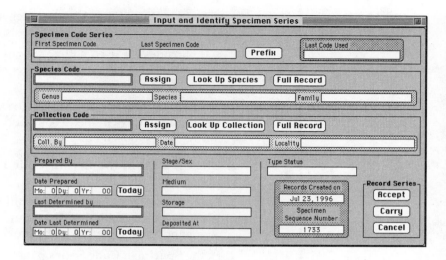

NOTE: Using the Input and Identify Specimen Series tool without entering a Species Code or determination information is exactly equivalent to using the Input Specimen Series tool for the same information. The latter is simply a subset of the former.

Specimen Series Input: Step by Step

Before using these tools, you should read "Specimen Input" (pp. 143–152) and Chapter 7, "Record Codes," especially "Using Alphanumeric Specimen Codes With the Series Tools" (pp. 105–106).

1. **From the Special menu, select Record Code Prefixes** to confirm that the Specimen Code settings are correct (pp. 106–108).

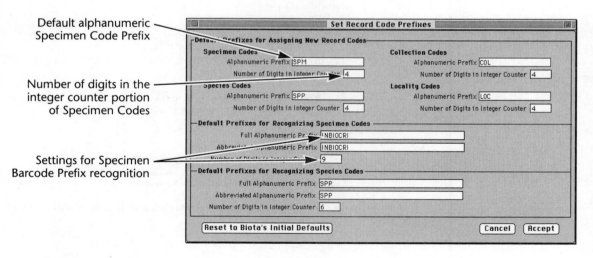

Default alphanumeric Specimen Code Prefix

Number of digits in the integer counter portion of Specimen Codes

Settings for Specimen Barcode Prefix recognition

In the illustration above:

- **The default alphanumeric prefix** for Specimen Codes produced by the Series tools will be *SPM*.

- **The number of digits in the integer counter** portion of Specimen Codes produced by the Series tools will be 4.
- **If barcodes are used** with the Series tools, the alphanumeric prefix must be *INBIOCRI* and the integer counter portion of the barcodes must contain 9 digits.

All these settings can be changed to suit your needs (pp. 105–108). Th rest of the settings in the Record Code Prefixes screen are irrelevant for the Series tools themselves.

2. **Click Accept in the Record Code Prefixes screen** to record any changes and dismiss the screen.
3. **From the Series menu, choose Input Specimen Series or Input and Identify Specimen Series.** The input screen appears (screens for both commands are illustrated in the previous section).
4. **In the Specimen Code Series panel:**
   - *Either:* **Manually enter the integer counter for the First Specimen Code** in the series for which you want to create Specimen records (illustrated below). *If the counter has any leading zeros, omit them.* Press TAB.

Biota will prefix leading zeros (below), if necessary to complete the number of digits you specified in the Record Code Prefixes screen.

◊ **If you entered leading zeros,** you will see this error message.

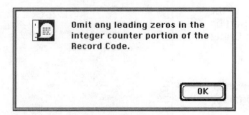

◊ **If you entered an alphabetic character,** you will see the error screen below.

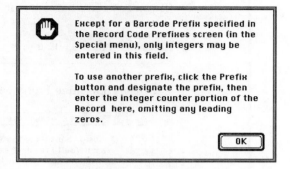

NOTE: If you want to continue with a Specimen Code series based on the *sequence number* for the Specimen table (p. 102), you should enter the next sequence number as the integer counter for the First Specimen Code. The next available sequence number is displayed and updated in a panel in the lower-right corner, below the Record Created date.

- *Or:* **Read in the barcode for the First Specimen Code** in the series for which you want to create Specimen records. (Then press TAB if your barcode reader is not set to enter an end-of-line character automatically.)

If the alphanumeric prefix correctly matches the barcode prefix setting in the Record Code Prefixes screen, the barcode will be accepted. Otherwise, you will see the alphabetic character error message illustrated in the previous paragraph.

5. **In the Specimen Code Series panel, repeat step 3 for the Last Specimen Code entry area.** If there is only one specimen in the series, you can skip this step. (Integer counter entry is illustrated first, then barcode entry, below.)

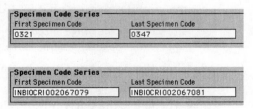

6. **If you entered an integer counter and wish to insert a Specimen Code Prefix** to complete the Specimen codes, click the Prefix button.

(If you entered a barcode, the Prefix button will have no effect.)

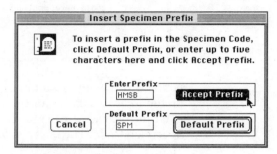

The Insert Specimen Prefix window appears.

7. **In the Prefix window:**

   - *Either:* **Enter an ad hoc prefix in the Enter Prefix area** (in the illustration above, the prefix *HMSB* has been entered) and click the Accept Prefix button.

   - *Or:* **Click the Default Prefix button** to enter the prefix displayed next to the button (*SPM* in the illustration; see step 1 to change the default prefix).

   Biota completes the First Specimen Code and Last Specimen Code entries by inserting the prefix.

   NOTE: If you prefer, you can use the Prefix button to set a prefix first, then enter the integer counters for the First and Last Specimen Codes. In this case, the codes will be completed as you type them.

8. **Complete as many of the rest of the entries in the Series input screen as you wish,** following the instructions for the standard Specimen input screen on pp. 143–152.

   All records in the range of codes you specified in the Specimen Code Series panel will be created with the entries you make.

9. **Click the Accept or Carry button to create the series of Specimen records.**

   - **If you use the Carry button,** all entries except for the First and Last Specimen Code will be carried forward. You can edit them, if necessary, for the next series.

- **If the records have been created successfully,** the Last Code Used display area (below) now shows the former Last Specimen Code entry, to guide your input for the next series. The Sequence Number display (step 4) is updated.

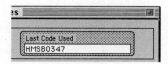

- **If you have requested the creation of more than 20 records,** a confirmation screen appears.

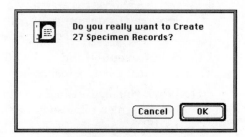

- **If the password system has been enabled** (pp. 409–415) and you have requested the creation of more records than the limit set in the Preferences screen (Appendix C), a message will appear explaining that you cannot complete the action. (A user with administration access privileges can change the limit.)

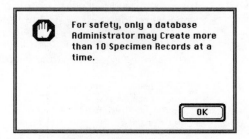

10. **Create additional series of records** if you wish, following steps 4 through 9.
11. **To examine the records created,** click the Cancel button in the Series input screen. The series you have created appear in a special version of the Specimen output screen. In place of the standard Add Specimen button, an Add New Series button is present.

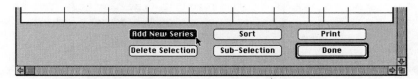

12. **To create another series of Specimen records,** click the Add New Series button (above). An option screen appears.

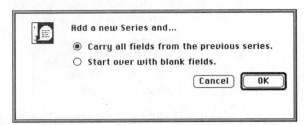

You can choose to carry forward all entries except the First and Last Specimen Code entries (exactly the equivalent of having clicked the Carry button after the last series you created), or start fresh.

13. **Select an option and click OK.** The Specimen Series input screen you were using before reappears.

14. **When you are finished creating series, click the Done button** in the output screen to present the standard Record Set option screen (pp. 13–15).

Using the Find Specimen Series and Find and Identify Specimen Series Tools

The tools of the Find and Tree menus offer many ways to find existing records, based on their content or their linkage to other records, and several ways to update, them if necessary, once found (Chapter 11). The Find Specimen Series and Find and Identify Specimen Series Tools have the specialized purposes of finding Specimen records based strictly on Specimen Code and updating them if you wish.

Specimen Codes may be entered in any order. Neither tool requires that Specimen Codes be sequential, although they may be.

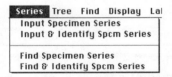

These two tools share a common interface for entering individual Specimen Codes or Specimen Code series.

The Find Specimen Series tool (below) simply finds a group of Specimen records, based on Specimen Code. One common use for the tool is to gather Specimen records into a Record Set to record a returned Specimen loan. (The New Loan screen has its own tools for gathering Specimen records to record a *new* loan.) See pp. 339–341.

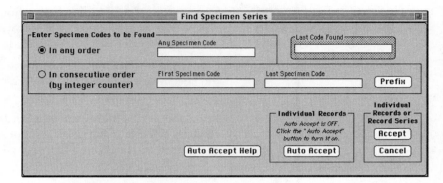

In addition to finding Specimen records based on Specimen Code, the Find and Identify Specimen Series tool (below) automatically updates the Species Code and other fields you specify, in all records found.

This tool is ideal for efficiently updating information in groups of existing Specimen records, whether or not their Specimen Codes are numerically sequential. Some common uses include:

- **Recording or changing the determination data** (Species Code, Last Determined By, Date of Determination) in the records for a group of specimens that have been grouped and identified.

- **Recording or changing Stage/Sex, Storage, or Type Status data** for a group of records. For example, the Storage field could be updated for a group of duplicate specimens or vouchers that have been moved to an auxiliary storage location. The Type Status field could be updated when a type series is designated. All three of these fields can may be renamed for other uses. See pp. 323–326.

Finding (or Finding and Identifying) Specimen Series: First Steps

Before using these tools, you should read "Specimen Input" (pp.143–152) and Chapter 7, "Record Codes," especially "Using Alphanumeric Specimen Codes With the Series Tools" (pp. 105–106).

1. **From the Special menu, select Record Code Prefixes** to confirm that the Specimen Code settings are correct (pp. 105-108). See steps 1 and 2 on pp. 229–230 for details.

2. **From the Series menu, choose Find Specimen Series or Find and Identify Specimen Series.** The input screen appears (screens for both commands are illustrated in the previous section).

3. **Choose a search option** in the upper panel:

- *Either:* **In any order.** With this option selected, you enter complete Specimen Codes, in any order. Continue with steps in the next section, "If You Choose 'In Any Order.'"

- *Or:* **In consecutive order.** If you select this option, you enter the integer counter portion of the First and Last Specimen Code of a consecutive series, then add a Prefix if necessary. Continue with the section entitled "If You Choose 'In Consecutive Order.'" (pp. 238–242)

If You Choose "In Any Order"

1. **Enter a complete Specimen Code** for an existing Specimen record, manually or using a barcode reader, in the Any Specimen Code entry area, illustrated below.

2. **Press TAB.** Biota searches immediately for the record.

NOTE: You need not press TAB if you are using a barcode reader that has been set to enter an end-of-line character (RETURN [Macintosh] or ENTER [Windows]) automatically after each read.

- **If the record is found,** the Specimen Code remains entered in the input area.

- **If the record is cannot be found,** an error message appears and the Specimen Code entry area is cleared.

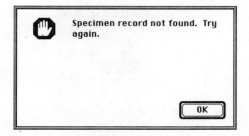

If you are using the Find and Identify Specimen Series tool, complete as many of the remaining entries in the screen as you wish, following the instructions for the standard Specimen input screen on pp. 143-152.

- **Values to enter:** Enter values for Species Code, Last Determined By, Date Last Determined, Stage/Sex, Storage, and/or Type Status. If you enter a value for Species Code, then Last Determined By and Date Last Determined are required entries, to make sure that changes in determination can be tracked.

- **Existing values that will be replaced:** All records in the range of Specimen Codes you just entered—and any additional records whose Specimen Codes you enter after using the Carry button—*will be updated with the entries you make in the Find and Identify Specimen Series screen for these fields, regardless of the existing values in the fields.*

- **Existing values will that will not be replaced:** Existing values in any field that you leave blank in the Find and Identify Specimen Series screen *will not be altered* in the Specimen records found.

3. **To accept the first Specimen Code** and clear the Specimen Code entry area for the next code:

    - *Either:* **Click the Accept button** if you want to make another manual entry (Find Specimen Series, below left; Find and Identify Specimen Series, below right),

     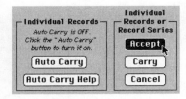

    - *Or:* **Click the Auto Accept button** (Find Specimen Series, below left) **or the Auto Carry button** (Find and Identify Specimen Series, below right), if you want to set up for automatic Specimen Code entries.

    These are toggle buttons. Click once to turn on Auto Accept or Auto Carry, click again to turn it off.

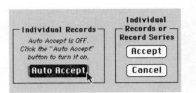 

- *Or:* **Click the Carry button** (Find and Identify Specimen Series only).

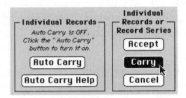

If you click Auto Carry or Carry in the Find and Identify Specimen Series screen, any entries you made in the update fields will carried over to the next record to be updated.

The Last Specimen Code from the previous series will be displayed in the Last Code Found panel (upper right corner of the screen).

4. **If you clicked the Auto Accept or Auto Carry button** to toggle it on, you can now proceed to enter additional Specimen Codes without having to click any buttons on the screen.

   Each time you enter a new Specimen Code and press TAB (or the barcode reader reads a Specimen barcode and enters an end-of-line character), the Specimen record is found automatically. If you are using the Find and Identify Specimen Series screen, the record is not only found automatically, but also updated automatically.

   With a barcode reader set to append the end-of-line character, the entire process requires neither keyboard nor mouse once the Auto Accept or Auto Carry button is toggled on.

5. **Follow the steps in the section entitled "Finding (or Finding and Identifying) Specimen Series: Final Steps."** (p. 242)

If You Choose "In Consecutive Order"

1. **In the "In consecutive order" panel:**
   - *Either:* **Manually enter the integer counter for the First Specimen Code** in the series want to find or find and update (illustrated below). *If the counter has any leading zeros, omit them.* Press TAB.

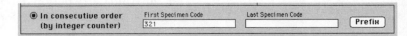

Biota will prefix leading zeros, if necessary to complete the number of digits you specified in the Record Code Prefixes screen (below).

◊ **If you entered leading zeros,** you will see this error message.

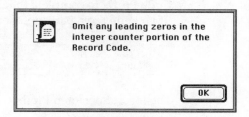

◊ **If you entered an alphabetic character,** you will see the error screen below.

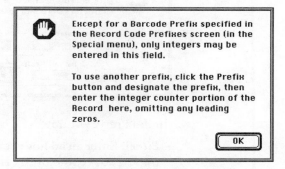

- *Or:* **Read in the barcode for the First Specimen Code** in the series for which you want to create Specimen records. (Then press TAB if your barcode reader is not set to enter an end-of-line character automatically.)

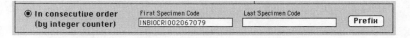

If the alphanumeric prefix correctly matches the barcode prefix setting in the Record Code Prefixes screen, the barcode will be accepted. Otherwise, you will see the alphabetic character error message illustrated in the previous paragraph.

2. **Repeat step 1 for the Last Specimen Code entry area in the "In consecutive order" panel.** If there is only one specimen in the series, you can skip this step. (Integer counter entry is illustrated first, then barcode entry, below.)

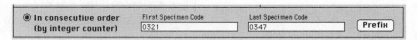

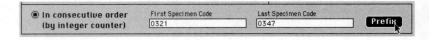

3. **If you entered an integer counter and wish to insert a Specimen Code Prefix** to complete the Specimen codes, click the Prefix button. (If you entered a barcode, the Prefix button will have no effect.)

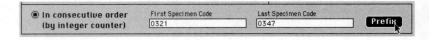

The Insert Specimen Prefix window appears.

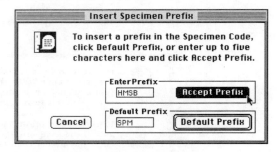

4. **In the Prefix window:**

   ♦ *Either:* **Enter an ad hoc prefix in the Enter Prefix area** (in the illustration above, the prefix *HMSB* has been entered) and click the Accept Prefix button.

   ♦ *Or:* **Click the Default Prefix button** to enter the prefix displayed next to the button (*SPM* in the illustration; see pp. 102–103, step 1, to change the default prefix).

   Biota completes the First Specimen Code and Last Specimen Code entries by inserting the prefix.

NOTE: If you prefer, you can use the Prefix button to set a prefix first, then enter the integer counters for the First and Last Specimen Codes. In this case, the Codes will be completed as you type them.

5. **If you are using the Find and Identify Specimen Series tool,** complete as many of the rest of the entries in the screen as you wish, following the instructions for the standard Specimen input screen on pp. 143–152.

   ♦ **Values to enter:** Enter values for Species Code, Last Determined By, Date Determined, Stage/Sex, Storage, and/or Type Status. If you enter a value for Species Code, then Last Determined By and Date Last Determined are required entries, to make sure that changes in determination can be tracked.

- **Existing values that will be replaced:** All records in the range of Specimen Codes you just entered—and any additional records whose Specimen Codes you enter after using the Carry button—*will be updated with the entries you make in the Find and Identify Specimen Series screen for these fields, regardless of the existing values in the fields.*

- **Existing values that will not be replaced:** Existing values in any field that you leave blank in the Find and Identify Specimen Series screen *will not be altered* in the Specimen records found.

6. **Click the Accept button** (either tool) **or the Carry** Button (Find and Identify Specimen Series only) to accept the first series of consecutive Specimen Codes.

The Last Specimen Code entry will be displayed in the Last Code Found panel (upper-right corner of the screen), and Specimen Code entry areas will be cleared for the next series.

- **If you have asked Biota to find** (left, below) **or find and update** (right, below) **more than 20 records,** a confirmation screen appears.

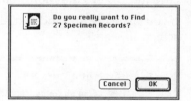 

- **If the password system has been enabled** (pp. 409–415) and you have requested the updating of more records than the limit set in the Preferences screen (Appendix C), a message will appear explaining that you cannot complete the action. (A user with Administration privileges can change the limit.)

NOTE: The Auto Accept or Auto Carry toggle button is disabled when you choose the "Consecutive order" option, since these buttons are used only for single Specimen Code entry.

7. **Follow the steps in the next section,** "Finding (or Finding and Identifying) Specimen Series: Final Steps."

Finding (or Finding and Identifying) Specimen Series: Final Steps

This section describes final steps for using the Find Specimen Series and Find and Identify Specimen Series tools. These steps are the same for both the "In any order" and "In consecutive order" options described in the previous two sections.

1. **Create additional series of records** if you wish, following the steps in the appropriate preceding section.

2. **To examine the records found (or found and updated),** click the Cancel button in either Find Series input screen. The records appear in a special version of the Specimen output screen. In place of the standard Add Specimen button, an "ID or Find More" button is present.

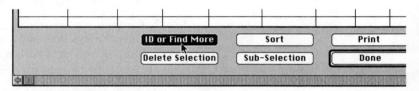

3. **To find (or find and update) another series of Specimen records,** click the "ID or Find More" button (above).

If you are using the Find and Identify Specimen Series tool, an option screen appears.

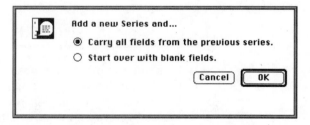

You can choose to carry forward all entries except the First and Last Specimen Code entries (exactly the equivalent of having clicked the Carry button after the last series you created), or start fresh. (There is nothing to "carry" with the Find Specimen Series tool, so this option screen does not appear.)

4. **If the option screen appears, select an option and click OK.** The Specimen Series screen you were using before reappears.

5. **When you are finished using the Series tool, click the Done button** in the output screen to present the standard Record Set option screen (pp. 13–15).

Chapter 13 Printing Labels

Biota provides several special tools (in the Labels menu) for producing labels for specimens, collections, and species. You can print labels on plain, acid-free paper or, for slide labels, adhesive-backed acid-free paper. Alternatively, you can export label data to text files, where you can format and duplicate the label text with a word processing or spreadsheet application to suit your personal or institutional preferences.

Warning: All Biota preformatted labels use the Helvetica font (Macintosh) or the Arial font (Windows). If you do not have the appropriate font installed (in the printer's memory as well as on your computer, for some printers), the layout of labels will probably not be correct, depending on what font is substituted for Helvetica or Arial.

Printing Procedures

For all printed labels and reports, Biota presents the same series of screens and options when you are ready to preview and print. This section outlines the steps to take once you have chosen the label or report options you want. See the section that applies to the labels (this chapter) or report (Chapter 14) you are printing, before using the steps below.

1. **Launch the printing process** by clicking a Print button or the OK button in a Label Option screen, after setting up any options for the labels or report. (See the appropriate section later in this chapter for help setting up options). A Page Setup window appears.

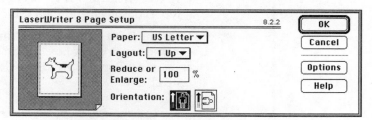

The window shown here is for the Macintosh operating system, with a LaserWriter selected. Your own printer and operating system (the Windows Page Setup window has the same basic options) will affect the exact appearance of the window.

2. **Set the reduction or enlargement in the Page Setup window,** if this option is available for your printer. With this setting:

    ♦ **You can change the apparent font size and total size** of Biota label formats if they do not suit your needs.

    ♦ **For very tiny labels,** such as pin labels, you can set the enlargement to 200% to 400% to allow you to study them in Preview mode (see below) before printing. If you do this, don't forget cancel the actual printing, then launch it again and choose the 100% default, once you are satisfied with the preview. (Otherwise, you will print some very large pin labels.) Since Biota remembers your previous label option settings, it is easy to relaunch the printing.

3. **Click the OK button in the Page Setup window** (illustrated above). The Print window appears.

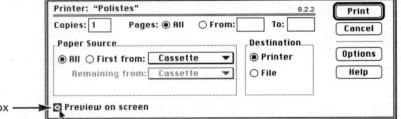

Preview on screen checkbox

Again, the window shown here is for the Macintosh operating system, with a LaserWriter. Your own printer and operating system will affect the exact appearance of the window.

4. **Choose a Preview option.** Notice the "Preview on screen" checkbox in the lower-left corner of the Print window (illustrated above). (The box is in the same position in Windows versions of the screen.)

    ♦ **If you leave the Preview checkbox unchecked,** labels or reports go directly to your printer.

    ♦ **If you click the Preview checkbox,** you can check the labels or report onscreen and then Print or Cancel.

5. **Click the Print button in the Print window** (illustrated above). If you checked the Preview box, the first page of labels or of the report will appear in a Preview window with a row of buttons at bottom, illustrated below.

The Zoom button produces a "floating marquee" that you can grab and drag with the mouse to enlarge different areas of the preview window. The other buttons are self-explanatory.

NOTE: If you do not have a printer attached to your computer, the computer may lock up if you click the Print button in this screen. You can avoid this possibility if you remember to click Stop Printing when you are through with the preview.

Label Option Windows

Each Label tool in Biota presents an option window, with which you can control certain features of the labels to be produced. Because most of the options appear in option windows for several different label tools, it makes sense to describe them just once, in this section of the chapter.

Options that apply to only one or two kinds of labels are discussed in the section on the particular tool or tools to which they apply.

Sort Options

- *Either:* **Sort selection by Record Code or Key field.** This is the default option for all label tools. For all Specimen labels (both locality and determination labels), you can choose to sort by Specimen Code before printing them. For Collection labels, you can sort by Collection Code. With the Custom Label tool, which prints labels for any Core table, you can choose to sort by Key field (see pp. 6–7 for a discussion of Key fields).

- *Or:* **Use the Sort Editor to sort the records.** For all label tools, you can choose to sort the records using the Sort Editor (pp. 131–139).

- *Or:* **Sort the selection taxonomically.** For all specimen determination labels, you have the option of sorting Specimen records taxonomically before printing the labels. The Specimen records are sorted by Specimen Code *within* species, groups of records for the same species are sorted alphabetically by specific name, these groups are sorted by genus, and so on. This option does not apply to locality labels or collection labels.

Data Options

- **Include Specimen Code on each label.** A checkbox for this option is offered for both locality and determination labels of all kinds. It is not exclusive of any other option, where offered.

- **Include Host Specimen name (Genus, specific name, and Family) on each label.** A checkbox for this option is offered for collection labels (Host Specimen Code is an attribute the Collection table; see pp. 393–396) and for pin, slide, and vial Specimen locality labels. It is not exclusive of any other option, where offered. If you choose this option, Host Specimen data are checked case by case and are included only if present in individual Collection records.

- **Include Family on each label.** A checkbox for this option is offered for slide and vial specimen determination labels (but not for pin determination labels, on which Family is not traditionally included) and for herbarium specimen labels. It is not exclusive of any other option, where offered.

- **Collection date options.**
  - *Either:* **Use three-letter English abbreviations for Months** (Jan, Feb, Mar, Apr, etc.). This is the default option for all label tools.
  - *Or:* **Use Roman numerals for Months** (I, II, III, IV, etc. for Jan, Feb, Mar, Apr, etc.). In some disciplines, this style is preferred.

  In either case, partial dates (Month-Year or Year-only) and "intelligent" collection date ranges are exported or printed (pp. 113–114).

Collection Labels

Labels produced from Biota Collection records (with fields from the parent Locality record) have several uses. When mass collections are made, such as with quadrats, traps, trawls, extractors, foggings, sweeps, and so on, many specimens share the same collection data. In whatever manner the collection (or lot) is stored until sorted and identified, it needs a collection label. Often, "residual" specimens will stay in mass storage even after focal groups have been removed and labeled individually or in smaller groups.

In entomology, locality labels attached to specimens traditionally contain no information about the individual specimen (such as a Specimen Code or the equivalent) and provide only approximate geographical information. They are usually printed in large sets and applied to all specimens from a broadly defined locality (broadly, that is, by GIS/GPS standards, pp. 115–116).

Using Biota's ability to export standard locality data (from Collection and parent Locality records) to disk file, entomologists (and others) can fine-tune the format, delete unwanted fields, and then produce as many copies of each label as needed for each set of data exported, using a word processing or spreadsheet application. The labels can then be printed on a laser printer or master sheets can be sent to a commercial printing service.

Information Included on Collection Labels

When you use the Collection Label tool, as described in the next section, Biota exports or prints the following fields from the Collection and Locality tables, for each record in the active Collection Record Set. The intention in choosing these fields for the tool was to be fairly inclusive, since you can delete extra information easily in a text file, whereas looking it up in the database to add it is time-consuming.

- Country
- State/Province

- Locality Name
- District
- Elevation (p. 161)
- Latitude and Longitude (in DMS, p. 116)
- Collected By
- Date Collected or collection date range (pp. 113–114)
- Method
- Host Genus, specific name (Species Name field), and Family (optional)
- Collection Code

Here are three examples. In the second—a collection record for hummingbird flower mites taken from a plant specimen—Host Specimen data are included. An entry for District is included in the third label (*Sarapiquí*).

ECUADOR: Galápagos Islands Floreana Island Elev 10 m 1°17'0"S90°26'0"W coll. C. Darwin 24 Sep 1832 Shotgun Collection Code: CD002	PERU: Cajamarca Cajamarca-Celendin rd. Elev 3080m 7°5'0"S78°25'0"W coll. J. L. Luteyn 12 Feb 1985 Search Ex Siphocampylus sanguineus (Campanulaceae) Collection Code: JLL11314H	COSTA RICA: Heredia Sarapiquí La Selva Elev 150 m 10°26'0"N84°1'0"W coll. J. T. Longino 9 Oct 1989 Search Collection Code: JTL03621

(The examples above are not actual labels, just the data that are exported for each Collection record.)

If you choose the print option, rather than exporting to a text file, the printed labels carry the information shown above. Here is the middle example in printed format, shown at actual size.

```
PERU: Cajamarca
Cajamarca-Celendin rd.
Elev 3080m
7°5'0"S78°25'0"W
coll. J. L. Luteyn
12 Feb 1985
Search Ex Siphocampylus sanguineus
(Campanulaceae)
Collection Code: JLL11314H
```

Exporting or Printing Collection Labels: Step by Step

1. **Find the Collection records for which you wish to print labels or export label data,** using any of the tools in the Find menu (Chapter 11).
2. **Declare these records the current Collection Record Set** (pp. 13–15).
3. **From the Labels menu, choose Collection Labels.** The Collection label option screen appears.

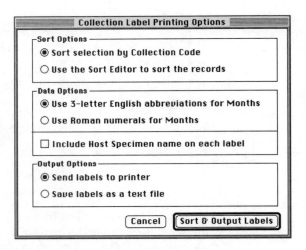

4. **Choose a Sort Option and Data Options.** See pp. 245–246.

5. **Choose an Output Option.**

   • *Either:* **Send labels to printer** (the default). If you choose this option, follow the instructions in the section on "Printing Procedures" earlier in this chapter. (pp. 243–245)

   • *Or:* **Save labels as a text file.** If you choose this option, the standard Save File window for your operating system will appear to let you name and place the file.

6. **Click the Sort & Output Labels button** in the options window.

Pin and Vial Specimen Labels

Using tools from the Labels menu, Biota can print standard locality and determination labels for pinned insect specimens as well as slightly larger ("vial") labels for fluid-preserved specimens of arthropods, plants, and other taxa.

With the advent of GPS technology (pp. 115–116), individual specimen barcodes (Appendix H), and high quality laser and ink-jet printers, *locality* labels for individual specimens are now practical, if still not yet widely used in most disciplines. Once determinations have been entered in Biota for individually identified specimens (identified by barcodes or by more traditional means), automatically printed determination labels are easy to produce.

Information Included on Pin and Vial Labels

Pin or Vial Locality Labels. When you use the Pin or Vial Locality Label tools, as described in the "Step by Step" section below, Biota prints the following fields from the Specimen, Collection, and Locality tables, for each record in the active Specimen Record Set.

- Country
- State/Province
- Locality Name
- Elevation (p. 161)
- Latitude and Longitude (in DMS, p. 116)
- Collected By
- Date Collected or collection date range (pp. 113–114)
- Host Specimen Genus, specific name (Species Name), and Family (optional)
- Specimen Code (optional)

Here are some examples of pin locality labels, at their normal printed size. They have no printed borders, so they can be trimmed if you want. To make them even smaller, use the Reduce or Enlarge setting in the Page Setup screen (pp. 243–244).

```
COSTA RICA: Heredia                    COSTA RICA: Heredia
La Selva                               La Selva
Elev 150 m 10°26'0"N84°1'0"W           Elev 150 m 10°26'0"N84°1'0"W
coll. J. T. Longino 22 Jan 1991        coll. J. T. Longino 9 Oct 1989
INBIOCRI001B459857                     INBIOCRI001459864
```

Below are vial locality labels, at their normal printed size, for the same collection data. They have printed borders.

COSTA RICA: Heredia La Selva Elev 150 m 10°26'0"N84°1'0"W coll. J. T. Longino 22 Jan 1991 INBIOCRI001B459857	COSTA RICA: Heredia La Selva Elev 150 m 10°26'0"N84°1'0"W coll. J. T. Longino 9 Oct 1989 INBIOCRI001459864

Pin or Vial Determination Labels. When you use the Pin or Vial Determination Label tools, as described in the "Step by Step" section below, Biota prints the following fields from the Specimen and Species tables for each record in the active Specimen Record Set:

- Genus
- Specific name
- Species Author
- Determined By
- Year determined
- Specimen Code (optional)
- Family (optional for vial labels, not available for pin labels)

Here are some examples of pin determination labels, at their normal printed size. They have no printed borders, so they can be trimmed if you want.

```
Strumigenys                            Procryptoceros
  godmani                                impressus
  Forel 1899                             Forel 1899
det. J. T. Longino 1992                det. J. T. Longino 1992
INBIOCRI001B459857                     INBIOCRI001459864
```

Below are vial determination labels, at their normal printed size, for the same specimen data. They have printed borders. The Family option has been included.

```
Strumigenys              Procryptoceros
  godmani                  impressus
Forel 1899               Forel 1899
det. J. T. Longino 1992  det. J. T. Longino 1992
INBIOCRI001B459857       INBIOCRI001459864
FORMICIDAE               FORMICIDAE
```

Printing Pin or Vial Specimen Labels: Step by Step

1. **Find the Specimen records for which you wish to print labels,** using any of the tools in the Find menu (Chapter 11).

2. **Declare these records the current Specimen Record Set** (pp. 13–15).

3. **To print Pin or Vial Locality labels, choose from the Labels menu:**
   - *Either:* **Pin Labels: Locality**
   - *Or:* **Vial Labels: Locality**

   In either case, the option screen that appears looks like the one below (with the appropriate window title).

   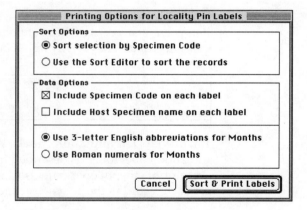

4. **To print Pin or Vial Determination labels, choose from the Labels menu:**
   - *Either:* **Pin Labels: Determination**
   - *Or:* **Vial Labels: Determination**

   In either case, the option screen that appears looks like the vial determination option screen below—except that the Family option is disabled for Pin labels.

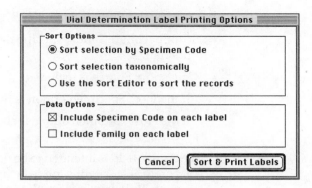

5. **Choose a Sort Option and Data Options.** See pp. 245–246.
6. **Click the Sort & Print Labels button** in the options window.
7. **Follow the instructions in the section "Printing Procedures"** (pp. 243–245) earlier in this chapter.

Slide Specimen Labels

Using tools from the Labels menu, Biota can print standard locality and determination labels for slide-mounted specimens. The precise format of these labels follows the acarological tradition (revealing Biota's ancestry), but this format may prove adequate for other small arthropods or parasites, and for organ or tissue specimens.

Individual locality labels have a long history for slide-mounted specimens in acarology and parasitology. With the advent of GPS technology (pp. 115–116), individual adhesive specimen barcodes (Appendix H), laser and ink-jet printers, and high quality paper with permanent adhesive backing, printing precise locality slide labels for individual specimens—automatically, using Biota—saves an enormous amount of time over traditional methods. Once determinations have been entered in the database for individually identified specimens, automatically printed determination labels are easy to produce and add to slides. (Barcoding the slides first makes it easy to match up labels and specimens by Specimen Code.)

Information Included on Slide Labels

Slide Locality Labels. When you use the Slide Label Locality tool, as described in the "Step by Step" section below, Biota prints the following fields from the Specimen, Collection, and Locality tables, for each record in the active Specimen Record Set:

- Specimen Code (optional)
- Country
- State/Province
- Locality Name
- Elevation (p. 161)

- Latitude and Longitude (in DMS, p. 116)
- Host Specimen Genus, specific name (Species Name), and Family (optional)
- [Collection] Source is included:
  - ◊ If the Host Specimen option is not checked.
  - ◊ If the Host Specimen option is checked, but a Specimen has no host record.

  The Source field is intended to be used for information such as "From soil core," "Blacklight sample," "In leaf litter." (There is not enough space on slide labels for both the Source field and the Host Specimen data.)
- Collected By
- Date Collected or collection date range (pp. 113–114)

Here are two pairs of slide locality labels, shown at normal printed size, that include Host Specimen information. The left pair are full-sized; the right pair leave room on the slide for a barcode label (see "Step by Step," below).

Slide Determination Labels. When you use the Slide Determination Label tool, as described in the "Step by Step" section below, Biota prints the following fields from the Specimen and Species tables, for each record in the active Specimen Record Set:

- Specimen Code (optional)
- Genus
- Specific name
- Species Author
- Year determined
- Determined By
- Stage/Sex
- Family (optional)

Here is a pair of slide determination labels, with the Specimen Code and Family options included, shown at normal printed size.

Printing Slide Specimen Labels: Step by Step

1. **Find the Specimen records for which you wish to print labels**, using any of the tools in the Find menu (Chapter 11).
2. **Declare these records the current Specimen Record Set** (pp. 13–15).
3. **To print slide labels, choose from the Labels menu:**
   - *Either:* **Slide Labels: Locality.** The option screen below appears.

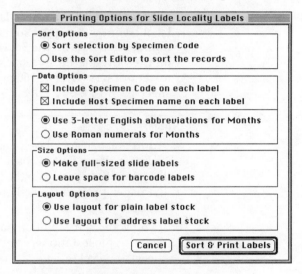

   - *Or:* **Slide Labels: Determination.** The option screen below appears.

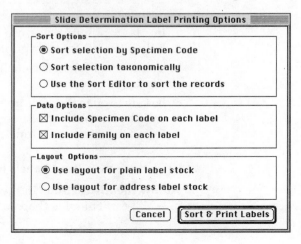

4. **Choose a Sort Option and Data Options.** See pp. 245–246.
5. **For slide Locality labels, choose a Size Option:**
   - *Either:* **Make full-sized labels** (22 by 22 mm).
   - *Or:* **Leave space for barcode labels.** In this case, the printed labels measure 22 mm by 16.5 mm, leaving room for a 22 mm by 8 mm barcode label (p. 252, Appendix H) spanning the end of the slide, with the specimen centered on the slide.
6. **Choose a Layout Option.**
   - *Either:* **Use layout for plain label stock.** With this option checked, label layout is edge-to-edge, with unscored paper. Acid-free, permanent-adhesive-backed laser or inkjet printer paper is ideal (Avery 5165 or the equivalent).
   - *Or:* **Use layout for address label stock.** With this option checked, label layout is designed so that groups of three or four labels (the number depends on the Size Option) fit within each address label on prescored, adhesive-backed, address label stock for laser or inkjet printers (Avery 5160 or 5260 or the equivalent, 30 labels per sheet, each 1 in. by $2\,5/8$ in., 3 columns by 10 rows).

   Notes:
   a. Address label stock is expensive. Print out a draft set of labels on ordinary paper first to check the text and layout and to make sure the labels fit properly on the label stock. (Hold the stock in front of the paper-printed label sheet against a strong light to check the alignment.)
   b. If you want to try out both layouts before buying special paper in quantity, most stationery and photocopy stores will sell you a few sheets of each kind.

   Warnings:
   a. In a laser printer, never use any adhesive label stock not specifically intended for laser printers.
   b. Use the single-sheet (manual) feed tray on your printer to feed label stock.
   c. Be sure to open the "face up tray"—the direct exit tray of the printer. Label stock can cause serious (and very sticky) paper jams if forced to exit by the normal, "face-down" route.

7. **Click the Sort & Print Labels button** in the options window.
8. **Follow the instructions in the section "Printing Procedures"** (pp. 243–245) earlier in this chapter.

Herbarium Specimen Labels

Biota can print standard herbarium specimen labels for dried, sheet-mounted specimens. Elements on the labels include both locality and determination data, as customary for herbarium labels. Options allow flexibility regarding which elements are included on the labels, and for some elements, the fields they represent in the database. Multiple copies of each label can be printed for duplicate specimens.

Information Included on Herbarium Specimen Labels

When you use the Herbarium Label tool, as described in the "Step by Step" section below, Biota prints the following fields from the Genus, Species, Specimen, Collection, and Locality tables, for each record in the active Specimen Record Set:

- Country
- Specimen Code (optional)
- Family (the [Genus] Family field) (optional)
- Genus
- Specific name (the Species Name field)
- Species Author
- State/Province
- District
- Locality Name
- Site (optional)
- Latitude and Longitude (in DMS, p. 116)
- Elevation (p. 161)
- Text of the first Specimen Note, containing a field description of the plant (optional). See details below
- Specimen Code (optional)
- Collector or collectors:
  - *An individual collector:* If the Short Name entered in the [Collection] Collected By field represents an *Individual* Personnel record (pp. 173–174), the Short Name itself is used for the collector element of the labels.
  - *A group of collectors:* If the Short Name entered in the [Collection] Collected By field represents a *Group* Personnel record (pp. 174–177), the Short Name for each member of the Group is included in the collector element of the labels. The order of individual names on the label follows the order displayed in the Group Editor when you click the Edit Group button in the Personnel input screen while the Group record is displayed. The names can be reordered using the Group Editor (p. 177).

- Date Collected or collection date range (pp. 113–114)
- Collector's field number from:
  - ◊ *Either:* The [Collection] Collection Code field (optional)
  - ◊ *Or:* The [Collection] Source field (optional)
- Collection (or herbarium) name, from the [Personnel] Notes field of the Project Name record (pp. 177–179)

Below, an herbarium label produced by Biota is shown, with all elements labeled. The actual size of the labels, when printed without reduction or enlargement, is 3 in. by 5 in. (7.6 by 12.5 cm.). They are printed four to a page, in landscape format.

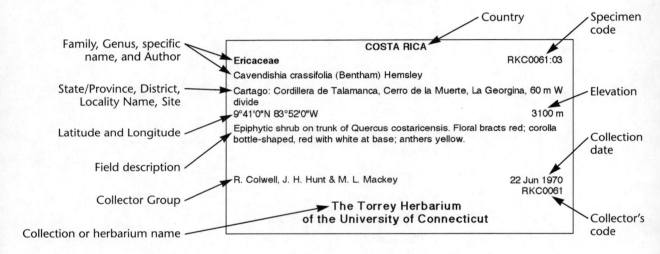

Printing Herbarium Specimen Labels: Step by Step

1. **Find the Specimen records for which you wish to print labels,** using any of the tools in the Find menu (Chapter 11).
2. **Declare these records the current Specimen Record Set** (pp. 13–15).
3. **Choose Herbarium Labels from the Labels menu.** The herbarium label option screen appears.

```
┌─────── Herbarium Label Printing Options ───────┐
│ ┌─Sort Options─────────────────────────────────┐│
│ │ ⦿ Sort selection by Specimen Code            ││
│ │ ○ Sort selection taxonomically               ││
│ │ ○ Use the Sort Editor to sort the records    ││
│ └──────────────────────────────────────────────┘│
│ ┌─Data Options─────────────────────────────────┐│
│ │ ☒ Include Specimen Code    ☒ Include Family  ││
│ │ ☐ Include [Collection]Site field             ││
│ │ ☐ Use first Specimen Note as field description of plant │
│ │ ☐ Use [Collection]CollectionCode as Collector's Number  │
│ │ ☐ Use [Collection]Source as Collector's Number          │
│ └──────────────────────────────────────────────┘│
│ ┌─Print Options────────────────────────────────┐│
│ │ ⦿ Print one label per Specimen record        ││
│ │ ○ Print [Specimen]Abundance labels per Specimen record │
│ └──────────────────────────────────────────────┘│
│              [ Cancel ]  [ Sort & Print Labels ]│
└─────────────────────────────────────────────────┘
```

4. **Choose a Sort Option.** See p. 245.

5. **Choose Data Options.** (Each is independent of the others, and none need be checked.)

    - **Include Specimen Code.** See p. 245.

    - **Include Family.** See p. 246.

    - **Include [Collection] Site field.** If you check this option, whatever has been entered in the Site field of the Collection record will be appended to the State/Province, District, Locality Name string on the label. See the sample label illustrated in the previous section (p. 256)—the Site field value was "La Georgina, 60 m W of divide" in the Collection record for this Specimen.

    - **Use the first Specimen Note as the field description of the plant.** To accommodate the inclusion of a field description of the living material for each Specimen (habit, colors of floral structures, etc.), you can record the description as a Specimen Note for that Specimen record (pp. 179–183). Since each Specimen record can have any number of attached Notes (e.g., comments on the specimen by different botanists who have studied it), Biota needs some way to tell which note contains the field description in its Note Text field.

    The convention Biota uses is a natural one: *the earliest Specimen Note, by Note Date, for each Specimen is assumed to contain a field description in its Note Text field.* Since field descriptions normally carry the date the specimen was collected, no earlier Note Date could be legitimate, in any case.

    If there are no Notes for a Specimen, or the earliest Note has a blank Note Text field, the field description of the plant is simply left blank on the label. If you need to print labels for a mixture of specimens, some of which have field descriptions in the earliest Specimen Note and some of which have other information in the earliest Note, you can either (1) print them in two sets, with this option checked for the set with field descriptions; or (2) enter a Note with the collection date in the Note Date field but no entry in the Note Text field, for the specimens with that lack field descriptions.

- **Collector's field number options.** The collector's field number is the link between the specimen and the collector's field notes. There are two alternative places Biota can look for the collector's number for the specimen, to include on the label.
  - ◊ **Use the [Collection] Collection Code as the Collector's Number.** This is a natural way to enter data from a field notebook. Each Collection Code in the database must be unique, however, so a large, multicollector database might run into ambiguities even with collector's initials or name prefixed to the field number to create a Collection Code.
  - ◊ **Use the [Collection] Source field as the Collector's Number.** If used, this alternative should be used consistently. You can change the name of the Source field to *Collector's Number* using a Field Alias (pp. 323–326).

NOTE: Once you select one of these two options, the other is disabled for this group of labels. (You can enable both again by unchecking the checked option). Neither is required.

6. **Choose a Print Option.**
   - *Either:* **Print one herbarium label per Specimen record** in the active Specimen Record Set.
   - *Or:* **Print [Specimen] Abundance duplicate labels per Specimen Record.** With this option selected, Biota looks at the Abundance field for each record in the current Specimen Record Set. If the Abundance value in the record is 1 (the default for all new Specimen records, p. 152), one label is printed for that Specimen record. If the Abundance value is the number n, then n identical, duplicate labels are printed for the record.

NOTE: If no label is printed for a record in the current Specimen Record Set, the record probably has a zero in the Abundance field. Change it to 1 for one label, or to the number n for n labels.

7. **Click the Sort & Print Labels button** in the options window.
8. **Follow the instructions in the section on "Printing Procedures"** (pp. 243–245) earlier in this chapter.

Species Labels

Species Labels are used to identify conspecific sets of specimens in collections—unit trays in entomological collections and species folders in herbaria, for example.

Information Included on Species Labels

When you use the Species Label tool, as described in the next section, Biota prints the following fields from the Species table.

- Genus
- Specific name (the Species Name field)
- Species Author
- Species Code

As an option, you can include an additional line to show who determined the particular set of specimens that you intend to label with a Species label:

- Personnel Short Name field for the determiner (optional)

Here are three examples of Species labels (determination option not included). To change the size, use the Reduce or Enlarge setting in the Page Setup screen for your printer (pp. 243–244).

Cactospiza heliobates (Snodgrass & Heller) 1901 Sp. Code: cactheli	Strumigenys godmani Forel 1899 Sp. Code:	Cavendishia crassifolia (Bentham) Hemsley Sp. Code:

Printing Species Labels: Step by Step

1. **From the Labels menu, choose Species Labels.** The Species Labels setup screen appears.

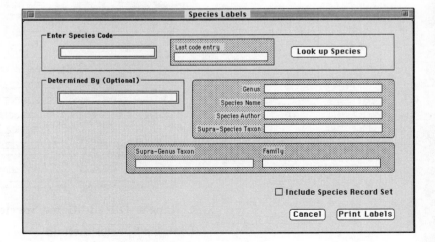

2. **Define the set of Species records for which to print labels.** There are three different ways to specify which species to include in a print run of Species Labels. *These methods can be used in any combination.*
   - **Species Record Set method.**
     a. **Find the Species records** for which you wish to print labels, using any of the tools in the Find menu (Chapter 11).
     b. **Declare these records the current Species Record Set** (pp. 13–15).
     c. **Click the Include Species Record Set checkbox,** to set this option, in the Species Labels setup screen.

   - **Species Code entry method.**
     a. **In the Enter Species Code box, enter a Species Code.** You can enter a full Species Code—manually or with a barcode reader (p. 164), or enter a partial code and use the wildcard entry method (p. 96).

     b. **Press tab.** The "Last code entry" display area will now show the Species Code you entered and the lower stippled display area will show information for the record found.

     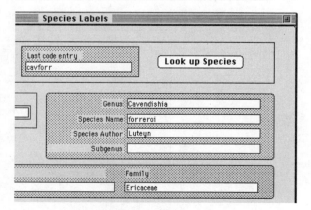

     c. **Repeat with additional Species Codes.**
   - **Look Up Species method.**
     a. **Click the Look Up Species button** (illustrated above). Use the techniques described for this tool on pp. 145–146 to find and enter a Species Code.

b. **Repeat with additional Species Codes.** After each entry, the previous Code will be displayed in the "Last code entry" box.

3. **Enter a determiner's name (optional).** Use the techniques described on pp. 149–150 to find and enter the Short Name of a determiner. The name will be included on every Species label in this print run.

4. **Click the Print Labels button**. The labels will be sorted taxonomically (p. 245).

5. **Follow the instructions in the section on "Printing Procedures"** (pp. 243–245) earlier in this chapter.

Designing and Printing Custom Labels

Inevitably, the format of the standard labels produced by the tools of the Labels menu will not meet everyone's specialized label needs. Alternatively, for Collection labels, you can automatically export data to a text file from the most commonly used fields, then edit the labels to your liking (pp. 246–248). More generally, you can use the Export Editor (pp. 434–443) to export data to a text file from any one table at a time, or the Quick Report Editor (pp. 269–278) to export data for fields from related tables. Once in a text file, you can edit data if necessary before printing.

For specialized labels that require fields from related (child or parent) tables, consider trying the Label Editor, the "off the shelf" 4th Dimension utility described in this section. It has some limitations, but is very easy to use.

You can use the Label Editor to make labels for records in any Core table (p. 5), specifying fields from that table and any of its *parent* tables. Fields in tables more than one link away ("grandparent" tables, etc.) cannot be accessed with the Label Editor, nor can child table fields (child values would be ambiguous anyway).

Using the Label Editor: Step by Step

1. **Find the records for which you want to print labels, and declare them the Record Set for that table** (pp. 13–15).

2. **From the Labels menu, choose Design & Print Custom Labels.** A table selection and option screen appears.

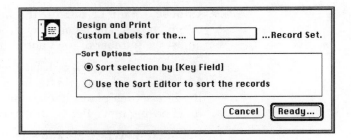

3. **Click and hold on the table selection popup list, then select the table for which want to print labels.**

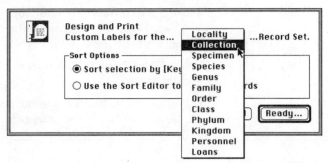

4. **Choose a Sort Option** (see p. 245).
5. **Click the Ready button in the table selection window.** The Page Setup window appears (see pp. 243–244).
6. **Click the OK button in the Page Setup screen.** The Label Editor appears.

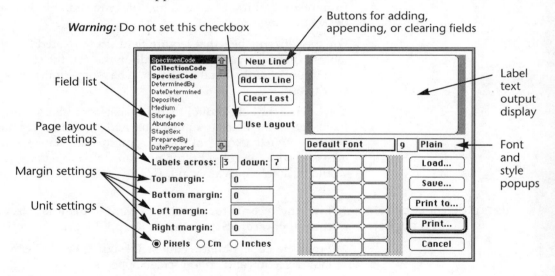

7. **Choose a page layout for the labels,** using the "Labels across/down" boxes. The layout image on the lower right changes as you adjust the number of rows and columns.
8. **Set the page margins,** or leave them at zero until you try printing a trial set. To pull a margin further towards the edge of the paper than the zero setting places it, use a negative number. Notice the buttons for choosing units: Pixels, Cm, or Inches.
9. **Set the font, type size, and style.** These settings apply to *all* text in the label—you cannot set each line separately.

10. **Create the label text layout.**

    - **To add a field, click the field** in the scrollable field list to select it, then:
        ◊ *Either:* **Click the New Line button** to create the next line in the text.
        ◊ **Or: Click the Add to Line button** to append the field to the end of the last line entered.

    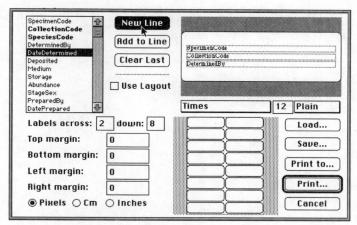

    - **To delete a field, click the Clear Last button.** If the field you want to delete is not the last field you added, you will have to delete other entries to get back to it.

    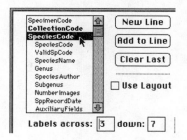

    - **To access fields from a parent table,** double-click the linking field in the table list (linking fields are boldfaced). In the illustration at the left, the Species Code field in the Specimen table was clicked to give access to the fields of the Species table.

    Note: Only parent tables of the table you selected in step 3 can be accessed. For Specimen labels (illustrated here), for example, you can access the Collection, Species, or Personnel tables, but no others.

11. **If you want to save the layout,** click the Save button, then name and place the layout file.

12. **To load a previously saved label layout,** click the Load button, find the file, and load it.

> **Warning:** *Do not check the Use Layout checkbox*—if you already clicked it, *uncheck* it now. If it is checked when you click the Print button, 4D tries to print using whatever screen layout you last displayed. (Sorry, the checkbox is not under programmer control.)

13. **To print the labels, read the warning above, then click the Print button.** See pp. 243–245 for print preview instructions.

> **Note:** The Print To button has no function in Biota.

Chapter 14 Printing Reports

Biota offers printed reports for individual records as well as groups of records.

- **For an individual record** from any Core table or an image from the Image Archive, you can print a replica of the record just as it is as displayed in the input screen, using the Print button on the input screen itself.

- **For the records displayed in an output screen (whether or not they have been declared a Record Set),** you can use the Print button on the output screen or the Print command from the File menu to print a list of the records. You can choose a preformatted Biota report or use the Quick Report Editor, instead, to design your own printed reports. You can save and reload the formats you create in the Quick Report Editor. With certain quick report designs, you can also use the built-in Graph Editor to display or print simple graphs (pp. 279–280).

Warning: All Biota preformatted reports use the Helvetica font (Macintosh) or the Arial font (Windows). If you do not have the appropriate font installed (in the printer's memory as well as on your computer, for some printers), the layout of reports will probably not be correct, depending on what font is substituted for Helvetica or Arial.

Printing Procedures

For all printed reports, Biota presents the same series of Page Setup and Print screens and options when you are ready to preview and print. Because these printing procedures are the same for both reports and labels and are described in full in Chapter 13, "Printing Labels," please consult pp. 243–245.

Printing an Individual Record

To print a copy of an individual record from a Core table or the Image Archive table, just as displayed in the input screen:

1. **Display the record you want to print in the input screen** for its table. For records from Core tables, use the techniques described in Chapter 9. For Image records, display the image in the Image input screen (pp. 355–357).

2. **Click the Print button in the input screen** to print a replica of the input screen showing the record currently displayed.

The Page Setup window appears.

3. **To preview the printed record,** see "Printing Procedures," pp. 243–245.

Printing a Report Based on a Selection of Records

Printing a Report Based on the Active Record Set for a Table

To print a report listing all the records in the active Record Set for a Core table or for the Determination History table:

1. **Display the Record Set** in the standard output screen for the table.
   - **For Core tables:** From the Display menu, choose the command for the table.
   - **For the Determination History table:** From the Special menu, choose Display Determination Histories. If you have enabled the password system, only the Administrator can use this command.

2. **Follow the directions in the section** "Printing a Standard or Custom Report Based on the Records in an Output Screen," (pp. 266–268).

Printing a Report Based on a Record Set Pointer File

To print a report listing all the records in a Record Set pointer file:

1. **Load the Record Set Pointer File** from disk to display the records in the output screen for the table (see pp. 18–19).

2. **Follow the directions in the next section,** "Printing a Standard or Custom Report Based on the Records in an Output Screen."

Printing a Standard or Custom Report Based on the Records in an Output Screen

For each Core table and the Determination History table, Biota offers a preformatted report that you can print, listing all records in the current Selection, including fields from related tables where appropriate. Each table has its own specially designed report. Rather than attempt to display them in this book, the best way to evaluate these report formats is to print some records from your own Data File or from the Biota Demo Data File (Chapter 3), or preview reports on the screen (pp. 243–245).

If the preformatted report for a table does not suit your needs, you can print the records using a custom report layout that you design yourself using the Quick Report Editor (pp. 269–278).

Here are the steps to follow for printing a standard report.

1. **Display the records to be printed in the output screen.** See the preceding two sections of this chapter and Chapters 9 and 11 for help, if you need it.

2. **Click the Print button at the bottom of the output screen.**

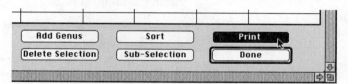

Alternatively, you can select Print from the File menu while the records are displayed in the output screen.

3. **Choose a print option.** A print option window appears offering two or three choices, depending on the table for which you are printing records. (The Determination History table has no options.) Three different option windows are illustrated below.

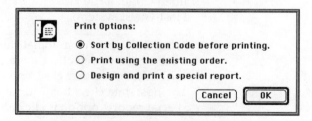

- **Sort by Record Code before printing.** Use this option to print a report with the records sorted by Record Code (Species Code, Specimen Code, Collection Code, or Locality Code). The Loans report can be sorted by Loan Code.

- **Print using the existing order.** If you have used the Sort Editor to create a special order for the records that you want maintained in the report, choose this option.

- **Design and print a special report.** If you choose this option, Biota will present the 4th Dimension Quick Report Editor, which you can use to design, save, and/or load custom report formats that you create yourself. The report will be based on the Selection of records in the output layout when you clicked the Print button or selected

Print from the File menu. See the next section.

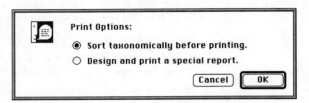

- **Sort taxonomically before printing.** For records from taxonomic tables (Species, Genus, Family, etc., and Specimen as well), this option prints the records for the table in alphabetical order within taxonomic levels (ranks) in a hierarchical report organized by rank (as in the table on p. 2).

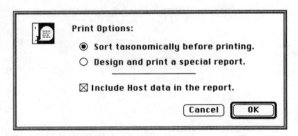

- **Include Host data in the report.** For the Specimen table, an additional option is offered. If the "Include Host data" checkbox is checked, the report will include, for each Specimen record, the Specimen Code of the Host for that specimen (see Chapter 21 on the Host-Guest system).

NOTE: This option is automatically checked if any Specimen records in the current selection have host data. Otherwise, it appears unchecked. In either case, you can override the setting manually. The host data checkbox is disabled when the "Design and print a special report" option is clicked.

4. **Click the OK button in the print option screen.**

    - **If you chose the standard Biota report,** see "Printing Procedures," pp. 243–245.
    - **If you chose to Design and Print a Special Report,** see the next section, "Designing and Printing Reports with the Quick Report Editor."

Printing a Specimen Count by Taxon Report

The Specimen Count by Taxon report provides quantitative information on the taxonomic breakdown of a Specimen Record Set. The report computes and displays the number of Specimen records for each taxon and rank.

To use this tool, choose "Specimen Count by Taxon" from the Special menu. You can use the screen preview option to display the report on the screen (pp. 243–245).

Designing and Printing Reports with the Quick Report Editor

The Quick Report Editor is an off-the-shelf, 4th Dimension tool for designing, using, saving, and loading custom formats for printing reports. (This tool can also be used to exporting text flatfiles. See "Exporting Custom Flatfiles" in Chapter 24, pp. 457–462.) With certain quick report designs, you can also use the built-in Graph Editor to display or print simple graphs (pp. 279–280).

To launch the Quick Report Editor for printing a report, see "Printing a Report Based on a Selection of Records," pp. 266–268.

The Quick Report Editor appears with its own menus. The upper left panel—the *field selection panel*—will initially display the fields of the table—called the *base table*—for which records were displayed in an output screen when you clicked the Print button in the output screen. The base table for the illustrations in this section is the Collection table.

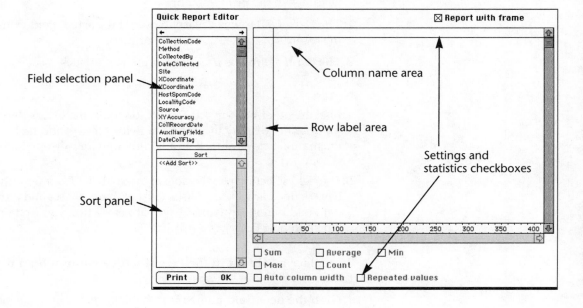

Setting Up Columns in the Quick Report Layout

The first step in designing a quick report is to lay out the data columns for the report. Generally, these columns will represent fields from the base table or its parent tables, but you can also include formulas that modify or combine field values (p. 272).

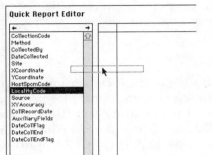 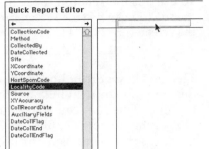 

1. **Drag each field name from the field selection panel to the column name area,** as shown here, then release the mouse button to place it.

    Once the first column has been inserted, three rows appear below the column name area (above right):

    * The *Header (H)* **row** contains an editable copy of the field name you dragged to the layout. Information in the H row is used for column headings in the printed report.

    * The *Detail (D)* **row** is a placeholder for the actual field values from the records in the selection to be printed.

    * The *Total (T)* **row** is used for summary statistics.

    NOTES:

    a. Like the other built-in 4th Dimension tools, the Quick Report Editor does not display the names of fields (above) and tables (center illustration below) alphabetically, although they do appear in a consistent order.

    b. The field selection panel displays Internal Field Names only. If you have defined any Field Aliases, a list of these Aliases and their Internal Field Name equivalents will appear when you first open the Quick Report Editor. See pp. 325–326.

2. **To replace a column in the layout with a column for a different field:**

    a. **Grab the new field name** from the field selection panel and drag it to highlight the column to be replaced.

    b. **Release the mouse button.** The new field replaces the old one.

3. **To delete a column:**

   a. **Click the name of the column to be deleted** in the column name area (the upper row of names) to highlight the column.

   b. **Select Delete Column from the Edit menu.**

4. **To insert a column:**

   a. **Click the name of an existing column** to highlight it.

   b. **Select Insert Column from the Edit menu.** A blank column is inserted to the left of the column you highlighted.

   c. **Drag a field name for the new column over the blank column** to highlight it.

   d. **Release the mouse button.**

5. **To include fields from parent tables for adding columns:**

   a. **Click and hold in the box at the top of the field selection panel**, as shown below, left.

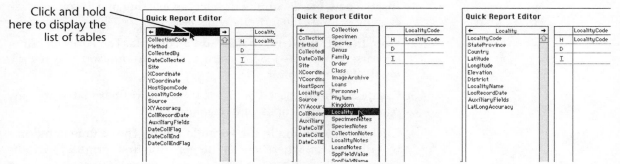

Click and hold here to display the list of tables

   b. **Select the parent table name from the popup list** that appears (above center). The field selection panel will display the field list for the table you selected (above right).

   c. **Add columns to the layout from the parent table** using the procedure in step 1.

   NOTE: Although the Quick Report Editor will permit you to create a column in the report layout using any field from any table, the only fields that will produce sensible results are fields from parent tables (pp. 7–8). In the illustration above, since the report is for Collection records, fields from the Locality and Personnel tables can be used in the layout as well as those from the Collection table. Parent tables *of any rank* may be included (unlike the Label Editor, pp. 261–264). For example, if you are creating a layout for Specimen records, you can use fields from any table in either the taxonomic or place hierarchies.

6. **To edit a column label,** double-click the label cell in the Header row (second row in the Quick Report screen). A text insertion point appears in the cell to allow editing.

Specifying a Formula in a Column Instead of a Field Name

In Biota, the two most likely conditions under which you might need to use a formula to specify the contents of a column in a Quick Report layout are:

- **To print partial dates** (Month-Year or Year only). See pp. 112–113.
- **To transform alphanumeric field values into numeric values** for computing statistics. For this purpose, you use the *Num* function (p. 139). For example, if you have used the Site field (an alphanumeric field, see Appendix A) to record air or water temperature at the time a collection was made, you would use the formula: Num ([Collection] Site) as the column name. You could then use the Quick Report Editor's statistical capabilities (all of which except for *Count* require numerical field values) to compute and display the minimum, average, and maximum temperature for the Collections from each Locality. (The Num function ignores any nonnumeric characters.)

Here is the procedure for entering a formula to specify a column in a quick report layout:

1. **Insert a new, blank column or select an existing one.** If you select an existing column, the formula will *replace* it.
2. **From the Other menu, choose Edit a Formula.** The Formula Editor appears. See pp. 136–138 for instructions *and a critical warning* regarding the Formula Editor.
3. **Create a formula** based on fields (from the primary table and any parent tables), functions, and operators.
4. **Click the OK button in the Formula Editor.** The column name area will display *Cn*, where *n* is an integer—the first formula is C1, the second C2, etc. You can use these formula names as variables in other formulas in the same layout.

Resizing a Column

1. **Select the column** by clicking its name in the column name area.
2. *Deselect* **the Auto column width box** at the bottom of the screen by clicking the box.

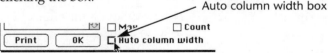

3. **Place the pointer over the column border in the column name area** (the upper row of column names). The pointer changes into a column width cursor (below).

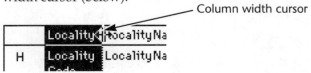

4. **Drag the column border left or right to resize the column.**

   NOTE: If you leave the Auto column width box checked for a column, the column width is adjusted automatically to accommodate the longest field value for the column. If you resize a column, any values that are too long for the column width are wrapped to the next line.

Adding or Removing Sort Criteria

1. **Drag the <<Add Sort>> marker (from the sort panel) to the right** until the column you want to use as a sort criterion is highlighted. You can sort on formulas (see p. 272) as well as fields.
2. **Release the mouse button.** The name of the column will be added to the list of sort criteria in the sort panel, with a small arrow indicating the sort direction.
3. **To change the sort direction** click the small arrow by the field name in the sort panel (up for ascending order, down for descending).

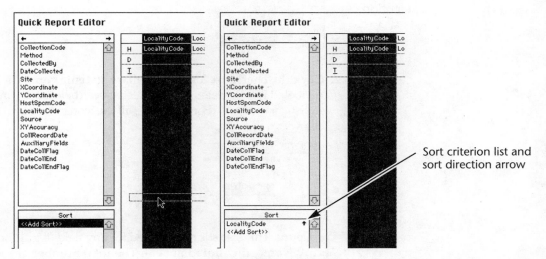

Sort criterion list and sort direction arrow

You can add as many sort levels to the Sort panel as you wish. The sort proceeds from the top of the list to the bottom.

4. **To remove a field (or a formula) from the sort list,** select Delete Last Sort from the Special menu. To remove an earlier addition from the sort list, you must delete all later additions first, then add them back in again.

Displaying All Values in a Sorted Column

The Quick Report Editor allows you to choose whether to print a report that includes all repeated values or, instead, an indented list, for a column in a quick report layout that you have added to the Sort list. (Examples of indented lists appear on pp. 278–279 in this chapter and on p. 2 in Chapter 1.)

To display the values from *all* the selected records in a sorted report column, regardless of whether a value repeats the one above it in the report:

1. **Select the column** by clicking in the column name area (the upper row of column names).

2. **Click the Repeated Values checkbox** (at the bottom of the screen) to check it.

With the Repeated Values checkbox *unchecked*—the default for all columns—repeated values are left blank in the printed report if the column is included in the Sort list, creating an indented table. If you do not include a column in the Sort list, all values appear, regardless of the setting of the Repeated Values box.

Computing and Displaying Summary Statistics for All Records in the Selection

1. **Select the cell where the Totals (*T*) row intersects the column to be used to compute statistics.** To select a cell, click it once.

2. **Click the checkbox for the summary statistic you wish to compute.** A labeled icon appears in the selected cell (below left), indicating the statistic you chose. (For reference, all the icons are shown on the right, below).

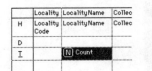

- **Count.** This statistic can be used for any field (e.g., [Locality] Locality Name, illustrated above). The total number of records in the selection will be printed in the cell.

- **Sum, Average, Min, and Max.** These statistics can be computed only for numeric fields or for alphanumeric fields that have been transformed using the *Num* function (see "Specifying a Formula in a Column Instead of a Field Name," p. 272). See Appendix A for field type information.

3. **Label the statistic if you wish.** Double-click a cell *adjacent* to the statistic. A text insertion point appears. Enter the label text. (You cannot put a label in the same cell with the statistic.)

4. **To delete a statistic or a cell label,** select it, then choose Clear from the Edit menu.

Computing and Displaying Summary Statistics for Sorted Groups of Records in the Selection

1. **Set up the sort criteria in the sort panel** (see "Adding or Removing Sort Criteria," p. 273) so that records will be grouped, when sorted, according to the field or fields for which you want to compute statistics.

   For example, Collection records will be sorted by Locality Code in the example below, creating groups of Collection records from each Locality. (See the sample output on pp. 278–279.)

   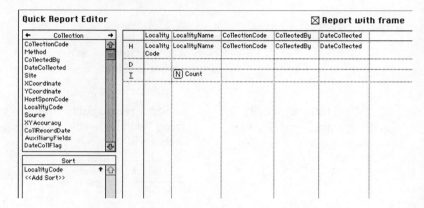

2. **Select the Totals row** by clicking on the underlined *T*.

   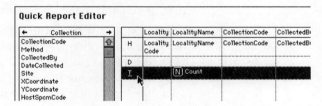

3. **From the Edit menu, select Insert Break.** A new row appears, labeled B1 (for *break level 1*).

	Locality	LocalityName	CollectionCode	Collecte
H	Locality Code	LocalityName	CollectionCode	Collecte
D				
B1		[N] Count		
T		[N] Count		

   If you had already inserted statistics icons in the Total row, the icons are copied to the Break line. If not, you can add statistics to the Break line using the method in the previous section.

4. **Label the statistics in the Break line,** using the technique of step 3 in the previous section.

   In Break line labels, you can include the *break field value* in the label by inserting a pound sign (#). In the example below, the label "Number From #:" has been entered next to the Count statistic in the B1 row.

The intermediate counts of collection records for each Locality will include the Locality Code in the label (e.g., "Number From Ansermanuevo:"; see the sample output on pp. 278–279).

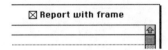

NOTE: You can add as many hierarchical break levels to the layout as you wish, but there must be at least as many sort levels as break levels.

Framing, Fonts, and Text Styles

- **To add a rectangular frame to the report,** leave the "Report with frame" checkbox checked. To eliminate the frame, click the box to uncheck it.

- **To specify font, font size, justification, and style,** select a column, row, or cell and make choices from the Font and Style menus.

   You can style not only column titles and cell labels, but the text of the field values themselves. Select the Detail row by clicking the "D" and choose from the Font and Style menus. You will not see any effect in the Quick Report Editor screen of styling in the Detail area, but the styling will appear in the printed report (and in the screen preview).

Hiding a Row or Column in the Printed Report

You might want to hide the rows of the Detail area if you were interested only in statistics in the break or total lines, or hide columns for some reports and show them for others using the same layout.

1. **Select the row or column** to be hidden in the Quick Report Editor. If you click the Detail (D) row, all field values from the actual records will be hidden.

2. **From the Edit menu, select Hide Line (row) or Hide Column.**

Adding Page Headers and Footers to the Report

1. **From the File menu, choose Print To** (while the Quick Report Editor is open).

2. **Make sure Normal Printer is selected.**

3. **Close the Print To window** by clicking the OK button.

4. **From the File menu, choose Page Setup.** The normal Page Setup screen for your operating system and printer appears (pp. 243–245).

5. **Click OK in the Page Setup screen** after setting any options you want. The Header and Footer setup screen appears.

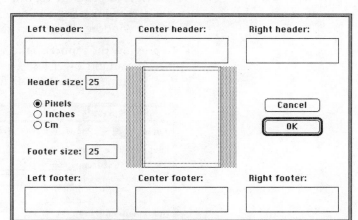

6. **Enter header and footer text** in any of the six text areas. The preview panel in the middle shows the layout.

7. **Include page numbers and the time and date of printing** if you wish, by imbedding the following codes in the header and footer text:
   - #P (adds a sequential page number)
   - #H (adds the time of printing)
   - #D (adds the date of printing)

8. **Style the text as you wish** by selecting text and choosing items from the Font and Style menus.

Saving and Loading a Quick Report Layout

When you have completed the report design, consider saving the layout as a text file so you can use the same design again later.

- **To save a quick report layout** to a disk file, choose Save As from the File menu while the Quick Report Editor is open, then name and place the file. As you are working on a saved filed, you can choose Save from the File menu to update the file.

- **To load a saved quick report layout,** choose Open from the File menu while the Quick Report Editor is open, then locate the file and open it.

Printing a Report or Dismissing the Quick Report Editor Without Printing

When you have completed the quick report design and saved it if you wish (see previous section), you can print the report, preview it on the screen, or dismiss the editor without printing.

1. **From the File menu, choose Print To** (while the Quick Report Editor is open).

2. **Make sure Normal Printer is selected.**

3. **Close the Print To window** by clicking the OK button.

4. **To preview or print the report,** click the Print button in the lower left corner of the Quick Report Editor, or choose Print from the File menu. The Print window for your operating system and printer appears.

   • **To preview the report on the screen,** click the "Preview on screen" checkbox in the lower-left corner of the Print screen, then click the Print button (see pp. 243–245 for details).

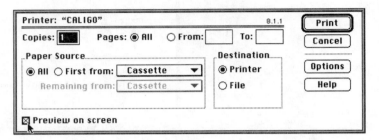

   • **To print the report,** make sure the "Preview on screen" checkbox in the lower-left corner of the Print screen is *not* checked, then click the Print button. Or you can preview first, then click the Print button in the Preview window. (See pp. 243–245 for details.)

5. **To dismiss the report without printing,** click the OK button in the lower-left corner of the Quick Report Editor. Be sure to save the layout first if you want to use it or modify it later (see the previous section).

An Example of a Quick Report

In the preceding sections, all the illustrations depicted the construction of a quick report layout for Collection records, including fields from both the Collection and Locality tables. A counter for the total number of records in the report was placed to the Total row, and a counter to show the number of Collection records linked to each Locality record was added by creating a Break row, after declaring the Locality Code column the primary Sort level. (The Locality Name column was later added as a second Sort level to suppress repetition in that column.)

Here is the final layout:

	LocalityCode	LocalityName	CollectionCode	CollectedBy	DateCollected
H	**LocalityCode**	**LocalityName**	**CollectionCode**	**CollectedBy**	**DateCollected**
D					
B1	**Number From ✱ :** [N] Count				
T	**Total Records** [N] Count				

The figure below shows a sample report using this layout.

Luteyn Collections				Page 1
LocalityCode	LocalityName	CollectionCode	CollectedBy	DateCollected
Alto de Galápago	Alto de Galápago	JLL12692H	J.L. Luteyn	4/27/89
		JLL12692	J.L. Luteyn	4/27/89
Number From Alto de Galápago:	2			
Ansermanuevo	Ansermanuevo-SanJosé del Palmar rd.	JLL7281H	J.L. Luteyn	4/19/79
		JLL7269H	J.L. Luteyn	4/19/79
		JLL10573H	J.L. Luteyn	5/15/84
		JLL7269	J.L. Luteyn	4/19/79
		JLL10573	J.L. Luteyn	5/15/84
Number From Ansermanuevo:	5			
Baeza-Tena rd.	Baeza-Tena rd.	JLL5677H	J.L. Luteyn	4/6/78
		JLL5674H	J.L. Luteyn	4/6/78
		JLL5674	J.L. Luteyn	4/6/78
Number From Baeza-Tena rd.:	3			
Cajamarca-Cel	Cajamarca-Celendin rd.	JLL11314H	J.L. Luteyn	2/12/85
		JLL11314	J.L. Luteyn	2/12/85
Number From Cajamarca-Cel:	2			
Cerro Horqueta	Cerro Horqueta	JLL3756H	J.L. Luteyn	5/24/73
		JLL3756	J.L. Luteyn	5/24/73
Number From Cerro Horqueta:	2			
Total Records	14			

Using the Quick Report Graph Editor

The Quick Report Editor has an integrated graphing tool—of very modest capability—called the Graph Editor. The Graph Editor plots data from the first break row (B1) of the layout. To use the Graph Editor with a quick report, you should include only one break level in the design (see "Computing and Displaying Summary Statistics for Sorted Groups of Records in the Selection," pp. 275–276).

1. **Create a quick report layout with the following options:**

   - **No more than one kind of summary statistics per column, in one Break row** (pp. 275–276).
   - **At least one sort level** (p. 273).
   - **The column to be used for the X-axis** of the graph placed in the leftmost column.

2. **In the Quick Report Editor, choose Print To from the File menu.** The "Print to" option window appears.

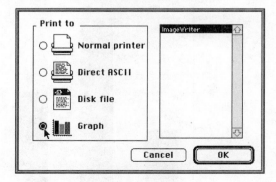

3. **Click the Graph button in the left panel, then click the OK button in the "Print to" option window (above).** (Ignore the right panel.)

4. **Click the Print button** in the lower left corner of the Quick Report Editor, or choose Print from the File menu. The Graph Editor window appears, showing a bar graph based on values in the first break row (B1).

5. **To change the graph type,** choose from the Graph Type menu. (Not all types will make sense with all quick report layouts.)

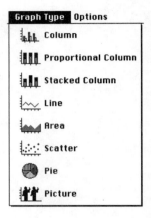

6. **To change the scaling,** select Scale from the Options menu.

7. **To insert an image from the Clipboard to be used in the Picture type graph,** select Paste to Y1 from the Pictures menu.

8. **To change the color and pattern scheme,** double click a legend box and choose a different color and pattern from the palettes.

9. **To preview the printed graph or print the graph,** choose Print from the File menu while the Graph Editor is open, then follow the directions in "Printing a Report or Dismissing the Quick Report Editor Without Printing," earlier in this chapter (pp. 277–278).

The bar graph corresponding to the Collection record report shown in the previous section is illustrated below. The height of each bar is the number of Collection records for the Locality listed on the abscissa.

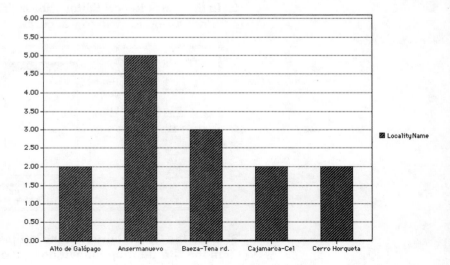

PART 4 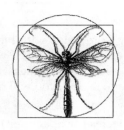 Customizing Biota

Chapter 15 User-Defined Auxiliary Fields

As your experience with software of other kinds will doubtless confirm, an application that is too narrow and inflexible is frustrating, whereas one that tries to do everything is ponderous and confusing. Biota is an attempt to strike the right balance between a lean, fast application that suits perfectly the needs of a few, and a jack-of-all trades application bloated with RAM-clogging features and customizable ornamentation. If Biota critics split equally between those that find it too limiting and those that find it too accommodating to special interests, the balance will prove to have been about right.

A relational database, by its nature, is confined within the limits of whatever relational structure the designer gives it. The challenge is to choose a structure that can accommodate a broad spectrum of uses, without greatly compromising the efficiency of the most common operations that users need to carry out and without in any way limiting the potential for exporting data to other applications.

This chapter and the next explain in detail how you can customize Biota to meet special needs by adding Auxiliary Fields (this chapter), renaming existing fields (pp. 323–326), setting default values for data entry (pp. 313–315), and setting up Entry Choice Lists (pick lists) for data entry. See also Chapter 7, which explains how to set up custom prefixes for Record Codes.

Core Fields and Auxiliary Fields

In designing Biota's relational structure (Appendix A), the goal was to place essential and commonly used fields, called Core fields, in the Core tables (Specimen, Collection, Locality, Species, and higher taxon tables, plus Personnel and Loans). Inevitably, however, different people with different needs will find some existing fields needless and other fields lacking. Biota's ability to rename many Core fields with Field Aliases can solve some of these problems (see pp. 323–326), but if you require more fields than exist in the Core tables (plus Notes, Images, and Determination Histories, which are in Peripheral tables), you will need to create Auxiliary Fields.

For example, systematists may use Species (or Specimen) Auxiliary Fields to create record values for a character matrix. Biota can export Auxiliary Fields in the NEXUS format for direct use in MacClade[1] or PAUP[2]. Ecologists may use Collection Auxiliary Fields to record the ecological features of collection sites or live-organism observation sites.

Auxiliary Fields, in contrast to Core fields, are special fields that you create and name as you like (see pp. 5–6). Core fields, however, are much faster to sort and search for, so *you should always use Core fields, renamed or not, in preference to Auxiliary Fields, wherever possible.*

You can create Auxiliary Fields for records in four Core tables in Biota: Specimen, Collection, Locality, and Species.

How Auxiliary Fields Work

Biota's Auxiliary Field system does not actually create "hard-wired" new fields in the Core tables. (There is no way to do that in a compiled database application.) Instead, data entered in an Auxiliary Field are kept in a special "triplet" format.

Suppose you create an Auxiliary Field called Size, for Specimen records. Each time you enter a Size value for a Specimen record, Biota creates a new record in a special table called Spcm Field Value in the format: Specimen Code, Field Name, Field Value. (The Spcm Field Value table has only these three fields, hence the term *triplet*.)

For example, for a large specimen whose Specimen Code was SPM00634, the new record in the Spcm Field Value table would be: SPM00634, Size, Large. For an Specimen Auxiliary Field called Color, the record in the Spcm Field Value Table for the same specimen might be: SPM00634, Color, Pink. The Spcm Field Value Color record for a different specimen might read: ABC01093, Color, Red.

If you use statistical applications, you are probably already familiar with this alternative way of representing tabular data: Row Index, Column Index, Value. Even though it requires three entries for each Value, this format is a more memory-efficient way than setting out a full table for storing "sparse matrices"—tables that are mostly blanks or zeros. For this reason, you need not hesitate to add a new Auxiliary Field that will be used for only a subset of the records in a Core table, for fear of adding unnecessarily to the size of the Data File. An Auxiliary Field "triplet" record is created only when you enter a value for an Auxiliary Field.

---

[1] Maddison, W. P., and D. R. Maddison. 1992. *MacClade: Analysis of Phylogeny and Character Evolution.* Sinauer Associates, Sunderland, MA.

[2] Swofford, D. L. 1993. *Phylogenetic Analysis Using Parsimony (PAUP), version 3.1.1.* Smithsonian Institution, Laboratory of Molecular Systematics, Washington, D. C.

Linked to each Auxiliary Field Value table (e.g., Spcm Field Value) is a separate table that simply keeps a list of Auxiliary Field Names (e.g., the Spcm Field Name table; see Appendix A) with an Index field to keep track of the Auxiliary Field order you specify.

Because Biota presents Auxiliary Fields for input and output as if they were in column-by-row tabular format, you may never see (or even have to understand) the triplet format or the separate tables for Names and Values. But you will need to understand the triplet format to use the Search Editor or Sort Editor on Auxiliary Field Values, or to import or export the Field Values and Field Names tables using the Import or Export by Tables and Fields tools. These special methods are described later in this chapter.

Creating, Editing, and Ordering Auxiliary Field Names

This section tells you how to create new Auxiliary Fields, change the order or name of existing Auxiliary Fields, and how to delete Auxiliary Field Names when you no longer want them. The next section explains how to enter, edit, and delete *values* in Auxiliary Fields.

Opening the Field Name Editor

You use the Field Name Editor to create, change, or delete Auxiliary Fields. You access the Field Name Editor directly from the input screen for each Core table that supports Auxiliary Fields (Specimen, Collection, Locality, and Species).

In this section, Species Auxiliary Fields will be used as an example, but the procedures are identical for Specimen, Collection, or Locality Auxiliary Fields.

1. **From the Input menu, choose Species.** The Species input screen appears.

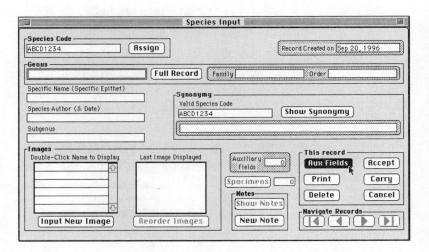

2. **Enter a Species Code.** Any value will do. You can even Cancel the Species record when you are finished modifying Species Auxiliary Field Names without losing the changes in Auxiliary Field Names.

3. **Click the Aux Fields (Auxiliary Fields) button in the Species input screen.** (The button is being clicked in the illustration above.) The Auxiliary Species Fields display screen appears in a window in front of the Species input window, showing all Species Auxiliary Fields currently defined, together with spaces on the right for recording their value for this record. (There will be no Auxiliary Fields already defined if you are working with your own Data File and have not yet defined any.)

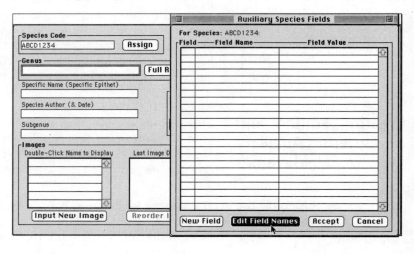

Hidden Window Cursor: Press any key to bring the active window to the front

NOTE: If you click in any portion of the Species input window that still shows to the left of the Auxiliary Species Fields display screen (above), the cursor turns into the "hidden window" cursor (the window-behind-a-window cursor, shown here), and the Species input screen moves "in front" of the Species Auxiliary Fields screen. *To return to the Auxiliary Fields window, press any key on the keyboard.* (The keystroke is not recorded.)

4. **Click the Edit Field Names button in the Auxiliary Species Fields display screen.** The Field Name Editor appears, showing the names of any Auxiliary Species Fields already defined. (There are none in this example.)

Creating New Auxiliary Fields (First Method)

1. **Open the Field Name Editor,** as explained in the previous section.
2. **Enter a new Field Name in the entry area,** as shown here.

Field Name entry area

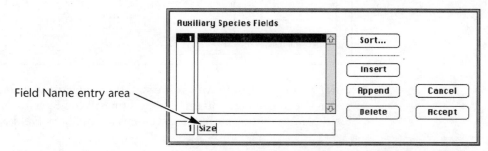

3. **To add the new Field Name to the list**, press the TAB key. Do not use any of the screen buttons to accept a single field if you intend to enter another.

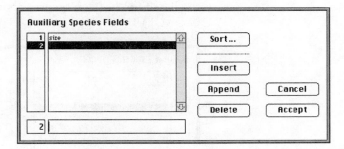

Your first field is moved into the list at the top of the Field Name Editor window.

4. **To add another Field Name,** repeat the process: enter the new Field Name and then press the TAB key.

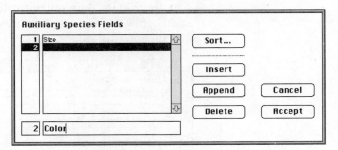

Auxiliary Field Names must be unique, within the set of Auxiliary Field Names for each Core table. If you try to enter a duplicate Auxiliary Field Name in the Field Name Editor, Biota displays the message shown here, and the name is not accepted.

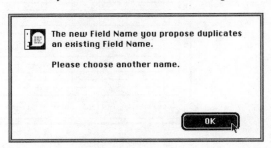

NOTE: Although you cannot create two Species Auxiliary Fields, both named Size, you could create a Species Auxiliary Field and a Specimen Auxiliary Field, both named Size. Of course, there is no restriction on repeated *values* for Auxiliary Fields in different records. For example, any number of Species can have the value Large for the Size field.

5. **Continue entering Field Names,** each followed by TAB, until you have entered all you want.

6. **Click the Accept button** in the Field Name Editor window to accept the list of fields. (Don't worry about the final, blank field. It will not be saved.)

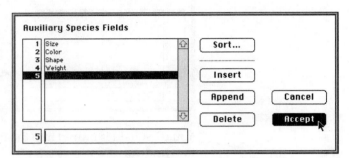

The Field Name Editor disappears, leaving the Auxiliary Species Fields display screen showing the new fields you created, ready for Auxiliary Field data entry for this Species record. (See "Entering and Displaying Data in Auxiliary Fields," pp. 295–304, for information on entering values.)

Creating New Auxiliary Fields (Second Method)

1. **With the Auxiliary Species Fields display screen displayed, click the New Field button.** (Alternatively, you can click on any blank line in the window.)

A dialog window appears, requesting a name for the new field.

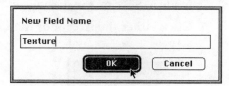

2. **Enter a Field Name and click OK.** A second dialog window appears immediately, requesting *for this Species record*, the value for the new field you have just created.

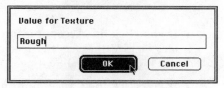

3. **Enter a value and click OK or click the Cancel button in the dialog window.** Either way, the new Field Name is added to the list of Species Auxiliary Fields.

    • **If you click the OK button (above),** the New Field Name dialog window appears again, ready for the next field, allowing rapid entry of a series of Species Field Names (which apply to all Species records) and their values for this particular Species.

    • **If you click Cancel in the Value dialog window,** shown above, control returns to the Auxiliary Species Fields display screen, where all Field Names and Values are then displayed.

4. **To accept the new Field Names,** click the Accept button in the Auxiliary Species Fields display screen. (If you click Cancel instead, no changes will be saved.)

Control now returns to the Species input screen. Any changes you made in Species Auxiliary Field Names have been accepted.

5. **Click either the Accept or the Cancel button on the Species input screen.**

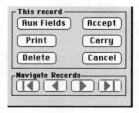

- **If you click the Accept button in the Species input screen,** Auxiliary Field Values for this Species record will be saved ("Rough" for the Texture field, above).

- **If you click the Cancel button in the Species input screen** to dismiss the Species record without changes, the changes you accepted in the previous step for Species Auxiliary Field *Names* will not be affected, but any *values* entered for this particular Species ("Rough" for the Texture field, above) will be lost.

6. **If you try to enter a duplicate Auxiliary Field Name using the New Field button,** Biota posts the following message, and the name is not not accepted. Auxiliary Field Names must be unique, within the set of Auxiliary Field Names for each Core table. (See the Note at step 4 in the previous section.)

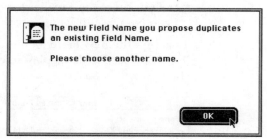

Editing the List of Auxiliary Field Names

1. **Open the Field Name Editor for the Core table whose Auxiliary Field Names you wish to edit** (Specimen, Collection, Locality, or Species), following the steps in the section "Opening the Field Name Editor" (pp. 283–284).

The list of existing fields appears in the Field Name Editor window, with the first field highlighted and displayed in the entry/editing area.

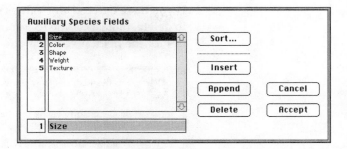

2. To add an Auxiliary Field to the end of the list:

   a. **Click the Append button** in the Field Name Editor.

   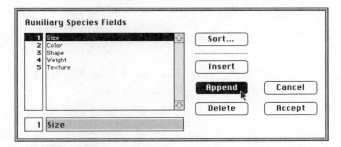

   b. **Fill in the new Field Name in the input area and press the TAB key.** The new field is added to the end of the list.

   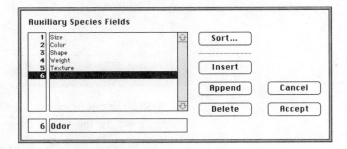

3. To delete an Auxiliary Field:

   a. **Click on the field, then click the Delete button** in the Field Name Editor.

   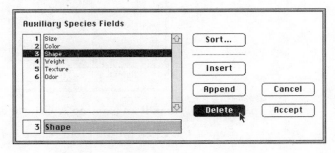

b. If you have entered any Values for the Auxiliary Field you intend to delete, Biota asks you to confirm the deletion. (If there are no Values for the field, no confirmation is requested.)

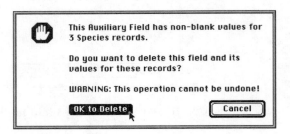

WARNING: Deleting an Auxiliary Field also deletes all Field Value records for that field. *Any values you have already entered for the deleted field will be lost.*

4. **To insert a new Auxiliary Field** at the beginning or in the middle of the list of fields:

   a. **Click on the Field Name** *before which* **you want the new item to be placed.**

   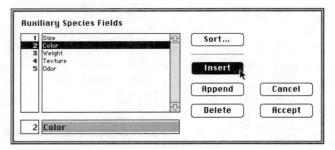

   b. **Click the Insert button** in the Field Name Editor. A blank Field Name appears in the list and in the entry area at the bottom of the window.

   c. **Enter the new Field Name and press the TAB key.**

   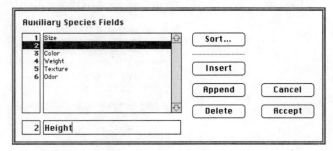

5. **To edit an existing Auxiliary Field Name:**

   a. **Click on the Field Name** in the list.

b. **Edit the Field Name** in the editing area at the bottom of the Field Name Editor window (below, "Weight" has been changed to "Mass").

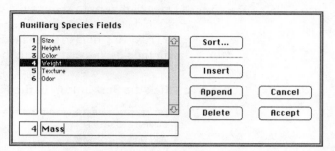

c. **Press the TAB key to register the change in the list.** Whenever you change an Auxiliary Field Name, Biota asks you to confirm the change.

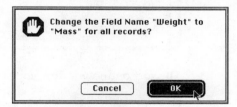

d. **Click OK to accept the changed Field Name.**

WARNING: If you want to change the *order* of Field Names in the list of Auxiliary Fields, see the next two sections. Do *not* attempt to change the order of Fields by changing the *names* of Fields (e.g., changing "Size" to "Color" and "Color" to "Size"). If you have any data in these fields and you interchange names, the data will then be associated with the wrong fields.

Reordering Existing Auxiliary Field Names Alphabetically

1. **Open the Field Name Editor for the Core table whose Auxiliary Field you wish to edit** (Specimen, Collection, Locality, or Species), following the steps in the section "Opening the Field Name Editor," earlier in this chapter (pp. 283–284).

   The list of existing fields appears in the Field Name Editor window, with the first field highlighted and appearing in the entry/editing area.

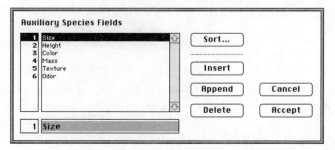

WARNING: If you want to change the *order* of Field Names in the list of Auxiliary Fields, follow the instructions in this section or the next. Do *not* attempt to change the order of Fields by changing the *names* of Fields (e.g., changing "Size" to "Color" and "Color" to "Size"). If you have any data in these fields and you interchange names, the data will then be associated with the wrong fields.

2. **Click the Sort button in the Field Name Editor.**

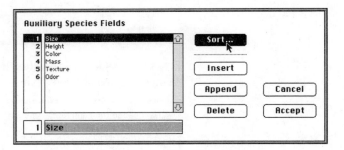

The sort option window appears.

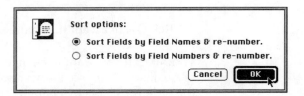

3. **Click the OK button to accept the default setting, "Sort Fields by Field Names and renumber."** The Field Names reappear in the Field Name Editor, sorted alphabetically, with corresponding Field Numbers in the narrow vertical window to the left of the list of Field Names.

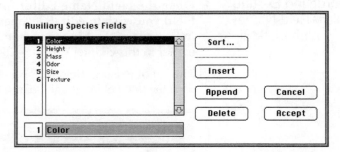

4. **Accept or Cancel the new ordering of Field Names.**
    - **If you like this arrangement of Auxiliary Fields,** click the Accept button in the Field Name Editor. The fields appear in this order in all contexts in Biota until you change the order again.

- **If you want to restore the original order,** click the Cancel button in the Field Name Editor.

Reordering Existing Auxiliary Field Names in a Specific Order

1. **Open the Field Name Editor for the Core table whose Auxiliary Field you wish to edit** (Specimen, Collection, Locality, or Species), following the steps in the section "Opening the Field Name Editor," earlier in this chapter (pp. 283–284). The list of existing fields appears in the Field Name Editor window, with the first field highlighted and displayed in the entry/editing area.

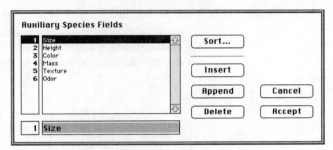

   WARNING: If you want to change the order of Field Names in the list of Auxiliary Fields, use the techniques in this section or the previous one. Do *not* attempt to change the order of Fields by changing the *names* of Fields (e.g., changing "Size" to "Color" and "Color" to "Size"). If you have any data in these fields and you interchange names, the data will then be associated with the wrong fields.

2. **Change the Field Numbers, one at a time, until they reflect the desired order.** In the example shown here, suppose you want to move the Color field so that it appears between that of Odor and Size. (You cannot use the alphabetical sort method of the previous section, since this order is not alphabetical.)

   - **Click on Color, then press the TAB key to move to the Field Number entry area** (the small box in the extreme lower-left corner of the Field Name Editor; illustrated above). Alternatively, you could select 1 with the mouse in the Field Number entry area.

   - **Type 4.5 or any other decimal number between 4 and 5** in the Field Number entry area, then press the TAB key. The Field Number for Color now appears as "4.5" in the Field Number column of the Field Name Editor.

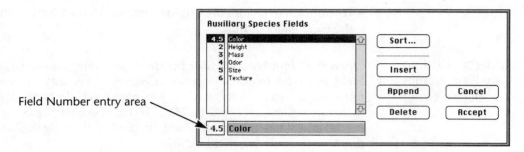

Field Number entry area

3. **Click the Sort button in the Field Name Editor.**

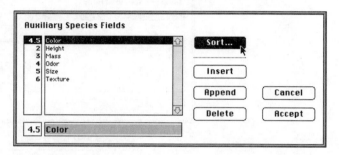

The sort option window appears.

4. **Click the lower option button in the sort option window, "Sort Fields by Field Numbers and renumber," then click OK.**

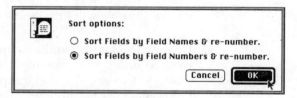

The Field Names reappear in the Field Name Editor, sorted in the new order that you set up, with corresponding new Field Numbers in the narrow vertical window to the left of the list of Field Names.

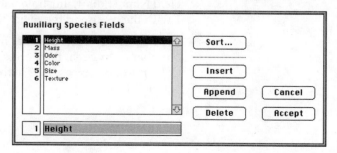

NOTES:

a. **You can reorder any number of Field Names at once** by changing Field Numbers to reflect the new order, using any combination of integer and decimal values, then following steps 3 and 4, above.

b. **To move a Field Name to the top of the list,** enter a decimal Field Number between zero and 1 (e.g., 0.5), then follow Steps 3 and 4, above.

c. **Duplicate Field Numbers** can arise if you change a Field Number, then click the Accept button in the Field Name Editor instead of the Sort button. Duplicate Field Numbers do no harm (Field *Numbers* are used by Biota only to order fields, and need not be unique, unlike Auxiliary Field *Names*), but two fields with the same Field Number will sort in unpredictable order. For this reason, you should correct the problem by opening the Field Name Editor, and following steps 2 through 4, above.

5. Accept or Cancel the new ordering of Field Names.

   - **If you like this arrangement of Auxiliary Fields,** click the Accept button in the Field Name Editor. The fields appear in this order in all contexts in Biota until you change the order again.

   - **If you want the original order back,** click the Cancel button in the Field Name Editor.

Entering and Displaying Data in Auxiliary Fields

Once you have defined the Auxiliary Field *Names* you want by using the Field Name Editor (see pp. 283–295), you are ready to enter Auxiliary Field *Values* (data) in your Auxiliary Fields. This section uses Species Auxiliary Fields as an example, but the procedures are identical for Specimen, Collection, or Locality Auxiliary Fields.

If you created the Auxiliary Fields by using the Field Name Editor (see "Creating New Auxiliary Fields [First Method]," pp. 284–286), there will be no values in Auxiliary Fields for any record. If, on the other hand, you used the New Field button on the Field Value entry screen (see "Creating New Auxiliary Fields [Second Method]," pp. 286–288), you may have entered values for one record as you created the fields. In either case, the procedures for adding new values, editing existing values, and working with Auxiliary Field records are the same.

Entering New Data in Auxiliary Fields

1. **Display a new or existing Species, Specimen, Collection, or Locality record in the input screen,** using any tools from the Input (Chapter 10), Find (Chapter 11), or Tree (pp. 186–192) menus. (Species will be used as an example, here.)

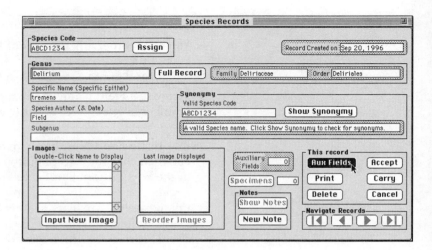

2. **Click the Aux Fields button at the lower right.** The Auxiliary (Species) Fields display screen appears.

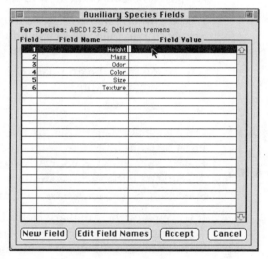

3. **Click anywhere on the line for the field for which you want to enter a value.** (The Height field has just been clicked in the example, above.) The Field Value entry window appears.

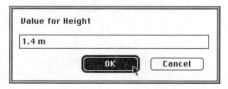

4. **Enter a value, then click OK** (or press the Return key). (The value 1.4m has been entered for the Height Auxiliary Field, in the example above.) The Field Value entry dialog reappears, ready for the Field Value for the next field in the Field Name list.

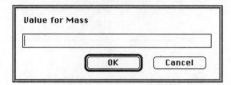

5. **If you want to enter a value for this field, do so and click OK.** This cycle continues until you reach the bottom of the Field Name list or until you click Cancel in the Field Value entry dialog.

6. **For nonsequential value entry,** click on the line you want, make the entry, then click Cancel when asked for a value for the next field.

7. **To edit a Field Value,** click on the value to be edited in the Auxiliary (Species) Fields display screen, then edit the value in Field Value entry window. Click OK, then Cancel the Field Value entry window for the following field, if it appears.

8. **When you have completed the entries** and they appear as you wish in the Field Value list, click the Accept button in the Auxiliary (Species) Fields display screen.

The Auxiliary Fields display screen disappears, and control returns to the Core table input screen (the Species input screen in the illustrations here). Notice that the number of Auxiliary Field Values for this Species record is displayed near the Aux Fields button.

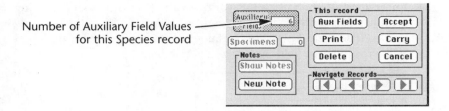

Number of Auxiliary Field Values for this Species record

Carrying Auxiliary Fields

The Carry button in Biota input screens allows you to use an existing record (or a record you have just entered) as a template for a new record (pp. 41–42, 127–128). As an option, you can choose to Carry (copy) all Auxiliary Field Values from the existing record to the new record (for the tables that support Auxiliary Fields).

To enable or disable this option:

1. **Choose Preferences from the Special menu.** The Preferences screen appears.

2. **Click the "Carry Auxiliary Field Values from the template records" checkbox** to enable (check) or disable (uncheck) the Carry Auxiliary Field Values option. The default setting is disabled (unchecked).

3. **Click Accept** in the Preferences screen.

Displaying Auxiliary Fields and Their Values for a Selection of Records

1. **Display a selection of records for the Specimen, Collection, Locality, or Species table in the output screen** for that table (Chapter 9).

   In this section, Species records will be used as an illustration, but the procedures are identical for the other three tables.

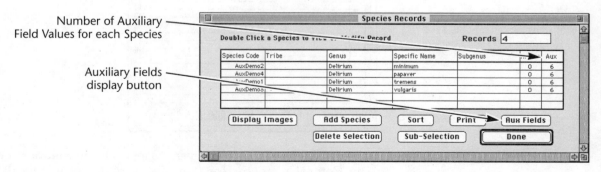

Number of Auxiliary Field Values for each Species

Auxiliary Fields display button

The numbers in the Aux column (illustrated above) show how many Auxiliary Field Values have been entered from each Core Table record (6 for each Species record, in the example).

2. **Click the Aux Fields button in the selection display screen.**

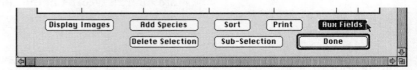

An option window appears, offering two alternative display formats for the Auxiliary Fields for these Species records.

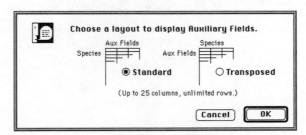

3. **Choose a display option.**

   ♦ **The Standard option** (the default) displays the Auxiliary Fields in the same orientation as ordinary (Core) fields in the output screen—in the example, Species as rows, Auxiliary Fields as columns.

   ♦ **The Transposed option** (right option button) displays the same information in a transposed format—in the example, it would display Auxiliary Fields as rows, Species as columns.

   Although these Auxiliary Field display formats are limited to 25 columns, you will be able to examine all your data, regardless the number of Auxiliary Fields you have defined and regardless of the number of records in the selection.

   ♦ **If you have defined 25 or fewer Auxiliary Fields and there 25 or fewer records in the current selection,** choose the option that makes more sense for the information you want to find or compare. (You may have to experiment a bit to know which is better for your purposes.) See the following section, "Displaying Auxiliary Fields in the Standard Format," or the section "Displaying Auxiliary Fields in the Transposed Format" (pp. 302–304).

   ♦ **If you have defined fewer than 25 Auxiliary Fields but there are more than 25 records in the current selection,** you will usually want to choose the *Standard* Auxiliary Fields display (the left option button). See the following section, "Displaying Auxiliary Fields in the Standard Format."

   ♦ **If you have defined more than 25 Auxiliary Fields but there are fewer than 25 records in the current selection,** you will usually want to choose the *Transposed* Auxiliary Fields display (the right-hand option button). See the section "Displaying Auxiliary Fields in the Transposed Format" (p. 304).

- **If you have defined more than 25 Auxiliary Fields and there are more than 25 records in the current selection,** choose the option that makes more sense for the information you want to find or compare.

Displaying Auxiliary Fields in the Standard Format: Records as Rows, Auxiliary Fields as Columns

1. **Follow the instructions in the previous section,** "Displaying Auxiliary Fields and Their Values for a Selection of Records" (pp. 298–300).
2. **In the display options window, make sure the Standard (left button) option is selected, and then click the OK button.**

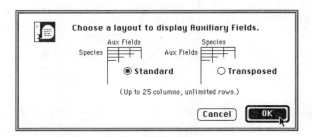

The selection of Core table records (Species, in the example) appears, with the corresponding values for Auxiliary Fields in labeled columns.

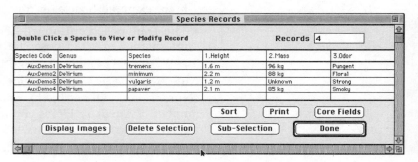

In the example above, notice that only the first three Auxiliary Fields are visible (for Height, Mass, and Odor).

3. **To see the next set of Auxiliary fields (columns), click in the *horizontal* scroll bar** (where the cursor arrow appears, above), or use the horizontal scroll arrows or "thumb-slider" to move to the right. The next set of fields appears, along with the Record Code (Species Code, in the example).

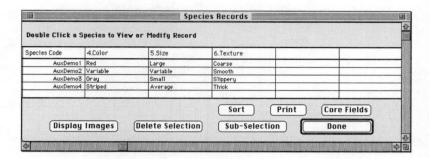

4. **If not all Auxiliary Fields have been revealed, continue to move to the right in the display,** using the horizontal scroll bar tools. Up to 25 Auxiliary Fields can be displayed.

   NOTE: If you have a monitor wider than the display window, you can expand the window horizontally to see more Auxiliary Fields all at once, as shown below. (The width of columns cannot be adjusted interactively on the screen—an unfortunate limitation of 4th Dimension.)

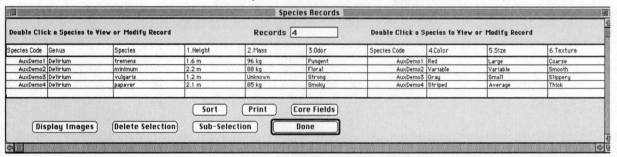

5. **To view all Auxiliary Fields for a record (or to edit them),** double-click a row (a Species record, in the example) in the output screen. The individual record appears in the standard input screen (the Species input screen, in the example).

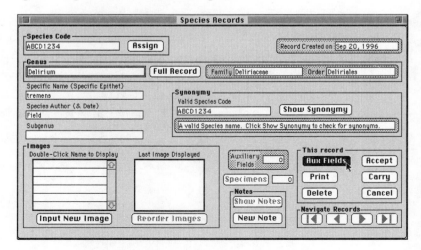

You can now use the input screen Auxiliary Fields button (clicked above) to display or edit all Auxiliary Field Values for that species, even if you have defined more than 25 Auxiliary Fields (see pp. 295–296).

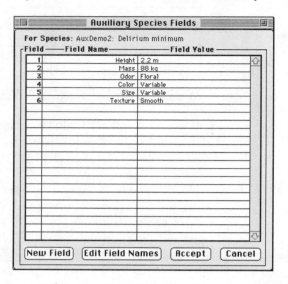

6. **When you are through using the Auxiliary Fields display screen,** you can either dismiss the selection by clicking the Done button (illustated below), or return to the Core Field display for the current selection of records by clicking the Core fields button on the Auxiliary Fields display screen.

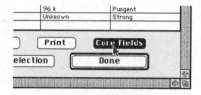

Displaying Auxiliary Fields in the Transposed Format: Auxiliary Fields as Rows, Records as Columns

1. **Follow the instructions in the section,** "Displaying Auxiliary Fields and Their Values for a Selection of Records" (pp. 298–300).
2. **In the display options window, click the button for the Transposed format option, then click the OK button.**

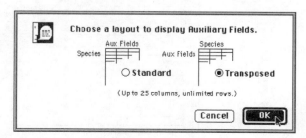

The selection of Core table records (Species, in the example) appears, with Auxiliary Field Names labeling the rows of the display and identifiers of the Core table records as column headings.

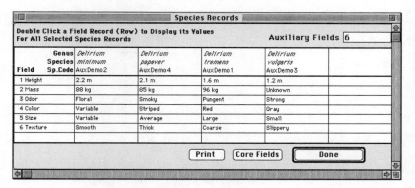

3. **If not all columns (records) are visible, use the horizontal scroll bar to see additional columns.** In the example above, all four Species (columns) in the current selection are visible, plus an empty column at the right. Had there been more than four Species, you would need to click in the horizontal scroll bar or use the horizontal scroll arrows or "thumb-slider" to move to the right. The next set of records (Species, in the example) would appear, along with the Auxiliary Field Names (rows), for reference. If all Species still had not been revealed, you would continue to move to the right in the display, using the horizontal scroll bar tools. Up to 25 Species (or Specimens, Collections, or Localities) can be displayed.

NOTE: If you have a monitor wider than the display window, you can open the window wider to see more Auxiliary Fields all at once, as shown for the Standard Auxiliary Fields display on page 301.

4. **To view the Values for all Core records in the selection, for a given Auxiliary Field**, double-click a row (a Species Auxiliary Field, in the example).

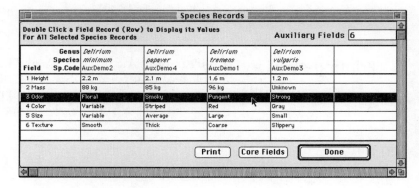

A special screen appears listing the Value for that Auxiliary Field for each Core record (each Species, in the example) in the current selection, even if there are more than 25 Core table records in the selection.

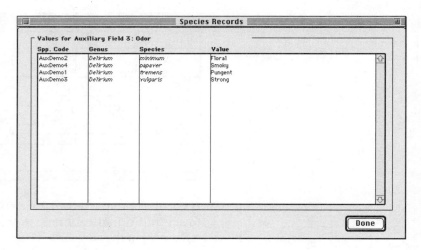

NOTE: This screen is available only with Transposed display format for Auxiliary Fields, not with the Standard format.

5. **Click Done at the bottom of the screen, above, when you are through examining the values.** Auxiliary Field Values and Auxiliary Field Names cannot be edited in this screen. (See pp. 286–288 and pp. 295–298 for two methods of editing Auxiliary Fields.)

6. **When you are through using the Auxiliary Fields display screen,** you can either dismiss the selection by clicking the Done button, or return to the Core Field display for the current selection of records by clicking the Core fields button on the Auxiliary Fields display screen.

Printing Auxiliary Fields

You can print Auxiliary Fields and their values for a selection of Core Table records (Specimen, Species, Collection, or Locality) in any of four formats: in matrix (row-by-column) format, either Standard or Transposed (see below); or in triplet format, either Standard (pp. 305–306) or Transposed (pp. 307–308).

Printing Auxiliary Fields in Matrix (Row-by-Column) Format

The most flexible method for printing Auxiliary Field data is to export the Auxiliary Field Values to a text file in full matrix format (either Standard or Transposed), open the text file using a spreadsheet application (such as Microsoft Excel) or a text processing application (such as Word or WordPerfect), then format and print the data to suit your own needs. This is the method to use if you want to print Auxiliary Field data and Core Field data together for the same records.

Biota does not provide a preformatted, printed report for Auxiliary Field data in matrix format. For help exporting Auxiliary Fields, see the section "Exporting Auxiliary Field Values," later in this chapter (pp. 308–311). For help exporting Core Field data, see Chapter 24.

Printing Auxiliary Fields in Standard Triplet Format (Record Code, Auxiliary Field Name, Auxiliary Field Value)

1. **Display Auxiliary Fields and their values** in the Standard Auxiliary Fields output screen, following the steps on pp. 300–302.

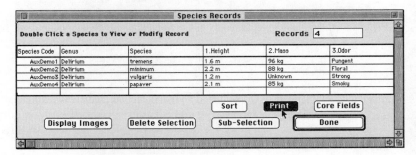

2. **Click the Print button at the bottom of the screen** (see above). A warning message appears, explaining that the data will be printed in triplet format.

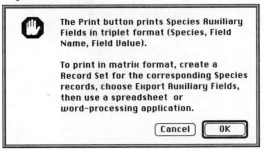

3. **Click the OK button in the warning window to proceed.** The Page Setup window appears. (The appearance and options available in the Page Setup window depends on the printer and operating system you are using.)

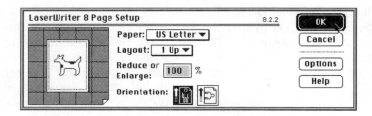

4. **Click the OK button in the Page Setup window.** The Print window appears. (The appearance and options available in the Print window depends on the printer and operating system you are using.)

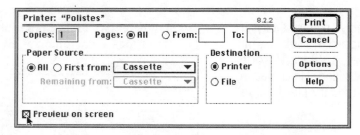

5. **If you want to preview the printed report,** click the "Preview on screen" checkbox in the lower-left corner of the Print window. The report appears in the preview screen or on paper.

Species Auxiliary Fields				
Species Code	Genus	Specific Name	FieldName	FieldValue
AuxDemo1	Delirium	tremens		
			Height	1.6 m
			Mass	96 kg
			Odor	Pungent
			Color	Red
			Size	Large
			Texture	Coarse
AuxDemo2	Delirium	minimum		
			Height	2.2 m
			Mass	88 kg
			Odor	Floral
			Color	Variable
			Size	Variable
			Texture	Smooth
AuxDemo3	Delirium	vulgaris		
			Height	1.2 m
			Mass	Unknown
			Odor	Strong
			Color	Gray
			Size	Small
			Texture	Slippery
AuxDemo4	Delirium	papaver		
			Height	2.1 m
			Mass	85 kg
			Odor	Smoky
			Color	Striped
			Size	Average
			Texture	Thick

Printing Auxiliary Fields in Transposed Triplet Format (Auxiliary Field Name, Record Code, Auxiliary Field Value)

1. **Display Auxiliary Fields and their values** in the Transposed Auxiliary Fields output screen, following the steps on pp. 302–304.

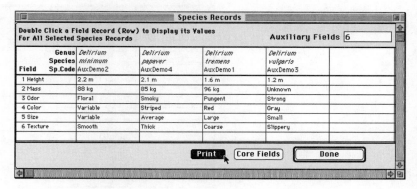

2. **Click the Print button at the bottom of the display** (see above). A warning window appears, explaining that the data will be printed in triplet format.

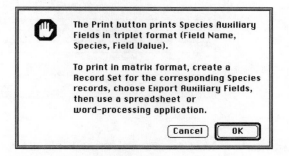

3. **Follow steps 3 through 5 in the previous section** (pp. 305–306). The report appears in the preview screen or on paper.

Species Auxiliary Fields				
FieldName	Species Code	Genus	Specific Name	FieldValue
Height				
	AuxDemo1	Delirium	tremens	1.6 m
	AuxDemo2	Delirium	minimum	2.2 m
	AuxDemo3	Delirium	vulgaris	1.2 m
	AuxDemo4	Delirium	papaver	2.1 m
Mass				
	AuxDemo1	Delirium	tremens	96 kg
	AuxDemo2	Delirium	minimum	88 kg
	AuxDemo3	Delirium	vulgaris	Unknown
	AuxDemo4	Delirium	papaver	85 kg
Odor				
	AuxDemo1	Delirium	tremens	Pungent
	AuxDemo2	Delirium	minimum	Floral
	AuxDemo3	Delirium	vulgaris	Strong
	AuxDemo4	Delirium	papaver	Smoky
Color				
	AuxDemo1	Delirium	tremens	Red
	AuxDemo2	Delirium	minimum	Variable
	AuxDemo3	Delirium	vulgaris	Gray
	AuxDemo4	Delirium	papaver	Striped
Size				
	AuxDemo1	Delirium	tremens	Large
	AuxDemo2	Delirium	minimum	Variable
	AuxDemo3	Delirium	vulgaris	Small
	AuxDemo4	Delirium	papaver	Average
Texture				
	AuxDemo1	Delirium	tremens	Coarse
	AuxDemo2	Delirium	minimum	Smooth
	AuxDemo3	Delirium	vulgaris	Slippery
	AuxDemo4	Delirium	papaver	Thick

Exporting Auxiliary Field Values

Like data in any other table in Biota, the data in the four Auxiliary Field Value tables (Spcm Field Value, Spp Field Value, Coll Field Value, and Loc Field Value) can be exported directly to a text file using the Export Editor (see pp. 434–443). In fact, if you are moving Auxiliary Fields data between Biota Data Files, you *must* use this method to create a text file in a format suitable for the Import Editor to import the text file into another Biota Data File.

Because the information in Auxiliary Fields appears internally in triplet format (see "How Auxiliary Fields Work," pp. 282–283, and Appendix A), however, rather than in columns and rows, Biota offers special tools for exporting Auxiliary Field Values in matrix format, just as they appear in the Auxiliary Field display screens (pp. 298–302) or in NEXUS format (see pp. 310–311).

1. **Establish a Record Set composed of the Core Table records whose Auxiliary Field Values you wish to export.** (See pp. 13–15 if you need help declaring Record Sets.) Suppose, for example, you want to export values for the Species Auxiliary Fields records used as an illustration earlier in this chapter and you have already declared the parent Species records the current Species Record Set.

2. **From the Im/Export menu, choose Export Auxiliary Fields.** An option screen appears.

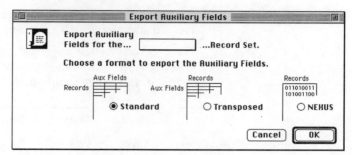

3. **In the option window, click and hold on the table name popup list.**

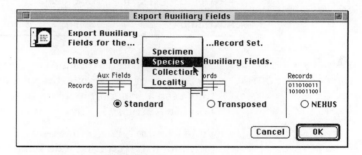

4. **Select the name of the Core table** for which you want to export Auxiliary Field Values. (Species is selected, above.)

5. **Select the output format you want,** using one of the three option buttons:

    ♦ **Standard format** (Records as rows, Auxiliary Fields as columns).

    ♦ **Transposed format** (Auxiliary Fields as rows, records as columns).

    ♦ **NEXUS format** (see "Exporting Auxiliary Fields in the NEXUS Format," pp. 310–311).

6. **Click the OK button in the option window.** The Save File window for your operating system appears.

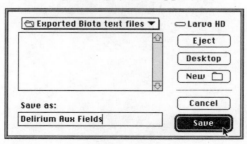

7. **Enter a name for the text file and click the Save button** (Macintosh) **or OK button** (Windows) in the Save File window.

   A progress indicator appears, then a notice that the text file has been created successfully. (If an error comment appears, try again.)

   Now you can view, edit, format, or print the text file using any spreadsheet or word processing application. Here is how the example data, in Standard format, would look in an exported text table:

	Height	Mass	Odor	Color	Size	Texture
AuxDemo1	1.6 m	96 k	Pungent	Red	Large	Coarse
AuxDemo2	2.2 m	88 kg	Floral	Variable	Variable	Smooth
AuxDemo3	1.2 m	Unknown	Strong	Gray	Small	Slippery
AuxDemo4	2.1 m	85 kg	Smoky	Striped	Average	Thick

Exporting Auxiliary Fields in the NEXUS Format

Systematists can use Biota Auxiliary Fields to create character matrices for Species or Specimens, then export the matrices in NEXUS format. (Biota can also export matrices in NEXUS format for the Collection and Locality Auxiliary Fields).

Biota exports NEXUS files using the "STANDARD" NEXUS datatype only (not to be confused with Biota's Standard Auxiliary Fields format). NEXUS files exported by Biota can be used without modification as input for MacClade[3] or PAUP[4]. You may want to edit the NEXUS documents that Biota exports, however, to modify the way they are used by these applications.

To export a NEXUS file, just follow the directions in the previous section (pp. 308–310), but in step 4, click the NEXUS button in the Export Auxiliary Fields option window.

Below is an example of a NEXUS file exported by Biota for several Species records for hummingbird flower mites, from the BiotaDemo Data File that is distributed with Biota. Notice that any internal space characters in Auxiliary Field Names are automatically filled with the underline character, as required in the NEXUS standards. Please consult the NEXUS data standards documentation for an explanation of the various components of the file.

---

[3] Maddison, W. P., and D. R. Maddison. 1992. *MacClade: Analysis of Phylogeny and Character Evolution.* Sinauer Associates, Sunderland, MA.

[4] Swofford, D. L. 1993. *Phylogenetic Analysis Using Parsimony (PAUP), version 3.1.1.* Smithsonian Institution, Laboratory of Molecular Systematics, Washington, D. C.

```
#NEXUS

BEGIN TAXA;
  DIMENSIONS NTAX=5;
  TAXLABELS rhinrich rhinhaplo troperio tropfuent tropkressi;
END;

BEGIN CHARACTERS;
  DIMENSIONS NCHAR=11;
  FORMAT
    DATATYPE=STANDARD
    SYMBOLS=" 0 1 2";
  CHARSTATELABELS
    1 metapodal_plates,
    2 opisthoventral_setae,
    3 exopodal_plates,
    4 genital_shield,
    5 genital_setae,
    6 anal_shield,
    7 dorsal_shield,
    8 peritrematic_plates,
    9 setae_z1,
    10 podonotal_setae,
    11 coxa_IV_spur;
  MATRIX
    rhinrich 1 1 1 1 1 0 2 1 1 1 1
    rhinhaplo 1 0 1 1 1 0 2 1 0 1 1
    troperio 1 1 0 0 0 0 2 0 0 0 0
    tropfuent 1 1 0 0 0 0 2 0 0 0 1
    tropkressi 1 0 0 0 0 0 1 0 0 0 0;
END;
```

NOTE: There is one important restriction. To export Auxiliary Field Values in NEXUS format, all Auxiliary Field Values in the set must be single-digit integers (0 through 9) or single alphabetic characters (A through Z).

If you attempt to export a longer value or a value with other characters in NEXUS format, Biota posts the error comment shown here. Please note that this restriction does not apply to Auxiliary Field values in general, only to those exported in NEXUS format.

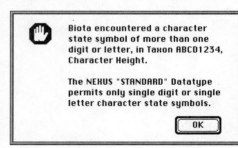

Chapter 16 Setting Default Entries, Using Entry Choice Lists, and Renaming Fields

This chapter and the previous one explain in detail how you can customize Biota to meet special needs by adding Auxiliary Fields (Chapter 15), renaming existing fields (pp. 323–326), setting default values for data entry (pp. 313–315), and setting up Entry Choice Lists (pick lists) for data entry. See also Chapter 7, which explains how to set up custom prefixes for Record Codes.

Setting Default Entries: Field Value Defaults

Some fields in your database may require the very same entry for many or even all the records in a Data File. For fields likely to require default entries in the Specimen, Collection, and Locality tables, you can assign a *Field Value Default* that is entered in the data entry area for that field automatically whenever you create a new record. (You can select or erase the entry and replace it with an ad hoc one if you need to.)

If you need to select from *several* repetitive values for data entry in a field, see the next section, on Entry Choice Lists (pp. 315–322).

To Set Field Value Defaults

1. **Choose Field Value Defaults from the Special menu.** Screen 1 of the Set Field Value Defaults tool appears, showing default entries for fields in the Specimen table.

![Set Field Value Defaults dialog showing Specimen Table fields: Prepared By "P. Naskrecki", Date Prepared Mo: 6 Dy: 24 Yr: 1995, Determined By, Date Determined, Stage/Sex, Medium, Storage, Deposited At, Type Status. Buttons: Clear Default Values, Cancel, Accept, Second Screen of Fields. Screen 1 of 2.]

The example above shows a default preparator's name and date, but any of the fields shown can receive a default entry. Please note these important points:

- **If you have renamed any fields,** using the Core field Alias editor (see pp. 323–326), then the Aliases, not the Internal Field Names, appear in the Set Field Value Defaults screens.

- **The Prepared By and Determined By entry areas,** with the double borders, are standard Biota wildcard lookup entries, linked to the Personnel file. (See p. 96 for information on wildcard entry techniques.)

- **The Date fields** accept either full (DD/MM/YYYY) or partial (MM/YYYY or YYYY) dates. (See pp. 111–114.)

> **WARNING:** Be careful how you use the Deposited At field. If you use the Loans system, this field is used to store the Loan Code for specimens you have lent or borrowed. Entering your own collection name or institution name as a default entry value often makes sense for the entry of original data, however. If you later lend the specimens, the Loans system will automatically replace your collection name with the Loan Code for the new loan.

2. **Display the second screen of fields if necessary.** If you click the Second Screen of Fields button, the screen for Collection and Locality default entries appears.

[Screenshot: Set Field Value Defaults dialog, Screen 2 of 2, showing Collection Table fields ([Collection] Locality Code: La Selva, Collected By, Date Collected (or Date Started), Date Collection Completed (Optional), Site, Method: Malaise Trap, Source) and Locality Table fields (District, State/Province, Country), with buttons Clear Default Values, Cancel, Accept, First Screen of Fields...]

The example above shows a default Locality Code for Collection records and a default collection Method, but any of the fields shown can receive a default entry. Please note these important points:

- **Entering a default Locality code for *Collection* records** (the [Collection] Locality Code field) links all Collection records created while this default is set to the Locality record for the default (La Selva, in the example).

- **The Collected By entry area,** with the double borders, is a standard Biota wildcard lookup entry, linked to the Personnel file. (See p. 96 for information on wildcard entry techniques.)

- **The Date Collected and Date Collection Completed entry areas** each accept either full (DD/MM/YYYY) or partial (MM/YYYY or YYYY) dates. (See pp. 111–114.)

3. **Click the Accept button** on either screen to record the defaults in your Data File for both screens, once you are done entering default values in both screens.

NOTE: Field Value Defaults are recorded in the Lists table in your Biota Data File. See Appendix B.

Using Entry Choice Lists

Some fields in your database will probably have a limited number of repetitive entries—a field for sex, life stage, or growth habit, for example. For most nonlinking fields in Biota you can create an *Entry Choice List* (usually called simply a *Choice List* in this book) that presents entry options in a floating window during data entry. Since Choice Lists eliminate misspellings, they are especially useful when someone else assists you in entering data.

The Choice List window appears automatically whenever you TAB or click into the entry area for the field during data input or revision. When the window appears, you can simply click an item on the List to enter it—or dismiss the List window (by pressing TAB or clicking the Cancel button in the Lists window) and then make a manual entry. You construct each List by entering items with the List Editor, which is accessed from the List window itself by clicking the Modify button, as detailed below.

Between Biota sessions, the Lists you have created are kept as part of your Biota Data File in the Lists table (see Appendix B), not in the Biota application file. Saved Lists are preserved whether currently activated or not, until you change or erase them. Each time you launch Biota, all activated Lists are loaded automatically from the Data File into the Biota application file. If you activate an existing List during a Biota session, the List is loaded at that time.

Activating or Deactivating Choice Lists

1. **Choose Entry Choice Lists from the Special menu.** The List settings screen appears.

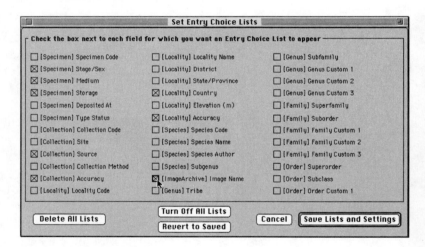

2. **Set the activation state for each List.**

    - **Click the empty checkbox next to the name of each field for which you want to activate an Entry Choice List.** Table names appear in square brackets, followed by the field name.

    - **Click checked boxes to uncheck any List you wish to deactivate.** *Deactivating a List does not erase or change its entries.* If you later reactivate the List, the entries appear just as they were last time the List was edited and saved.

3. **Click the Save Lists and Settings button to register the settings you have checked**. Newly activated Lists will appear (and newly deactivated Lists stop appearing) the next time you use the appropriate input screen. *Input screens already open are not affected.* Close any open input screens and reopen them to activate changes in List settings. (It is not necessary to quit and restart Biota).

Adding, Deleting, or Modifying Items in a Choice List

Open the input screen for the table that includes the List (field) you want to work on. You can do this by selecting the appropriate command from the Input menu (e.g., Input Specimens) or opening an existing record from an output screen (pp. 121–123).

1. **Click in the entry area for the field.** The existing List, or an empty one, appears in a Choices window.

NOTE: If no List appears, either the List has not been activated (see p. 316), or the field does not support Entry Choice Lists.

2. **To open the List Editor,** click the Modify button in the Choices window (above). The List Editor appears (below).

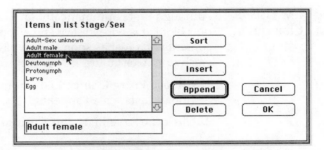

You use the buttons and List item entry/editing area in the List Editor to modify, delete, or add items to the Choice List for this field. Each field for which a List can be activated in the Entry Choice List screen (step 1) has it own separate List (see Appendix B).

3. **To add a new item at the end of the existing List,** click the Append button and enter the new item in the entry area at the bottom of the window. You will see the new item being entered in a new line at the end of the List.

4. **To insert a new item between to existing items,** click on the item you want to *follow* the new item, then type in the new item in the entry area. You will see the new item being entered in a new line just above the item you selected.

5. **To modify an existing item**, click on the item to be modified, then edit it in the entry area.

6. **To delete an existing item, a new item you don't want, or a blank item,** click on the item to be deleted to select it, then click the Delete button.

7. **To sort the items alphabetically,** click the Sort button.

8. **To undo changes made in an individual list** and dismiss the List Editor, without recording any changes you have made, click the Cancel button in the List Editor. The List Editor disappears and returns you to the Choices window. Now click Cancel in the Choices window to dismiss it.

9. **When a List is set as you wish,** click the Accept button in the List Editor window to record the changes. The List Editor disappears and returns you to the Choices window.

10. **Click Cancel in the Choices window to dismiss it.** This action does *not* undo any changes already accepted in the Lists Editor. Now, each time the cursor enters the field for this List, the List will automatically appear in a Choices window.

11. *Important:* **Save the Lists.** (See the next section for information on saving new Lists.)

Saving New Lists or Recording Changes in Existing Lists

Although Biota automatically saves any changes in Lists each time you quit Biota, it is wise to save any extensive changes you have made to the contents of a List as soon as you have made them, in case of a system crash or power failure.

1. **Choose Entry Choice Lists from the Special menu.** The Set Entry Choice Lists screen appears.

2. **Click the Save Lists and Settings button.**

Undoing All Changes Made in Choice Lists during the Current Biota Session

If you have made changes in Choice Lists during this Biota session, but you have not used the Entry Choice Lists screen to save them (see the previous section), you can undo all changes in Choice Lists made in the current Biota session. Once you quit Biota, all changes are automatically saved, and this method will not undo them when you launch Biota again.

1. **Choose Entry Choice Lists from the Special menu.** The Set Entry Choice Lists screen appears.

2. **Click the Revert to Saved button.**

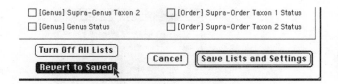

Using an Entry Choice List to Enter Data in a Record

1. **To enter an item from the Choices window, click on the item.** Biota enters the item and the cursor moves to the next entry area.

2. **To enter, manually, an item not shown in the Choices window,** press TAB or click the Cancel button in the Choices window (not the Cancel button of the input layout!) to dismiss the Choices window, then enter the special item manually in the field.

3. **To stop a Choices window from appearing,** you must deactivate the corresponding Choice List. See "Activating or Deactivating Choice Lists," earlier in this chapter (p. 316).

Transferring Choice Lists to a Different Biota Data File

If you have only a few short Lists, it will not take long to recreate them in a new or existing Biota Data File. Long or complex Lists, however, are worth transferring all at once. Biota keeps both Lists and List Settings (information on whether a List is currently activated or not) in a special table called Lists (Appendix B). You cannot view or edit the Lists table directly, but the Lists table can easily be exported and imported between Biota Data Files.

NOTE: Core field Aliases, which are also registered in the Biota's Lists table, will be transferred along Choice Lists. (See pp. 323–326).

Here is how to copy Lists from one Biota Data File to another.

1. **Launch Biota and open the "donor file"**—the existing Biota data from which you want to transfer the Lists (and Aliases). Press and hold the OPTION key (Macintosh) or the ALT key (Windows), while launching Biota. If the user password system has been activated, press and hold the key while clicking the Connect button in the Password window, to display the Open File window for your operating system.

2. **From the Im/Export menu, choose Export by Tables and Fields.** The Export Editor appears. (See Chapter 24, pp. 434–443, for a detailed introduction to the Export Editor.)

3. **Choose Lists from the Table Name popup list.**

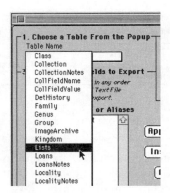

The fields for the Lists table appear the "Biota Field Names or Aliases" list in the lower-left panel of the Export Editor.

4. **Click the All>>> button.**

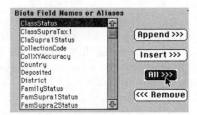

All fields appear in the Text File Fields list on the lower right.

5. **Click the Export button to launch the export.** The Save File window for your operating system appears.

6. **Place and name the text file** that Biota will create for the Lists table records. Click the Save button (Macintosh) or the OK button (Windows). A progress indicator appears. When the Export is complete, Biota posts this message:

7. **Edit the text file if necessary.** If you wish, you can now edit the exported Lists records using a text processing or spreadsheet application, but it is safer not to. (If you do edit the text file, make sure you understand its structure—see Appendix B, and *be sure to save it as plain text*.)

8. **Open the "receiving file"**—the new or existing Biota Data File into which you want to import the Lists records. If you are using BiotaApp (not Biota4D) and you have sufficient available memory, you can open the receiving file using a duplicate copy of BiotaApp, if you have made one. Otherwise, close down the donor file and relaunch Biota (pp. 8–9).

   NOTE: Be sure to press and hold the OPTION key (Macintosh) or the ALT key (Windows) while launching Biota, to bring up the Open File window. (Otherwise, Biota will open the donor file again, if you have relaunched Biota.)

   - **If you want to create a new, blank Biota Data File,** click the New button in the Open File window (see pp. 9–10).
   - **Otherwise, find the receiving file and open it.**

9. **From the Special menu, choose Import by Tables and Fields.** An option screen appears:

10. **Click OK to accept the default option** "Import new records." The Import Editor appears. See Chapter 26 for full details on using the Import Editor.

    WARNING: Biota will replace any existing Lists records with the ones you are about to import—an action unique to Lists importation. For all other tables, existing records are never deleted using either import option.

11. **Choose Lists from the Table Name popup list.** The fields for the Lists table appear the "Biota Field Names or Aliases" list in the lower-left panel of the Export Editor.

12. **Click the All>>> button** to enter all fields into the Text File Fields list on the lower right.

13. **Click the Import button to launch the import.** A warning message appears. As long as you import the same file you just exported, the warning need not concern you.

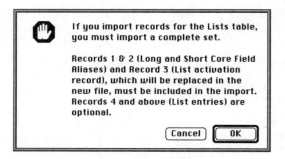

14. Click OK. The Open File window appears.

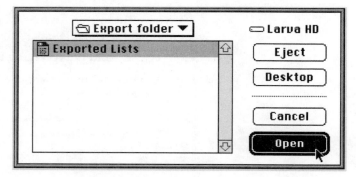

15. Find the text file you created for exported Lists records and open it. The Import progress indicator appears.

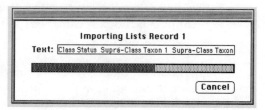

When the Import is complete, Biota posts this message:

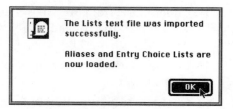

16. If problems arise or Biota posts an error message, see "Displays and Error Messages During Record Importing or Updating," pp. 495–502. If all else fails, repeat the steps in this section.

Changing the Names of Fields: The Core Field Alias System

Core fields in Biota include all fields that belong to Biota's Core tables: Specimen, Collection, Locality, Species, and all higher Taxon tables—plus Personnel and Loans (see Chapter 2 and Appendix A).

Auxiliary Fields, in contrast to Core fields, are special fields that you create and name as you like (see Chapter 15). Core fields, however, are much faster to sort and search for, so you should always use Core fields, renamed or not, in preference to Auxiliary Fields, wherever possible.

Each Core field has an *Internal Field Name*. An easily recognized version of each Internal Field Name is used to label input areas on input screens and output columns on output screens, as well as in printed reports. For example, the Internal Field Name [Specimen] Specimen Code appears on the screen as "Specimen Code." The internal field [Locality] State Province appears on the screen in two forms. Where space permits, Biota displays "State/Province." Where space is tight, you will see "State/Prov." (Notice the convention used for writing Internal Field Names: [Tablename] Fieldname.)

If your data entry, display, and reporting needs would be better served by renaming certain Core fields, you can use Biota's Core field Alias system to do so very easily. For example, a systematist (or a museum or herbarium) might want to rename a field in the Specimen table (say, [Specimen] Storage) as Specimen Condition, to accept information for each specimen regarding quality of preservation. The new name, Specimen Condition, is then called a *Core field Alias* (or just an *Alias*). It is an alias because Biota must still use the Internal Field Name behind the scenes.

Advantages and Disadvantages of Using Aliases

If you need to use a Core field for some special purpose, the obvious advantage of using an Alias to rename the field is that you can make the Alias (as displayed in input and output screens) precisely reflect the contents of the field.

There are, however, some relatively minor drawbacks that you should know about, in using Aliases for field names. The Search Editor (see pp. 192–200), the Sort Editor (see pp. 131–139), the Quick Report Editor (see pp. 269–278), and the Custom Label Editor (see pp. 261–264) all display Internal Field Names. Because these are off-the-shelf, standard 4th Dimension tools, (4D is the development environment for Biota) their behavior is not under the programmer's control, so there is no way to make these tools display Aliases instead.

But you do not need to memorize the Aliases you have defined, in order to use these tools. *Biota automatically displays a floating window with a list of all Alias = Internal Name equivalencies, whenever one of the four 4D tools is active.* (See "Checking Field Aliases," pp. 325–326).

All of Biota's other tools for searching, reporting, importing, and exporting—developed specifically for Biota—use Aliases, if you have defined any. (The Import and Export by Tables and Fields tools have a button to toggle between Aliases and Internal Field Names.)

Renaming Core Fields: Setting Aliases

1. **From the Special menu, select Core field Aliases.** The Set Core field Aliases window appears.

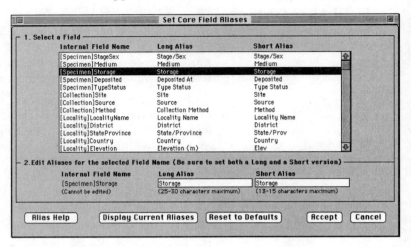

2. **Locate the field you want to rename.** Notice that the Internal Field Names (the left-hand column in the screen) follow the format: (Tablename) Fieldname.

   WARNING (Macintosh only): To move through the list of Internal Field Names and Aliases, *use the arrows at the top and bottom of the vertical scroll bar for the list*—do *not* click in the gray area above or below the "thumb slider." Due to a bug in the 4D scrollable-area tool, the list moves all the way to the top or bottom, skipping the middle part, if you click in the gray area. The thumbslider itself works correctly. (Check the ReadMe file on the Biota distribution disk to see if this 4D bug has been fixed by now—or just try it.)

   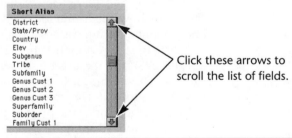

   Click these arrows to scroll the list of fields.

3. **Click once on the field you want to rename in the list of fields.** The field name now appears in the entry areas in the lower panel of the window.

4. **Edit the Long Alias and Short Alias entries in the lower panel.** In

Biota's screens and reports, the Long Alias (25–30 characters) appears where space permits, and the Short Alias (13–15 characters) otherwise. Be sure to define both.

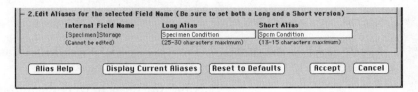

NOTE: You may have to experiment with the length of the Aliases after you see them in context in other screens. The number of characters that will fit depends on the exact letters used—e.g., *Trillium* and *Mammalia* are each eight letters long, but differ substantially in the space they occupy in proportional fonts.

5. **To check the Aliases you have defined,** click the Display Current Aliases button at the bottom of the Set Core Field Aliases screen. (See the next section.)

6. **When you have finished setting up all the Aliases you want, click the Accept button** in the Set Core field Aliases window.

Checking Field Aliases

1. **From the Special menu, select Core field Aliases.** The Set Core field Aliases window appears.

2. **Click the Display Current Aliases button.**

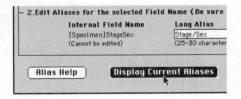

If any Aliases have been defined and saved, a floating window appears, listing all current Core field Aliases and their Internal Field Name equivalents. (If no Aliases have been defined and saved, a message appears instead.)

Depending on your screen size, the Core field Alias floating window has one of two formats shown below. The narrow window appears on screens less than 680 pixels wide.

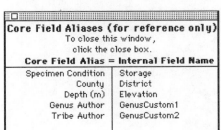

3. **To close the floating window,** either click in the close box in the upper left corner or click the Accept or Cancel button in the Set Core field Aliases window.

Clearing All Aliases and Resetting to Defaults

1. **From the Special menu, select Core Field Aliases.** The Set Core Field Aliases window appears.
2. **Click the Reset to Defaults button.**

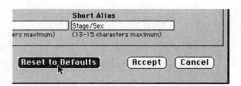

A confirmation dialog appears.

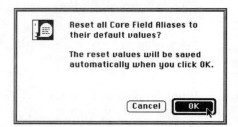

3. **Click OK if you want to proceed.** All Aliases will be reset to the values that Biota uses when you first create a new Data File. These default values correspond closely with the Internal Field Names.

PART 5   Special Tools and Features

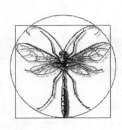

Chapter 17   The Specimen Loan System

The Loans menu contains the primary tools for managing specimen loans. Using these tools, you can keep track of loans you make to others as well as specimens you borrow from others. You can review existing loans and print or reprint loan forms and lists of specimens loaned.

How Biota Keeps Track of Specimen Loans

To use the Loan system, it is essential to understand how Biota keeps track of Loans. Each Specimen record in your Data File has a field called Deposited. Biota uses this field to identify each group of Specimen records that constitute a Loan. You can view the Deposited field in the standard Specimen output screen, as well as in individual Specimen records in the Specimen input screen.

- **When you create new Specimen records** for your own specimens, you should enter a name or code in the Deposited field (up to 20 alphanumeric characters), indicating that the physical specimens are presently in your custody. (You may wish to set up this code as the default entry for the Deposited field, by using the Field Value Defaults utility from the Special menu. See pp. 313–315.)

- **When you lend a group of specimens** from your collection to another institution or individual, you create a new record in the Loans table to record the loan, with a unique Loan Code to identify the Loan. Once you have indicated which Specimen records will be included in the Loan, Biota changes the Deposited field of each Specimen record in the Loan, replacing the existing value with the Loan Code.

- **When loaned specimens are returned** to your collection, you use the Record Returns tool to change the Deposited field of the corresponding Specimen records back to your institutional or personal collection name or to some other meaningful code.

- **When you borrow a group of specimens** from another institution or individual, you can create a Loan Code for the group and use the New Loan screen to record the specimens as a loan to your own institution or to your own collection.

- **When you return borrowed specimens,** you can use the Record Returns tool to change the Deposited field in the corresponding Specimen records of your Data File to a new code, indicating that the physical specimens have been returned.

Recording a New Loan

To create a Loan record for a new loan from your collection, or to record a loan you have received from another collection, take the following steps.

1. **From the Loans menu, choose New Loan.** The New Loan input screen appears.

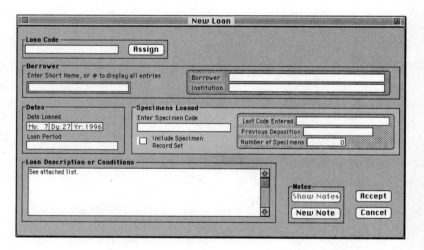

2. **Enter a Loan Code.** You can enter the Code either by typing it in the Loan Code entry area or by using the adjacent Assign button (pp. 102–103). Loan Code, the Key field for the Loan table, is an obligatory entry, which must be made first. All remaining entry areas are optional. The Loan Code you enter must not duplicate the Loan Code of any existing Loan Record, or you will receive an error message (p. 7).

   See the previous section of this chapter for ideas on what to use for a Loan Code.

   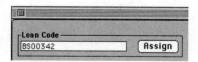

   If you use the Assign button, the Assign Loan Code window offers two options.

- **You can enter a prefix of your own choice,** then click the Accept Prefix button.

- **You can click the Default Prefix button** to use the prefix "Loan." (This default is fixed and cannot be altered in the Record Code Prefix screen, unlike other Record Code prefixes.)

In either case, Biota will format the Loan Code with the prefix (up to 14 characters in the case of an ad hoc prefix) plus a four-digit integer counter derived from the Loans table sequence number (see p. 102 for information on sequence numbers).

3. **Enter the Borrower's name.** The [Loans] Borrower field is linked to the [Personnel] Short Name field. You can use all the usual techniques for entering data in linking fields described in Chapter 6 (pp. 92–95).

    - **If you know that the Borrower has no existing record in the Personnel table,** enter a Short Name value for the Borrower (e.g., "A. B. Jones" or "Erewhon Museum NH") and press the TAB key. The Short Name may have up to 25 characters.

        NOTE: On Pin, Slide, Vial, and many Herbarium labels, Biota uses the Short Name field for the Collector's name (Chapter 13). Be sure to choose a version of the Borrower's name that is appropriate for this purpose if the Borrower is likely to be a Collector in the future— e.g., initials and last name, or first name and last name. You can also use a Group Name in the Borrower field, e.g., "A. B. Smith & C. D. Jones." Group names link records for individuals in the Personnel table. See pp. 174–177 for information on how to use Group Names.

    - **If you think the Borrower may already have a Personnel record, but you are not sure what Short Name the record carries,** enter the @ character and press the TAB key. A scrollable list of all existing Personnel records will appear, from which you can choose the correct record by clicking it. You can sort the records by Short Name, Last Name, or Institution by clicking the buttons above the columns.

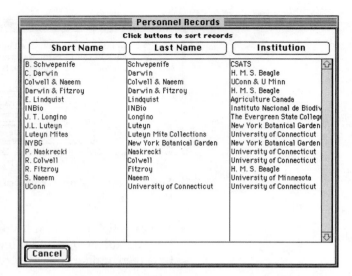

- **If you know the Short Name for the Borrower's existing Personnel record,** enter the first letter or the first few letters of the Short Name, or the full Short Name, and press the TAB key. If a single match is found, the name will be completed and the cursor will move into the next entry area. If more than one match is found, a choice list like the one above will appear, with all matches shown.

If the Borrower entry you make does not match the Short Name field of any existing Personnel record, Biota will offer you a choice between creating a new Personnel record for that name, "on the fly," or accepting the Loan record without creating a link to the Personnel table (pp. 92–95) (an orphan Loan record).

NOTE: *Creating an orphan Loan record is not recommended.* If you do so, none of the information on the Borrower's institution or contact information can be entered in the loan forms.

4. **Enter the Date Loaned.** When the New Loan screen opens, today's date (the date in your computer's internal clock) is entered in the Date Loaned entry area. If you need to change it, enter a full date (see pp. 112–113). If you want to keep today's date, just "TAB through" the date entry areas.

5. **Enter the Loan Period.** This alphanumeric entry can be up to 20 characters.

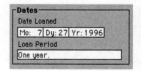

6. **Indicate the Specimens Loaned.** Although you can create a Loan record without linking it to any Specimen records (e.g., for unmounted material not in the Data File), normally a Loan record is linked to

the Specimen records for the physical specimens loaned (see "How Biota Keeps Track of Specimen Loans," pp.327–328).

There are two, nonexclusive methods of indicating which Specimen Records are included in the Loan.

- **Click the Include Specimen Record Set checkbox** in the Specimens Loaned panel.

To use this option, you must first have defined a Specimen Record Set (pp. 13–15). You can find the Specimen records using any of the tools of the Find, Series, or Tree menus (Chapter 11).

Once you click the checkbox, the Specimen Code for the last record in the Record Set appears in the Last Code Entered display area, showing the current value of the [Specimen] Deposited field for that record in the Previous Deposition display area (below).

The Number of Specimens display area (above) counts the number of Specimen Records you have entered so far—including the number in the Specimen Record Set plus any records entered using the "one by one" option described below.

- **Enter the Specimen Codes, one by one,** in the Enter Specimen Code entry area of the Specimens Loaned panel.

You can enter the Specimen Codes manually, pressing TAB after each entry, enter them with a barcode reader, or use both methods for the same Loan. If the barcode reader is set to enter an end-of-line character after each read, you can enter all Specimen Codes without touching the keyboard or mouse.

Once entered, each Specimen Code appears in the Last Code Entered display area, showing the current value of the [Specimen] Deposited field for that record in the Previous Deposition display area.

The Number of Specimens display area counts the number of Specimen Records you have entered so far—including any in the current Specimen Record Set if the Include Record Set checkbox has been checked.

7. **Make an entry in the Loan Description or Conditions entry area.** This optional entry (up to 32,000 characters) will appear on the printed Loan Invoice (p. 334). If the Loan includes any specimens or other material not represented by Specimen records, this is the place to describe that material.

8. **Add a Loan Note if you wish** (pp. 179–183). Loan Notes should be used for any comments on the loan that are *not* intended to appear on the Loan Invoice itself. Any number of additional Loan Notes can be added to the Loan record later, to document notices sent to the Borrower, condition of returned specimens, and so on.

9. **Click the Accept button** to save the Loan record (or the Cancel button to cancel it). If you Accept the record, the next step depends on whether or not any Specimen records have been included in the Loan in step 6, above.

   • **If *no* Specimen Records are linked to the Loan record,** the message below will appear.

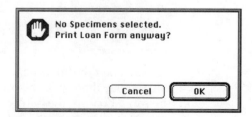

   If you click OK, the Loan form printing sequence begins. See "Previewing, Printing, and Exporting Loan Records," pp. 333–337.

   • **If Specimen Records are linked to the Loan record,** the Specimen Records will appear in a special version of the Specimen output screen entitled "Specimens in Loan." Notice that the Deposited field of each record in the new Loan has been changed to the Loan Code.

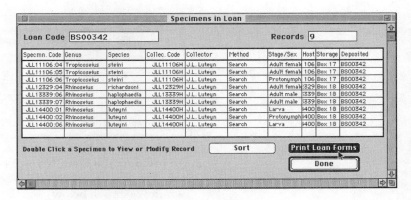

The purpose of this screen to provide an opportunity to confirm that no errors have been made in specifying the content of the loan. You can use the Sort button to sort the records. To examine a full Specimen record in the standard Specimen input screen, double click the record in the output screen.

10. **If the Specimens have been displayed:**

    ♦ *Either:* **Click the Print Loan Forms button** (shown above) to initiate the printing of a Loan Invoice and (as options) a list of the Specimen records in the Loan and a text file.

    ♦ *Or:* **Click the Done button** to leave the New Loan procedure without printing or exporting.

    This is the place to stop if you are recording a loan to you from another individual or institution (see "How Biota Keeps Track of Specimen Loans," pp. 327–328).

    The Record Set option screen appears to allow you to declare the Specimen records in the Loan a Record Set if you wish (pp. 13–15). (It is not necessary to do so to complete the Loan process.)

11. **Choose a Record Set option and click OK.** The Page Setup window for your operating system and printer appears.

12. **If you clicked the Print Loan button,** please continue with "Previewing, Printing, and Exporting Loan Records," below.

Previewing, Printing, and Exporting Loan Records

This section describes the procedures and options for printing Loan invoices, printing a list of the Specimen records in a Loan, and exporting text files to disk for the Specimen records in a Loan.

1. **Follow the instructions for "Recording a New Loan"** (pp. 328–333) **or for "Displaying an Existing Loan"** (pp. 337–339) until instructed to proceed with this section.

2. **Click the OK button in the Page Setup window.** The Print window for your operating system and printer appears.

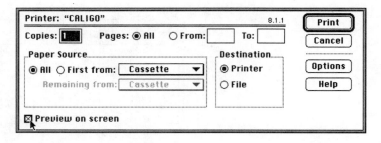

3. **To preview the Loan Invoice before (or instead of) printing,** click the "Preview on screen" checkbox in the lower-left corner of the Print window (shown below).

4. **Click the Print button in the Print window.**

   • **If you requested print preview,** the Loan Invoice for the active Loan record appears in the preview screen.

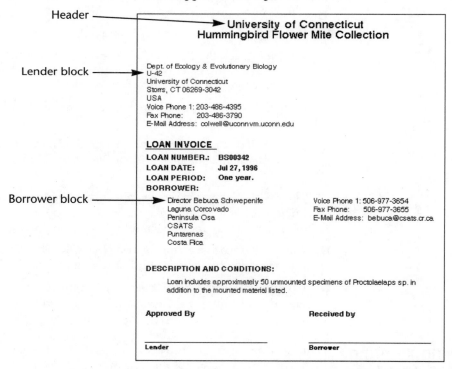

◊ The *header* at the top of the Loan Invoice shows the information in the [Personnel] Notes field for the special Project Name record (pp. 177–179) in the Personnel table, unless the field is blank—if so, the header in the Loan Invoice shows the [Personnel] Last Name field for the Project Name record, instead.

◊ The *lender block* displays address and contact information from the Project Name record of the Personnel table.

◊ **The remaining information on the Loan Invoice** comes from the corresponding fields of the Loan record, or from the Personnel record for the Borrower (*borrower block*, above). (*Loan Number* on the Invoice is the Loan Code for the record.)

- **To print the Invoice after previewing it,** click the Print button at the bottom of the preview screen (see pp. 243–245 for more details about the preview screen).

- **To proceed *without* printing the Invoice,** after viewing the Invoice in the preview screen, click the Stop Printing button at the bottom the preview screen.

- **If you choose to Print without previewing the invoice, or you clicked the Print button on the preview screen,** the Loan Invoice is printed and the "Print and Export Options" screen appears.

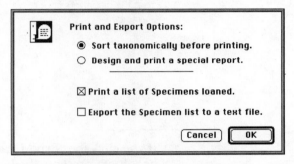

5. **Choose Print and Export options and click the OK button,** or Cancel to dismiss the rest of the procedure.

    - *Either:* **Choose "Sort taxonomically before printing."** This option sorts Specimen records taxonomically (p. 245) before printing if you select the "Print a list of Specimens loaned" option in the lower half of the Print and Export Options screen.

    - *Or:* **Choose "Design and print a special report."** This option launches the Quick Report Editor (pp. 269–278), with which you can create your own layout for a report on the Specimen records in the Loan. Fields from the Loan record are not available, however, except for the Loan Code, which is reproduced in the [Specimen] Deposited field.

    - **If you chose "Sort taxonomically before printing,"** the two checkboxes in the Print and Export Options screen are enabled. You can check either or both.

        ◊ **If you click the "Print a list of Specimens loaned" checkbox,** Biota will print a special version of the standard Specimen report (pp. 266–268) that includes Loan information and lists all Specimen records in the Loan (below).

Luteyn Mite Collections
Specimen Loan

Dept. of Ecology & Evolutionary Biology
U-42
University of Connecticut
Storrs, CT 06269-3042
USA
Voice Phone 1: 203-486-4395
Fax Phone: 203-486-3790
E-Mail Address:

Loan Code
BS00342

9 Specimens
Lent to:

Loan Date
Jul 27, 1996

Director Bebuca Schwepenife
Laguna Corcovado
Peninsula Osa
CSATS
Puntarenas
Costa Rica

Voice Phone 1: 506-977-3654
Fax Phone: 506-977-3655
E-Mail Address:
bebuca@csats.cr.ca

Spcmn. Code	Collec. Code	Method	Date Coll.	Locality	Source	GIS1	GIS2	Stage/Sex	Vou.	Medium	Deposited
Class Arachnida											
Acari											
Order Gamasida											
Family Ascidae											
Melicharini											
Genus *Rhinoseius*											
luteyni sp. n. (Species Code: lute)											
JLL14400:01	JLL14400H	Search	Apr 4, 199	Latacunga-Quevedo rd.		0	0	Larva	1	Hoyers	BS00342
JLL14400:02	JLL14400H	Search	Apr 4, 199	Latacunga-Quevedo rd.		0	0	Protonymph	1	Hoyers	BS00342
JLL14400:06	JLL14400H	Search	Apr 4, 199	Latacunga-Quevedo rd.		0	0	Larva	1	Hoyers	BS00342
Group richardsoni											
haplophaedia Ohmer et al. 1991 (Species Code: rhinhaplo)											
JLL13339:06	JLL13339H	Search	Nov 15, 198	Nono-Pto. Quito Rd.		0	0	Adult male	1	Hoyers	BS00342
JLL13339:07	JLL13339H	Search	Nov 15, 198	Nono-Pto. Quito Rd.		0	0	Adult male	1	Hoyers	BS00342
richardsoni Hunter 1972 (Species Code: rhinrich)											
JLL12329:04	JLL12329H	Search	May 14, 198	Mpio.ElCairo,Correg. Boqu		0	0	Adult female	1	Hoyers	BS00342
Genus *Tropicoseius*											
steini sp. n. (Species Code: stein)											
JLL11106:04	JLL11106H	Search	Jan 19, 198	7-9km NE Pindilig tow.	Fvie In inllorescences	0	0	Adult female	1	Hoyers	BS00342
JLL11106:05	JLL11106H	Search	Jan 19, 198	7-9km NE Pindilig tow.	Fvie In inllorescences	0	0	Adult female	1	Hoyers	BS00342
JLL11106:06	JLL11106H	Search	Jan 19, 198	7-9km NE Pindilig tow.	Fvie In inllorescences	0	0	Protonymph	1	Hoyers	BS00342

July 28, 1996 Page 1

◊ **If you click the "Export the Specimen list to a text field" check-box,** Biota presents the Export Specimen Flatfile tool. See pp. 454–457 for instructions on using this tool. The text Specimen flatfile exported for the Specimen records in the Loan includes a special header with Loan information:

LOAN NUMBER: BS00342
LOAN DATE: Jul 27, 1996

LENDER:
Dept. of Ecology & Evolutionary Biology
U-42
University of Connecticut
Storrs, CT 06269-3042
USA
Voice Phone 1: 203-486-4395
Fax Phone: 203-486-3790
E-Mail Address: colwell@uconnvm.uconn.edu

BORROWER:
Bebuca Schwepenife
Laguna Corcovado
Peninsula Osa
CSATS
Puntarenas
Costa Rica
Voice Phone 1: 506-977-3654
Fax Phone: 506-977-3655
E-Mail Address: bebuca@csats.cr.ca

Specimen Code	Determined By	Date Determined	Stage/Sex	Deposited	Genus	Species Name
JLL14400:01	P. Naskrecki	12/16/93	Larva	BS00342	Rhinoseius	luteyni
JLL14400:02	P. Naskrecki	12/16/93	Protonymph	BS00342	Rhinoseius	luteyni
JLL14400:06	P. Naskrecki	12/16/93	Larva	BS00342	Rhinoseius	luteyni
JLL13339:06	P. Naskrecki	12/16/93	Adult male	BS00342	Rhinoseius	haplophaedia
JLL13339:07	P. Naskrecki	12/16/93	Adult male	BS00342	Rhinoseius	haplophaedia
JLL12329:04	P. Naskrecki	12/16/93	Adult female	BS00342	Rhinoseius	richardsoni
JLL11106:04	P. Naskrecki	12/16/93	Adult female	BS00342	Tropicoseius	steini
JLL11106:05	P. Naskrecki	12/16/93	Adult female	BS00342	Tropicoseius	steini
JLL11106:06	P. Naskrecki	12/16/93	Protonymph	BS00342	Tropicoseius	steini

The exported text table (above) can include virtually any selection of fields from the Core tables that you wish (p. 5).

The text file export option allows you to provide the Borrower with a complete archival electronic record of the loan (with or without a printed version). If the Borrower determines specimens in the Loan, he or she can add the new determination to the same text file and return it to you for updating the Specimen records (see Chapter 26).

Displaying an Existing Loan

Records in the Loans table, like other Core tables in Biota, can be displayed in an output layout and grouped as Record Sets, and Record Sets can be saved as Record Set Pointer Files (Chapter 4). The commands "Display All Loan Records" and "Display Loan Record Set," however, are located in the Loans menu, rather than the Find and Display menus, respectively, where the analogous commands for other Core tables can be found.

To display an existing Loan record, follow these steps.

1. **From the Loans menu, choose Display All Loans.** A list of all Loan records appears in the Loans output screen.

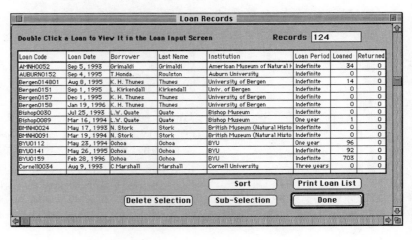

The Loans output screen differs from the standard output screens for other Core tables only in lacking a New Record button. (To create a new Loan record, choose New Loan from the Loans menu; see pp. 328–333). The Sort, Delete Selection, and Sub-Selection buttons work in the same way as the analogous buttons for other Core tables (Chapter 9).

2. **To print a list of the Loans displayed in the output screen,** click the Print Loan List button in the output screen, or select Print from the File menu while the screen is displayed. A print options window appears.

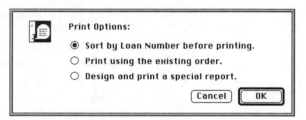

See Chapter 14, "Printing Reports," for help using these options.

3. **To display an individual Loan record,** double-click the record in the output screen. The record appears in a special version of the Loan Records input screen (below).

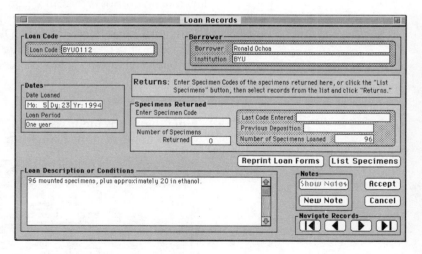

This screen resembles the New Loan screen (p. 328) but has a different set of buttons and functions. This is the same screen used for recording Returns (see "Recording Specimen Returns," below).

In this screen you can edit the Date Loaned, Loan Period, Loan Description or Conditions, or add a Loan Note. You can also *manually* change the Number of Specimens Returned and Number of Specimens Loaned, if necessary, but these numbers are normally (and preferably) adjusted *automatically* when you record returns by the methods described in "Recording Specimen Returns," below.

4. **To display the Specimen records linked to this Loan record,** click the List Specimens button in the screen above. The Specimen records will be displayed in a special Specimen output screen entitled "Specimens in Loan." (If the Loan is not linked to any Specimen records, a message will inform you.)

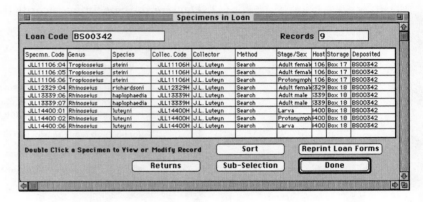

This screen has two special buttons, Returns (see "Recording Specimen Returns," below) and Reprint Loan Forms (see the next step).

5. **To reprint the Loan Invoice** (p. 334), **reprint the list of Specimens in the Loan** (pp. 335–336), **or export a text file** for the Specimen records in the Loan (p. 336–337):

   ♦ *Either:* **Click the Reprint Loan Forms button in the Loan Records input screen** shown in step 3, above.

   ♦ *Or:* **Click the Reprint Loan Forms button in the Specimens in Loan output screen** shown in step 4, above.

   Biota will offer the same sequence of options and functions available for New Loans.

6. **If you clicked either Reprint Loan button,** please continue with "Previewing, Printing, and Exporting Loan Records," earlier in this chapter (pp. 333–337).

Recording Specimen Returns

Biota offers three alternative methods of recording the return of physical specimens represented by Specimen records in the Data File.

Using the Specimen Record Set to Record Returns

This is usually the most efficient method when a set of returned specimens represents several different Loan records. For example, you may have made several Loans to the same individual or institution, who then returns some or all of the specimens, determined and sorted by species, but each conspecific group is a mixture of specimens from several Loans.

1. **Gather the Specimen records for the returned specimens** using any of the tools of the Find, Series, or Tree menus (Chapter 11).

   Probably the most useful tool for this purpose, especially when the specimens have barcodes, is the Find Specimen Series tool (Series menu; the same tool is called By Specimen Code Series in the Find menu). Use this tool with the "In any order" option and Auto Accept button. See pp. 236–238.

2. **Declare the Specimen records the current Specimen Record Set** (pp. 13–15).

3. **From the Loans menu, select Record Returns.** The Loan returns option screen appears.

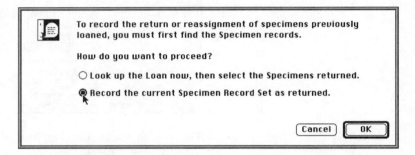

4. **Choose the second option, "Record the current Specimen Record Set as returned," then click the OK button** in the option window. (If the Specimen Record Set is empty, this option is disabled and the window displays a help message; see step 1 in the next section, for an illustration.)

   A window appears requesting a value to be entered in the Deposited field of the Specimen records returned.

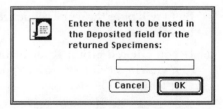

5. **Type in the value and click OK.**

   - **If you are recording returns to your own collection,** you would normally enter its code or name.

   - **If you are recording the return of specimens loaned to you,** you would normally enter the code or name for the collection to which they have been returned.

   A confirmation message appears.

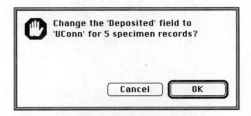

6. **Click the OK button to confirm the changes,** or Cancel to leave the records unchanged.

   ♦ **If you click OK,** Biota makes the changes in the Specimen records.

   ♦ **If any of the Specimen records in the Record Set is not part of an existing Loan,** Biota posts a message, for each such record.

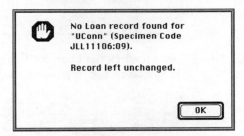

All records in the Record Set are then displayed in the Specimen output screen, whether changed or not.

NOTE: When you use this method to record returns, the Number Returned field in the Loan record for each Loan involved in the return is automatically updated to reflect the number of specimens returned.

Two Ways to Use the Loan Records Screens to Record Returns

With either of the methods described in this section, you record returns for specimens from one loan at time. If you need to record returned mixed specimens from several loans, see "Using the Specimen Record Set to Record Returns," earlier in this chapter (pp. 339–341).

1. **From the Loans menu, select Record Returns.** The Loan Returns option screen appears.

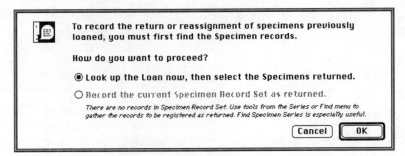

2. **Choose the first (default) option, "Look up the Loan now, then select the Specimens returned," then click the OK button** in the option window. (If the Specimen Record Set is empty, the second option is disabled and the window displays a help message, as illustrated above.)

    The Loans output screen appears, showing all Loan records.

    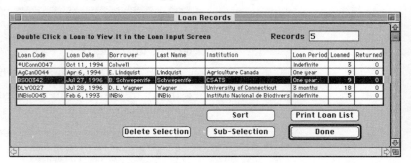

3. **Double-click the Loan for which you need to record returns** (above). The Loan record appears in the Loan Records input screen.

    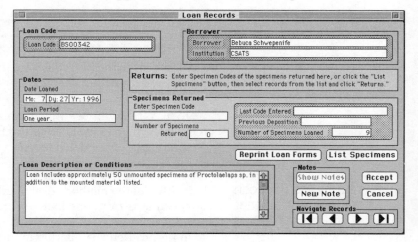

4. **To record returns by Specimen Code:**

    a. **Enter the Specimen Code for a returned specimen** in the Enter Specimen Code area of the Specimens Returned panel.

    b. **Press TAB.** (If you are using a barcode reader and Specimen barcodes, and the reader is set to enter an end-of-line character automatically after each read, this step is not necessary.)

◊ **If Biota determines that the Specimen Code is linked to this Loan record,** the Specimen Code appears in the Last Code Entered display area, and the Enter Specimen Code area is cleared for the next Specimen Code. The Number Returned field is incremented by 1.

◊ **If Biota finds the Specimen Code, but it is not linked to this Loan,** the message below appears. The Specimen Code may be linked to a different Loan record, or to no Loan record.

◊ **If Biota determines that the Specimen record is not in the Data File,** the message below appears.

c. **Enter the remaining Specimen Codes for the returned specimens** in the same way.

d. **Click the Accept button** in the Loan Records input screen.

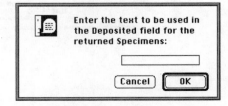

A window appears requesting a value to be entered in the Deposited field of the Specimen records returned.

e. **Go to step 6, below.**

5. **To record returns by selecting them from a Specimen list:**

   a. **Click the List Specimens button** in the Loan Records input screen

   The Specimen records will be displayed in the "Specimens in Loan" output screen. (If the Loan is not linked to any Specimen records, a message will inform you.)

   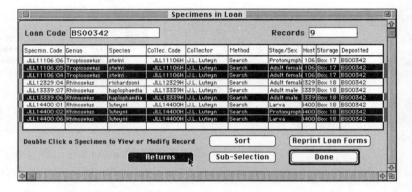

   b. **Select from the list all records for the returned Specimens,** using the mouse (see p. 128 for selection techniques).

   c. **Click the Returns button** at the bottom of the "Specimens in Loan" screen (shown above).

   A window appears requesting a value to be entered in the Deposited field of the Specimen records returned.

   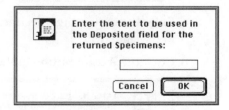

NOTES:

a. You can select a few Specimen records at a time, then click the Returns button for each group, if more convenient. Each time, you will need to fill in the request window, above.

b. As with all methods for recording Returns, when you use this method, the Number Returned field in the Loan record for each Loan involved in the return is automatically updated to reflect the number of specimens returned.

6. **Type in the value for the Deposited field and click OK.**

   ♦ **If you are recording returns to your own collection,** you would normally enter its code or name.

   ♦ **If you are recording the return of specimens loaned to you,** you would normally enter the code or name for the collection to which they have been returned.

   A confirmation message appears.

   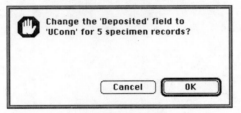

7. **Click the OK button to confirm the changes,** or Cancel to leave the records unchanged. When you click OK, Biota make the changes in the Specimen records and displays them again in the Specimens in Loan output screen.

8. **Click the Done button in the Specimens in Loan output screen.** The Record Set option screen appears (pp. 13–15).

9. **Select a Record Set option and click the OK button.** The Loan Records output screen reappears. (It has been open all along, but hidden by the Specimens in Loan screen.) The screen allows you to work with another Loan, if you wish.

10. **When you are through, click the Done button in the Loan Records screen** and choose a Record Set option for the Loans records displayed.

Chapter 18 Images

Biota records images in a table called Image Archive, which is linked to the Species table (Appendix A). Each Species record (parent record) can have any number of Image Archive records (child records) linked to it.

An Image Archive record (hereafter usually called an *Image record,* for short) has four fields: Species Code (the link to the Species record), Image Number (used to order Image records within Species), Image Name, and the Image itself.

Because each Image record is linked to a Species record, most input and display operations with Images are done through the Species input and output screens. (The exceptions are the Export Images and Create Web Pages commands in the Im/Export menu.)

Image Characteristics in Biota

Image records in Biota can come from any source of digitized images, although you may have to convert the image file format (see the section "Image File Formats," below).

Image Sources

You can scan drawings, photographs, illustrations in books, herbarium sheets, live plant specimens, or text documents using a flatbed scanner. If you have video signal digitizing hardware in your computer, you can capture video frames directly from a video camera, from videotape played on a VCR/VTR, or from other video storage technologies. You can use images from a digital camera or download image files from Web pages or from ftp sites.

Some graphics and image editing applications store supplementary information with image files that provide special instructions for output devices such as a PostScript printer. This information "tags along" when you paste or import an image into Biota and is used when printing the picture to an appropriate output device.

NOTE: The insect images illustrated in this chapter were captured directly from an S-Video input signal, using the built-in digitizing hardware in a PowerMacintosh 8100AV computer. The signal came from a lensless video camera (Hitachi KP-110) mounted on a dissecting microscope, or from a camcorder (Sony CCD-TR400) mounted on a copy stand, depending on the size of the object, using fiberoptic illumination. The mite drawings illustrated in Chapter 5 (p. 73) were scanned with a flatbed scanner.

Image Size and Shape

Images in Biota are never enlarged, cropped, or changed in proportion.

Enlargement generally reduces image quality, cropping might remove an important feature you intended to illustrate, and "stretch to fit" changes in proportion are a surrealistic touch not usually approved by biologists attempting to document living things. If you need to enlarge, crop, proportion, or change the palette for an image, you will need to use an application designed for those purposes before storing the image in a Biota Date File.

Image areas in Biota are uniformly 3 by 4 in proportion, but they vary in absolute size from thumbnail display areas (84 by 112 or 90 by 120 pixels) to the image display/input area in the Image input screen, which is 333 by 444 pixels in size.

If the original image is smaller than the dimensions of a display area, the entire image is shown at its original size. If the original image is larger than a display area, it is *proportionally reduced* to the largest size that can be displayed in the display area, for display only. The digital archive retains the original resolution and size for the image.

Image Color

Biota Image records accept images of any bit depth, including black and white (1 bit), gray scale (8 bit), or color (8, 16, 24, or 32 bit). As you know if you work with color image files, the bit depth of a color image greatly affects the amount of memory required to store the image on disk. Bear this in mind when adding image records to a Biota Data File. Archiving a single high-depth color image may enlarge the Data File more than saving hundreds of Specimen or Species records enlarges it.

NOTE: As an example of the memory requirement for gray-scale images, each of the insect images used to illustrate this chapter occupies about 30–40 KB, uncompressed.

Image File Formats

The intricacies of digital images are a world in themselves that this book will not attempt to penetrate very deeply. Moreover, it is a rapidly evolving world, especially in regard to images and the Internet.

Be sure to check the ReadMe file that you received on one of the Biota distribution diskettes and consult the Biota Web site (http://viceroy.eeb.uconn.edu/biota) for any updated information on image formats in Biota.

An image file format specifies the algorithm used to translate the pattern and color of pixels in a screen image into digital form in a disk file. Many different image formats exist, and there will surely be more. Three Image formats are currently relevant to Biota.

- **PICT.** The PICT format is the native image format of the Macintosh operating system, although Windows recognizes and displays images in this format. (In the Windows operating system, images in the PICT format have the .pic file extension.) When you place an image on the Macintosh Clipboard (using a video frame capture, screen capture, or graphics application), it is converted to a PICT image if was not already in the PICT format.

  Biota (all flavors, see pp. xxiii–xxiv) currently saves, loads, and exports images only in the PICT image format.

- **TIFF.** TIFF (Tagged Image File Format) images are widely used on both Windows and Macintosh platforms and in the publishing industry. (All the illustrations in this book are TIFF images.) To place a TIFF image in a Biota Image record, you must first convert it to the PICT format (Macintosh) or .pic format (Windows).

- **GIF.** Originated by CompuServe, GIF (Graphics Interchange Format) is currently the standard image format for World Wide Web pages. At the time of this writing, Biota cannot export images directly to the GIF format and most Web browsers cannot display the PICT format. (But check the ReadMe file to see if either limitation has changed.) Thus, the PICT images exported (as an option) by the Create Web Pages command (in the Biota Im/Export menu) must currently be converted to GIF format using an image conversion utility before being posted on a Web server. See p. 481.

Image Compression

Using Biota in the Macintosh operating system, you can compress images pasted into the Image input screen from the Clipboard, "on the fly," using any of the compression formats available in Macintosh QuickTime (see step 5 in the next section).

NOTE: Image compression is *not* currently available for 4th Dimension applications (including Biota) in the Windows operating system (but check the ReadMe file that you received with Biota). For this reason, if you intend to "transport" your Biota Data File from Biota for Macintosh to Biota for Windows (Appendix G), or use it under 4D Server (Appendix E) with Windows clients, you should not use the QuickTime compression option.

Creating a New Image Record by Pasting or Importing an Image

When you create a new Image Archive record in Biota, you can place the Image itself in the record in either of two ways—by pasting it in from the Clipboard or by importing it from a disk file. Here are the steps required.

1. **Display a Species record in the Species input screen** by selecting Species from the Input menu ("Species Input," pp. 164–168), by creating a new Species record "on the fly" from the Specimen input screen (pp. 144–146), or by displaying an existing Species record (Chapter 9).

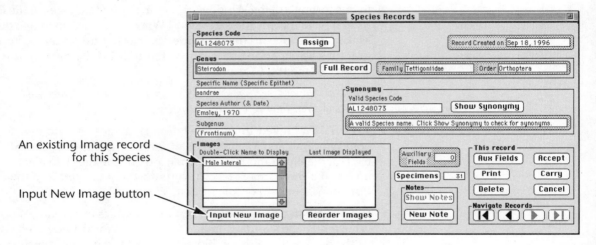

An existing Image record for this Species

Input New Image button

If the Species record is an existing record, any Images already linked to it appear by Image Name in the scrolling list in the lower-left corner of the Species input screen. The record illustrated above, for the katydid *Steirodon sandrae*, already has one image linked to it, with the Image Name "Male lateral."

2. **In the Species input screen, click the Input New Image button** in the lower-left corner (above) or double-click a blank line in the scrolling list of Image Names. The Image input screen appears, with information about the parent Species record displayed and the insertion point (cursor) in the Image Name input area.

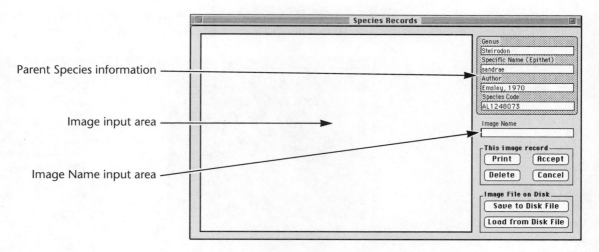

Parent Species information

Image input area

Image Name input area

3. **Enter an Image Name** (up to 30 characters) **and press the TAB key.** The name must be unique among the Image records for this Species. (The same Image Name can be used for Image records linked to other Species.)

- **If you enter an Image Name that has already been used** for another Image record for this Species, Biota posts an error message.

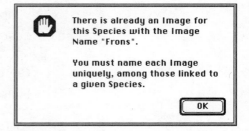

When you click the OK button in the error window, the cursor returns to the Image Name input area.

- **If you enter an acceptable Image Name,** the Image input area turns black (below), indicating that it has been activated (selected).

4. **To enter an image in the record by pasting from the Clipboard:**

   a. **Copy or capture an image to the Clipboard.** You can copy an image from a graphics document, a text document, or (in Macintosh) from the Scrapbook. Most video frame digitizing applications and screen capture utilities can usually capture directly to the Clipboard, as well.

b. **From the Edit menu in Biota, with the Image input area selected as shown above, select Paste.** The image appears in the input area.

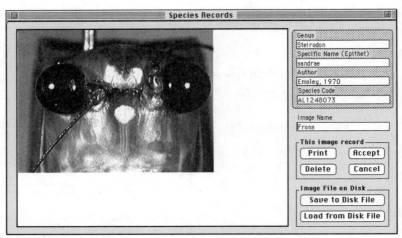

Notice in the example above (the *frons* or face of the katydid *Steirodon sandrae*) that the image does not fill the entire input Image input area. See "Image Size and Shape," p. 348.

c. **Go to step 7, below,** to complete the procedure.

5. **To compress an Image (Macintosh only) while Pasting from the Clipboard.**

   In the Macintosh operating system, if you have QuickTime installed and activated, you can compress the image "on the fly" as it is pasted. (In the Macintosh operating system, QuickTime is a Macintosh Extension that is component of System 7 and later.) To compress an Image using QuickTime:

   a. **Copy or capture an image to the Clipboard.** See step 4a, above.

   b. **Hold down the OPTION key while selecting Paste from the Edit menu.** (Do *not* use the keyboard equivalent for Paste, COMMAND-V, or this feature will not work!)

   The QuickTime settings screen appears.

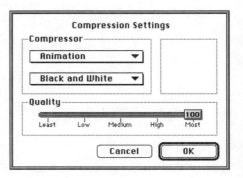

c. **Click and hold on the upper popup list** (showing the option "Animation") **and select a compression type.**

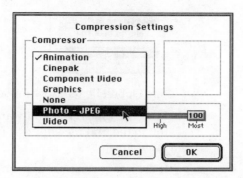

Experience indicates that the JPEG option is usually a good option for images in Biota, but you may wish to experiment with others.

d. **Click and hold on the lower popup list** (showing the option "Color") **and select a color option.**

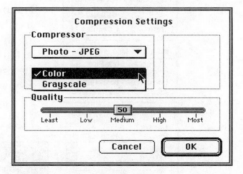

e. **Set the Quality slider** in the lower panel. The higher the Quality setting, the greater the memory requirement for the image. You will need to experiment with this setting.

For the katydid face in the illustration in step 4, the memory requirement for the uncompressed image was 36 KB. Using the "Least" setting for the Quality slider, the memory requirement was reduced to 12 KB, with no detectable loss of quality. (One way to check the memory requirement for a Biota image is to export it to a disk file, then check the size of the file.)

NOTE: Image compression is not currently available for 4th Dimension applications (including Biota) in the Windows operating system (but check the ReadMe file that you received with Biota). For this reason, if you intend to "transport" your Biota Data File from Biota for Macintosh to Biota for Windows (Appendix G), or use it under 4D Server (Appendix E), with Windows clients, you should not use the QuickTime compression option.

6. **To import an Image from a disk file.**

   a. Click the **Load from Disk File** button in the Image input screen.

   The Open File window for your operating system appears.

   b. **Find the image file you want to import and click the Open button** in the find file window. The image should appear in the Image input area.

   If the image does not appear, check to make sure the image format is appropriate (see pp. 348–349).

7. **Click the Accept button in the Image input screen.**

   The Image record is saved to the Data File, and the Species input layout reappears, showing the Name of the new Image record listed last in the scrolling list of Image Names, with the image itself displayed in the thumbnail display area.

   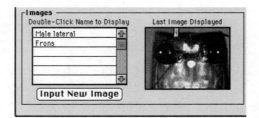

   NOTE: If you want to change the order of the Image records in the scrolling display (which controls the order in other contexts as well), see "Changing the Order of Image Records for a Species," pp. 359–363.

8. **Accept or Cancel the parent Species record.** If you Cancel the parent Species record after creating a new Image record for that Species:

   ♦ **If the parent Species record has previously been saved** (Accepted at some previous time), the new Image record is saved, without comment, when you Cancel the parent Species record. To delete the new Image record, you must open it in the Image input screen and click the Delete button (see "Deleting an Image Record," pp. 358–359).

- **If the parent Species record is a new record** that has not been saved (Accepted), the message below appears.

Since an Image record must always be linked to a particular Species record, Biota does not let you create an orphan Image record.

Displaying, Printing, Exporting, or Copying an Image Record

All operations on individual Image records are carried out in the Image input screen, accessible from the parent Species record. Here are the steps to take to display, export, or copy to the Clipboard an individual Image record. See the next section for instructions on deleting an Image record.

1. **Display a Species record in the Species input screen.**

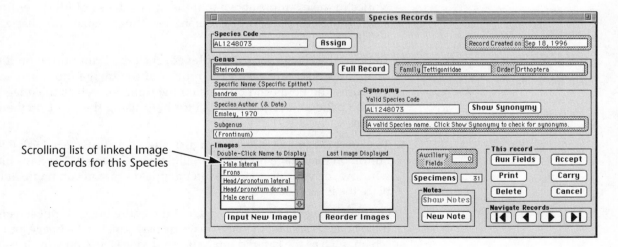

Scrolling list of linked Image records for this Species

If the Species record is an existing record, any Images already linked to it appear by Image Name in the scrolling list in the lower-left corner of the Species input screen. The record illustrated above, for the katydid *Steirodon sandrae*, already has five Image records linked to it, named in the list.

NOTE: If you want to change the order of the Image records in the scrolling display (which controls the order in other contexts as well), see "Changing the Order of Image Records for a Species," pp. 359–363.

2. **To display an existing Image** from the Species input screen, double-click the Image Name in the scrolling list. The Image appears in the Image input screen together with the Image Name and information about the parent Species record.

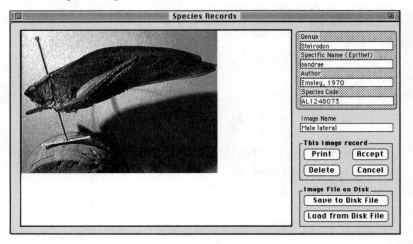

Notice in the example above that the image does not fill the entire input Image display/input screen. See "Image Size and Shape," p. 348.

3. **To print the Image record displayed,** click the Print button on the Image input screen (above). A replica of the Image input screen, including the fields and buttons on the right, as well as the image, will be printed. See pp. 243–245 for help using the print preview option.

4. **To export the image directly to a disk file,** click the Save to Disk File button on the Image input screen illustrated in step 2. The Save file window for your operating system appears so you can name and place the image file.

Only the image itself is exported, not the entire Image input screen. If the image is smaller than the image display panel in the Image input screen, such as the katydid image illustrated in step 2, only the image itself is exported, not the extra white space in the panel.

NOTE: If you intend to use the image file as a Web document, see "Image File Formats," pp. 348–349.

5. **To copy an image to the Clipboard and paste it in a document:**

   a. **Click the image to select it.**

   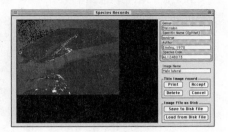

   The image changes to a "color negative," whether or not in color, and any white space in the panel reverses to black.

   b. **Copy the selected image to the Clipboard** by choosing the Copy command from the Edit menu (or the keyboard equivalent). You can check it on the Clipboard by choosing Show Clipboard from the Edit menu.

   c. **Paste the image from the Clipboard into a text document** (e.g., a Word or WordPerfect document) or an image document in an image editing or archiving application (e.g., Photoshop).

6. **To return to the parent Species record,** click the Cancel button in the Image input screen.

   The Species input layout reappears, showing the image last displayed in the Image input screen in the thumbnail display area.

   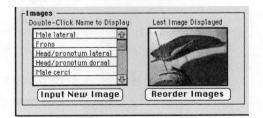

7. **Accept or Cancel the parent Species record.**

Deleting an Image Record

There are two ways to delete an Image record in Biota. You can delete an individual Image record, or you can delete the parent Species record along with all its linked Image records.

1. **Display a new or existing Species record in the Species input screen**. Image Names for the Image records already linked to the Species record appear in the scrolling list in the lower-left corner of the Species input screen (see the previous section, step 1).

2. **To delete a particular Image record:**

   a. **Double-click the Image Name in the scrolling list.** The Image appears in the Image input screen together with the Image Name and information about the parent Species record.

   b. **Click the Delete button on the Image input screen**.

   A confirmation message appears.

   c. **Click the OK button** to confirm the deletion.

   The Species input layout reappears. The name of the Image you deleted no longer appears in the scrolling list of Images.

   d. **Accept or Cancel the parent Species record.**

   NOTE: If you Cancel the parent Species record after deleting an Image record for that Species, the Image record is *not* restored.

3. **To delete a Species Record and all linked Image records:**

   a. **Click the Delete button in the Species input screen** (below).

Biota posts a warning message.

b. **Click the OK button** to confirm the deletion of the Species record and all linked Image records, or Cancel to prevent the deletions.

Changing the Order of Image Records for a Species

When you create a new Image record for a Species (pp. 350–355), the Image Name is added to the *end* of the scrolling list of Image records for that Species in the Species input screen. This section explains how to change the order of the Images in the scrolling list for an individual Species.

The order of the Image Names in the scrolling Image list of the Species input screen is important in three contexts in Biota.

- **The thumbnail image output screen** uses this order to display the first four images for each Species (see "Displaying Thumbnail Images in the Species Output Screen," pp. 363–365). To allow easy comparison of analogous views of related species by scanning "columns" of the thumbnail image screen, it is best to order images by view (and thus by Image Name) identically for related species. The images for two closely related katydid species on p. 364 is a good example of this technique.

- **The Export Images tool** can export either the first image for each selected Species, as ordered in the scrolling list in the Species input screen, or all images in the list for each selected Species. See "Using the Export Images Tool: Step by Step," p. 450.

- **The Export Web Pages tool** exports the first image for each Species, as ordered in the scrolling list in the Species input screen, and creates a hyperlink to the image from the Species record in a Species Web page. See Chapter 25, pp. 480–481.

If you want to change the order of the Images in the scrolling list for a Species, follow these steps.

1. **Display a new or existing Species record in the Species input screen**. Image Names for the Image records already linked to the Species record appear in the scrolling list in the lower-left corner of the Species input screen.

2. **Click the Reorder Images button** in the Image panel of the Species input screen. (If there are no Image records linked to this Species, the Reorder Images button is disabled—dimmed.)

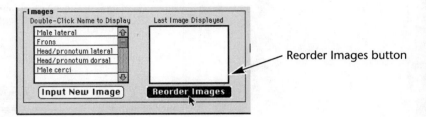

Reorder Images button

The Image Order Editor appears (below), displaying the Image Names for the Species named in the window title. The Image Names are ordered by Image Number, which is displayed in the left column. The first Image Name is displayed at the bottom of the window, with its Image Number highlighted in the Image Number editing area.

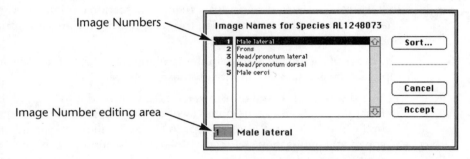

Image Numbers

Image Number editing area

3. **To sort Images alphabetically by Image Name:**

   a. **Click the Sort button** in the Image Order Editor window. A Sort option window appears.

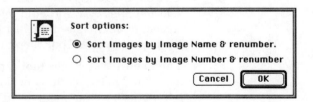

   b. **Click the OK button** in the option window to accept the default option, "Sort Images by Image Name and renumber." The Image Names reappear in the Image Order Editor, sorted alphabetically.

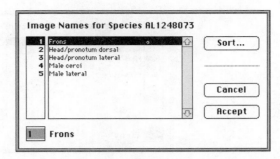

c. Click the Accept button in the Image Order Editor. The list reappears in alphabetical order in the Species input screen.

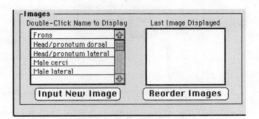

4. **To reorder Images as you wish (non-alphabetically).** To reorder Images, you edit the Image Numbers in the Image Order Editor to reflect the order you wish. You can skip numbers and use decimal or negative numbers if you wish.

   a. **Click the Image Name of the Image you want to relocate** in the order. "Male cerci" is clicked, below. The Image Name you clicked appears at the bottom of the editor.

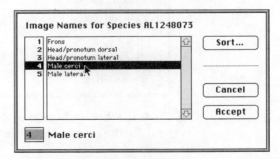

   b. **Change the Image Number** (in the Image Number editing area at the lower left) to place the Image in the desired position among the remaining Images in the list. In this example, the Image Number for "Male cerci" has been changed to 6, so this Image will sort numerically at the end of the list.

   c. **To move an Image to the beginning of the list,** enter a fractional or negative Image number. The Image called Frons will be moved to the beginning of the list in the example below.

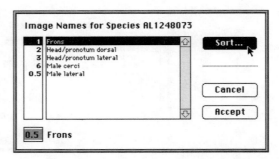

d. **Click the Sort button** in the Image Order Editor window. A Sort option window appears.

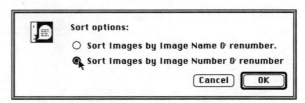

e. **Choose the second option,** "Sort Images by Image Number and renumber."

f. **Click the OK button** in the option window. The Image Names reappear in the Image Order Editor, sorted numerically according the edited Image Numbers, but renumbered sequentially by integers.

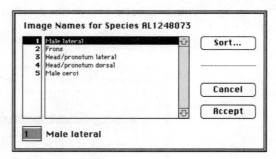

g. **Click the Accept button** in the Image Order Editor. The list reappears in the new order in the Species input screen.

NOTES:

a. Image Numbers need not be *sequential* within the set of Image records linked to a Species. Nonsequential Image Numbers, which may arise if you delete an Image record, nonetheless sort numerically. Image Numbers appear explicitly nowhere else in Biota, so their actual numerical values do not really matter as long as you are satisfied with the Image order. (You can always renumber them sequentially by using the Sort button in the Image Order Editor.)

b. Image *Numbers* need not be *unique* within the set of Image records linked to a Species, although Image records for the same Species that share the same Image Number appear in arbitrary order. Image *Names*, however, must be unique within the set of Image records linked to a Species (see p. 351).

Displaying Thumbnail Images in the Species Output Screen

If you want to compare images for several species at once, you have two options. You can launch separate processes and open the images simultaneously in the Image input screen, as described in the next section ("Comparing Images in the Image Input Screen," p. 365), or you can use the thumbnail Image output screen, accessed from the Species output screen. Here are the steps to take for the latter method.

1. **Find the Species records** for which you want to display and compare images.

   ♦ *Either:* **Using any of the tools of the Find menu,** display the Species records in the standard Species output screen (Chapters 9 and 11).

   ♦ *Or:* **Using tools of the Tree menu,** display the records for all the Species of a Genus in the Species Lookup screen (see pp. 187–192).

   The standard Species output screen is illustrated below. See p. 189 for an illustration of the Species Lookup screen.

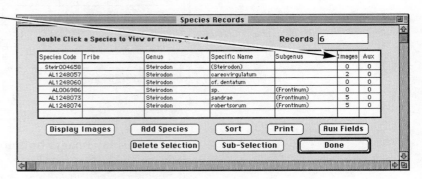

Number of linked Image records for each Species

Notice that the Images column in the Species output screen shows the number of linked Image records for each Species.

2. **If you wish, sort the Species records by decreasing number of linked images** (in the standard output screen). Sometimes, you may want to keep a group of Species records displayed, even though only some of them have linked images. (This is the case for the illustration above.)

   To sort the Species records by number of images, follow these steps.

   a. Click the Sort button at the bottom of the Species output screen. The Sort Species window appears.

   b. Click the Number Images field in the field names panel at the left.

c. **Click the sort order arrow** at the upper right to point it downwards, for descending order.

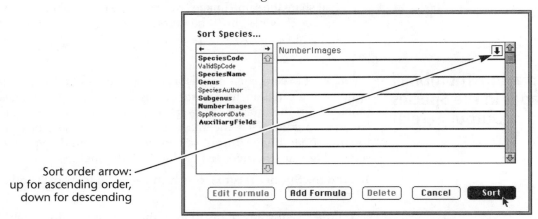

Sort order arrow: up for ascending order, down for descending

d. **Click the Sort button at the bottom of the Sort Species window** (above). The Species records reappear, sorted by *decreasing* number of linked images.

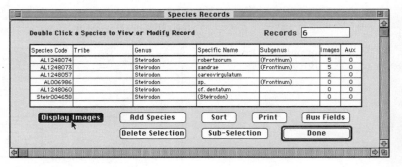

3. **Click the Display Images button** at the bottom of the screen (illustrated above). The first four Image records for each species in the current Species Selection appear in a row of thumbnail images, one Species per row.

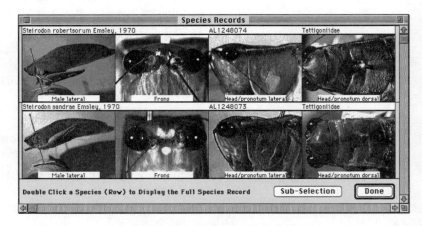

The Genus, specific name (Species Name), Species Author, Species Code, and Family appear in the white band above the each row of images. The Image Name for each Image appears in a small inset at the base of each image.

- **The order of the images, within rows,** follows the order you have set for the Image records in the Species input screen. See "Changing the Order of Image Records for a Species," pp. 359–363.
- **To display a Species record in the Species input screen from the thumbnail output display,** double-click the row of images for the Species.
- **To create a Sub-Selection of Species (thumbnail image rows),** select the rows you want to keep in the Sub-Selection and click the Sub-Selection button.
- **To return to the standard Species input screen,** click the Done button in the thumbnail display.

Comparing Images in the Image Input Screen

Sometimes, thumbnail images (see the previous section) are too small to show details that you want to compare in two or more images. To display two or more Image records simultaneously in the Image input screen, you can open the same or a different parent Species record simultaneously in separate Biota processes.

1. **Find the Species record you want,** using tools from the Find or Tree menus (Chapter 11), and display it in the Species input screen.
2. **Display an image** in the Image input screen and leave it displayed.
3. **Find the same or a different Species record,** using tools from the Find or Tree menus (every menu item launches a new process), and display it in the Species input screen.
4. **Display another image** in a second Image input screen.
5. **Repeat steps 3 and 4** for additional images.

NOTE: The practicality of this approach is limited by the amount of memory you allocate to Biota and by the size of your monitor screen.

Exporting Groups of Images to Disk Files

The Export Images tool in the Im/Export menu lets you export many images, each to its own file, with a single command. The tool is described in detail in the section "Exporting Images" in Chapter 24 (pp. 449–451).

In brief, the Export Images tool lets you choose between exporting *all* images for each Species in the active Species Record Set, or *the first* image for each Species in the active Species Record Set, as ordered in the Species input screen. (See "Changing the Order of Image Records for a Species," pp. 359–363.) In either case, each Image is saved in its own disk file, and the file is named automatically.

Chapter 19   Determination Histories

In the management of biological collections and in systematic revisions, the determination of individual specimens often changes as knowledge increases, nomenclature is corrected, or opinions shift. Botanists often record such changes right on the herbarium sheet, vertebrate biologists annotate tags or add secondary tags, whereas entomologists generally update determinations by adding secondary labels to specimens.

In biotic surveys and inventories, information on the taxonomic identity of a specimen is useful even when imprecise. A trained nonspecialist assistant can correctly identify most insects to order or family, for example. (See Chapter 22 for Biota's support for recording temporary determinations.) A specialist may at first determine the same specimens to genus, and later determine some of the same specimens to species while others await description of new taxa.

If you enable the Determination History system (in the Preferences screen from the Special menu), Biota will keep track of all sources of change in specimen determination automatically.

Determination History Records

Determination History records have a child-parent (many-to-one) relation with Specimen records (Appendix A). Each Determination History record (in the Determination History table) records the following information:

- **The Specimen Code field of the parent record** (the linking field for the child-parent relation to the Specimen table). If the parent record's Specimen Code is changed, this field in the Determination History table is automatically updated to keep the records linked.

- **The Determined By and Date Determined fields of the parent Specimen record,** at the time of the determination change was recorded in the Determination History record.

- **The Species Code, Species Name, and Species Author fields of the Species record** to which the Specimen record was linked at the time of the determination change was recorded in the Determination History record.
- **The Genus field of the Genus record** to which the Species record was linked, at the time of the determination change was recorded in the Determination History record.
- **A record of who entered the change** (the Changed By field). The person entering the change may or may not be the same person who changed the determination. This entry is made automatically based on who signed on when Biota was launched. If the user password system has not been activated (Chapter 23), Biota will record "Administrator" in the Changed By field of the Determination History record.

NOTE: If it is important to know who entered each change in specimen determination, be sure to set up signon names and passwords for each user, with appropriate privileges (Chapter 23).

- **The date the change was entered** (the Date Changed field), entered automatically from the computer's internal clock.
- **A record of where the change was made:**
  ◊ **In an individual Specimen record,** by changing the [Specimen] Species Code linking field.
  ◊ **In a series of Specimen records,** by changing the [Specimen] Species Code linking field by means of the Find and Identify Specimen Series tool (pp. 234–242).
  ◊ **In the linked Species record,** by changing the Species Name field, the Species Author field, or the [Species] Genus linking field and choosing to update linked Specimen records (pp. 218–220).
  ◊ **In the linked Genus record,** by changing the [Genus] Genus field and choosing to update linked Species records (pp. 218–220).
  ◊ **In the Synonymy screen,** by reassigning Specimen records to a Species that has been declared a senior synonym (Chapter 20).
- **The Sequence Number of the record** in the Determination History table. The only purpose for this field is to sort the records in the reverse order they were created (see illustrations below).

Note: The sequence number is the number of records entered in the table since the Data File was created. The number increases by one every time a record is created, no matter how many records are later deleted. (The Date Changed field cannot be used to sort by entry order since more than one change can occur on the same date.)

Enabling or Disabling the Determination History System

NOTE: If the user password system has been activated (Chapter 23), you must have Administration access privileges to enable or disable the Determination History system.

1. **From the Special menu, select Preferences.**
2. **Check to enable, or uncheck to disable** the option "Automatically record Determination Histories."

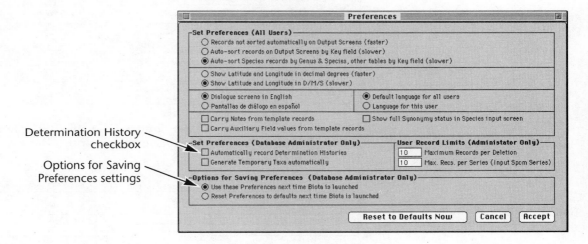

Determination History checkbox

Options for Saving Preferences settings

3. **Choose an option for saving any changes** in Preference settings made during this Biota session (above).

   - **If you want the changes to remain in force until altered** regardless of how many times Biota is launched in the meantime, choose the option "Use these Preferences next time Biota is launched."
   - **If you want the changes to be in force only for this session**, choose the option "Reset Preferences to defaults next time Biota is launched.

   The default setting for Determination Histories is unchecked *(off)*.

4. **Click the Accept button** on the Preferences screen.

Displaying the Determination History for a Specimen Record

You can display previously recorded Determination History records, whether or not the system is currently enabled (see the previous section to enable or disable the Determination History system).

1. **Display a Specimen record in the Specimen input screen** (Chapter 9). If any Determination History records exist for this record, the Determination History button is enabled (below). Otherwise, it is disabled (dimmed).

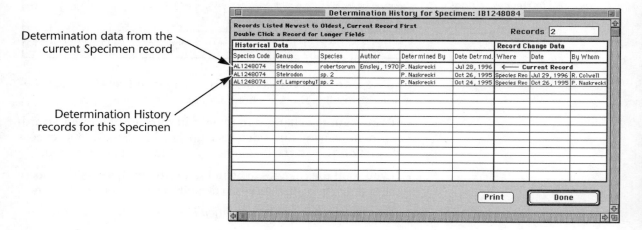

Determination History button

2. **Click the Determination History button** (if enabled). The Determination History records for this Specimen appear in a Determination History output screen.

Determination data from the current Specimen record

Determination History records for this Specimen

The screen requires some explanation.

- **The Specimen Code** appears in the window title. All the Determination History records shown are for the Specimen indicated in the title.

- **The screen has two groups of columns:**

  ◊ **The Historical Data panel,** on the left, shows the history of changes in Species Code, Genus, Species (specific name), Author, Determined by, and Date Determined.

  ◊ **The Record Change Data panel,** on the right, shows, for each line of Historical Data, the table in which the determination in the left panel was changed (Where), when the change was made (Date), and who made it (By Whom).

Keep in mind the convention used in this screen: Record Change Data in the right panel refer to changes made to the Historical Data *on the same line* in the left panel. See the summary, below, for an illustration.

- **The first data row of the screen,** labeled Current Record in the Record Change Date panel, is not a Determination History record. It shows fields from the parent Specimen record.

- **The rest of the data rows,** which are Determination History records, are listed from most recent to the earliest.

In the illustration above, the Determination History of Specimen *IB1248084* can be fully reconstructed, as follows:

- **On October 24, 1995,** the Specimen was tentatively identified as "cf. *Lamrophyllum* sp. 2" (Species Code *AL1248074*) by P. Naskrecki.

- **On October 26, 1995,** Naskecki determined the correct genus to be *Steirodon*. He himself entered this information in Genus field in the Species record (Species Code *AL1248074*) on the same date (shown on the *last* line of the display in the Record Change Date panel).

- **On July 28, 1996,** Naskrecki updated the determination again, by identifying the species as *Steirodon robertsorum* Emsley, 1970. On July 29, 1996, R. Colwell entered this change in the Species record, replacing "sp. 2" in the Species Name field with the definitive determination, "*robertsorum*," and adding the Author (and date), "Emsley, 1970."

3. **To display an individual Determination History record,** double-click the row. (Clicking the Current Record row has no effect.) The record appears in read-only mode in a Determination History input screen. You can use the navigation buttons to look at the sequence of Determination History records for this Specimen.

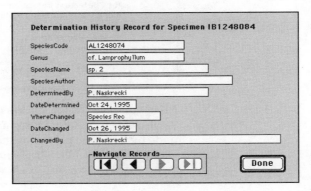

In the illustration, for example, "cf. *Lamprophyllum*" is too long to fit in the Genus column of the Determination History output screen, but can be read in the input screen.

Note: If you need to edit or delete a Determination History record, you must do so using commands from the Special menu. See "Displaying, Editing, or Deleting Determination History Records," pp. 375–377.

4. **To print Determination History records displayed** (plus the Current Record row), click the Print button in the Determination History output screen. See pp. 243–245 for help using the print preview screen.

5. **To return to the Specimen record,** click the Done button in the Determination History output screen.

How and When Changes in Determination Are Recorded

Changing a Determination in an Individual Specimen Record

Because Biota is a relational database, it records changes in Specimen determination made not only in Specimen records but also in the parent Species and Genus records.

Make sure you have enabled the Determination History system (see "Enabling or Disabling the Determination History System," p. 369.)

1. **Display the existing Specimen record** in the Specimen input screen (pp. 121–123).

2. **Change the existing Species Code in the Specimen record to a new value.** See pp. 144–146 for help in finding and entering a Species Code in a Specimen record.

   ♦ **Biota enters today's date** automatically in the Date Last Determined area. You can change the date if necessary. (See p. 151.)

   ♦ **Biota prompts you to enter a new value for "Last Determined by."**

When you click the OK button in this window, Biota highlights the "Last Determined by" entry area. If the new determination was made by the same person as the previous one, you need not make an entry. See pp. 149–151 for help in finding and entering a "Last Determined by" value.

NOTE: If there was no previous Species Code entry in the Specimen record, no Determination History record is created when you enter one, and the window above does not appear.

3. **Accept the revised Specimen record.**

- **If you did not update the "Last Determined by" area** after changing the determination, Biota prompts you again.

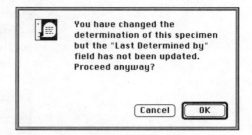

- **If you updated the "Last Determined by" area or you click OK in the window above,** Biota creates a linked record automatically in the Determination History based on information from the *old* Specimen determination. The new information then replaces the old in the Specimen record.

4. **To check that the Determination History record was created,** display the Specimen record in the input screen and click the Determination History button (see the previous section).

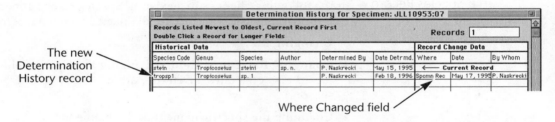

The new Determination History record

Where Changed field

The Where Changed field for the new record will read "Spcmn Rec," to indicate where the change was made.

Changing a Determination in a Specimen Record Series

Changing the determination for groups of specimens efficiently is the special function of the Find and Identify Specimen Series tool. The group of specimens may have sequential or nonsequential Specimen Codes. See Chapter 12.

Make sure you have enabled the Determination History system (see p. 369.)

1. **From the Series menu, choose Find and Identify Specimen Series.** Following the instructions in Chapter 12, set up the screen to find a group of Specimen records.

2. **Set up the Species Code for the new determination.** Follow the instructions in step 2 of the previous section of this chapter.

3. **Accept the revisions to the Specimen series,** following the guidelines in step 3 of the previous section. Biota automatically creates a Determination History record for each Specimen record in the Series.

4. **To check that the Determination History records were created:**

    a. **Click the Cancel button** in the Find and Identify Specimen Series screen to dismiss it.

    b. **Display a Specimen record** in the input screen and click the Determination History button.

The new Determination History record

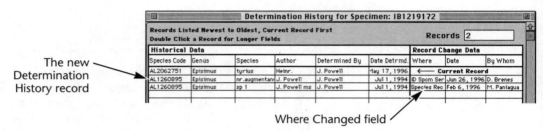

Where Changed field

The Where Changed field for the new record will read "ID Spcm Ser," to indicate where the change was made.

Changing a Determination in a Species Record

If you have enabled the Determination History system (see p. 369), any of the following changes in a existing Species record generates a new Determination History record for all linked Specimen records:

- **Changing the Species Code**.
- **Changing the Genus field**. (Changing the Genus field in a *Genus* record also generates a Determination History record. See the next section.)
- **Changing the specific name** (Species Name field).
- **Changing the Species Author field**.

Biota posts the following confirmation message:

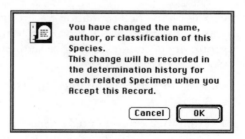

The Where Changed field in the Determination History record will read "Species Rec," to indicate where the change was made. An example is illustrated at the top of this page.

Changing a Determination in a Genus Record

If you have enabled the Determination History system (see p. 369), the following action generates a Determination History record for each Specimen linked to a Genus record through any Species record:

- **Changing the Genus field** in a Genus record.

Biota posts the following confirmation message:

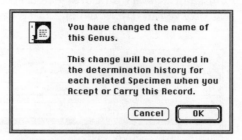

The Where Changed field in the Determination History record will read "Genus Rec."

Changing a Determination Using the Synonymy Tool

Using the Synonymy tool, you can automatically reassign Specimen records to a Species that has been declared a senior synonym (Chapter 20), unlinking those records from the junior synonym you specify.

If you have enabled the Determination History system (see p. 369), this action generates a Determination History record for each Specimen affected.

The Where Changed field in the Determination History record will read "Species Syn."

Displaying, Editing, or Deleting Determination History Records

Generally, the purpose of Determination History records is to keep a complete "audit trail" for all actions that affect the identification of Specimen records, including errors and corrections of errors.

Nonetheless, circumstances may arise that require editing or deletion of Determination History records. This section describes the tools available in Biota for working with existing Determination History records directly. The creation of *new* Determination History records occurs only through the mechanisms described in the previous section.

Note: If the user password system has been activated (pp. 409–412), you must have Administration access privileges to use the tools described in this section. If the user password system has not been activated, any user can use them.

- **To display all Determination History records,** choose All Determination Histories from the Special menu.
- **To display the Determination History records linked to the Specimen records in the active Specimen Record Set,** choose Determination Histories for the Specimen Record Set from the Special menu (abbreviated Det Hists for the Spcm Rec Set in the menu).

In either case, the records are displayed in a standard Biota output screen, with columns arranged in the same way as in the special Determination History output screen used to display Determination History records for an individual Specimen Record.

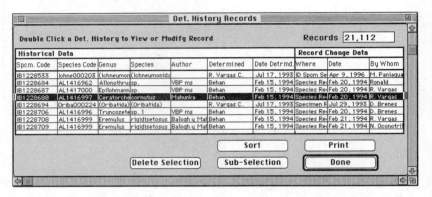

The screen includes all the standard output screen buttons except for the Add Record button (see Chapter 9). You can Sort the records, Print them, Delete a Selection, or create a Sub-Selection.

♦ **To display an individual Determination History record, double-click the record.** The record appears in the Determination History input screen.

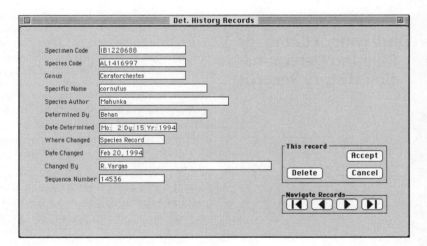

In this screen, you can edit any field except Specimen Code. (Editing the Specimen Code field in a Determination History record could create an orphan Determination History record, an action that would render the record useless.)

- **To create a Determination History Record Set or Record Set Pointer File,** click the Done button in the Determination History output screen and choose an option in the standard Record Set option screen (pp. 13–15).

- **To display the current Determination History Record Set,** choose Display Determination Histories from the Special menu.

- **To Save or Load a Record Set Pointer File** for Determination History records, use the appropriate command from the File menu. See the note below.

- **To use the Search Editor to find Determination History records by their content,** choose By Using the Search Editor from the Find menu (pp. 192–200). See the note below.

NOTE: If the user password system has been activated (Chapter 23), the Determination History table is accessible to the Save and Load Record Set Pointer File tools and the Search Editor only for users with Administration access privileges.

Exporting and Importing Determination History Records

Determination History records can be exported to text files using the Export by Tables and Fields tool from the Im/Export menu. There are no special considerations; just follow the general instructions in Chapter 24.

Determination History records can be imported from text files using the Import by Tables and Fields tool from the Im/Export menu. Be sure to consult the section "Importing Data to Notes Tables or the Determination History Table" (pp. 502–503) in Chapter 26.

Chapter 20 The Synonymy System

The Synonymy system can help you keep track of synonymies among Species records.

If your work is in systematics, you could use the system to record historical synonymies from the literature. But more usefully, you can update Species and linked Specimen records, in your own active database, when you conclude that two groups of specimens are conspecific. Biota keeps a record of the junior synonym by *synonymizing*—but retaining—its original Species record. With your confirmation, Biota will automatically switch the links for Specimen records from the junior synonym to the senior synonym. If you have activated the Determination History system (Chapter 19), the switch will be recorded for each Specimen.

In biotic surveys and inventories, Biota's Synonymy system can be used to resolve and document the "synonymies" that often arise in pooling two or more temporarily determined groups of Specimens, when each group was initially identified as representing a different Species (which may well have only a temporary name in the Species Name field; see Chapter 22). If you determine that both groups of Specimens represent the same Species, the Synonymy system can "move" all Specimens from one of the two Species to the other instantaneously. If you have enabled the Determination History system (Chapter 19) Biota will document the move for each Specimen affected.

Biota's Synonymy system is not designed to handle all the complexities that arise under the rules of nomenclature, although designs for such tools exist.[1] But it does well with the commonest case: simple synonymies at the Species level.

---

[1] Beach, J. H., S. Pramanik, and J. H. Beaman. 1993. Hierarchic taxonomic databases. In Fortuner, R. (ed.), *Advances in Computer Methods for Systematic Biology*. The Johns Hopkins University Press, Baltimore pp. 241–273.

How Biota Keeps Track of Species Synonymies

Chapter 7, "Record Codes," explains how and why Biota uses Species Codes to uniquely identify each record in the Species table. In addition to the Species Code field, however, each Species record has a separate field called *Valid Sp Code* (for Valid Species Code; see Appendix A).

In each new Species record, the Valid Sp Code field is automatically initialized with the same value as the Species Code—each Species record thus begins life as a *valid Species record*, defined in Biota as any record in which the Valid Sp Code is the same as the Species Code.

When you *synonymize Species Z with Species Y*, Species Z becomes a *junior synonym record* and Species Y becomes the *senior synonym record*. Biota replaces the Valid Sp Code in the record for the junior synonym (Species Z) with the Species Code of the senior synonym (Species Y). No change is made in the record of the senior synonym, which remains a valid Species record.

Thus Biota recognizes junior synonyms by the presence of a Valid Sp Code that does not match the Species Code. Of course, the Valid Sp Code need not be unique; several Species records may share the same Valid Sp Code when there are several junior synonyms of the same senior synonym.

Biota recognizes senior synonyms by searching for junior synonyms—that is, searching for other Species records that carry the Species Code of the senior synonym in their Valid Sp Code field.

To summarize, there are three *proper* kinds of Species records, with regard to the Synonymy system:

- **Species X is a valid Species with no synonyms in the database** if its Valid Sp Code matches its own Species Code, but the Valid Sp Code of no other Species in the database matches Species X's Species Code.

- **Species Y is a valid Species and a senior synonym** if its Valid Sp Code matches its own Species Code, and the Valid Sp Code for at least one other Species in the database matches Species Y's Species Code.

- **Species Z is a junior synonym** if its Valid Sp Code does not match its own Species Code, but instead matches the Species Code of a valid Species in the database.

In addition, there are two *improper* kinds of Species records, with regard to the Synonymy system. Biota safeguards against the inadvertent creation of either of these categories of junior synonyms, while permitting you to do so anyway, if you wish.

- **Species A is an** *orphan junior synonym* if its Valid Sp Code matches neither its own Species Code nor the Species Code of any other Species in the database. When you change the Species Code of a senior synonym, the Valid Sp Code of all its junior synonyms are automatically updated to prevent this anomaly.

- **Species B is an *compound junior synonym*** if its Valid Sp Code does not match its own Species Code, but instead matches the Species Code of a junior synonym in the database. When you synonymize a Species record that was previously a senior synonym using the Synonymize button, Biota automatically updates the Valid Sp Code of its former junior synonyms to the Species Code of the new senior synonym. (See p. 389, lowe illustration for an example.) If you wish to create compound synonyms, intentionally, to keep track of historical changes in nomenclature, you can do so by entering manual changes in Valid Sp Code.

Displaying the Synonymy Status of a Species Record

Each Species record falls into one of the five categories outlined in the previous section. To display complete information for a Species record, follow these steps.

1. **Display the Species record** in the Species input screen (see Chapter 9).

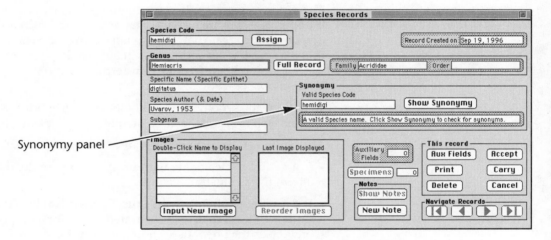

Synonymy panel

The Synonymy panel has three elements.

- **The Valid Species Code entry area.** Although the Synonymy system normally takes care of assigning values to the Valid Sp Code field, you can change the value manually in this entry area if necessary (for example, to correct an orphan or compound synonym anomaly).
- **The Show Synonymy button.** See the next numbered step.
- **The synonymy status display.** This line summarizes the synonymy status of the Species record. The degree of detail depends on the setting of the "Show full Synonymy status in Species input screen" checkbox in the Preferences screen. See the section "The Synonymy Status Display," later in this chapter (pp. 385–387).

2. **To display full synonymy information for the current Species record,** click the Show Synonymy button in the Synonymy panel of the Species input screen (above). The Synonymy screen appears.

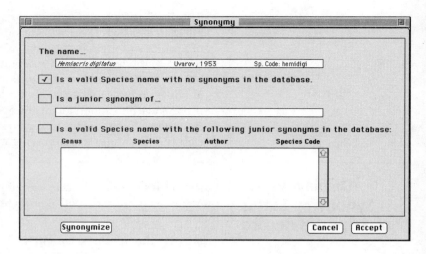

The synonymy screen displays complete synonymy information for the Species. There are no data entry areas—the screen is used only for display and to create new synonymies using the Synonymize button in the lower-left corner of the screen (see the section "Declaring a Species Junior Synonym and Transferring Its Specimens," later in this chapter, pp. 388–390).

To indicate the synonymy status of the current Species record, a check appears next to one of the three alternatives, completing the statement:

The name…[Genus, Specific Name, Author, Species Code]…

- **Is a valid Species name with no synonyms in the database.** This alternative is illustrated above.

- **Is a junior synonym of…[Genus, Specific Name, Author, Species Code].** This alternative is illustrated below for a proper junior synonym (see the previous section for definitions).

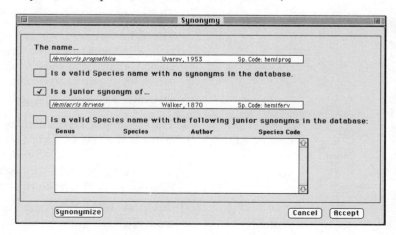

- **Is a valid Species name with the following junior synonyms in the database,** followed by a scrollable list of junior synonyms. This alternative is illustrated below.

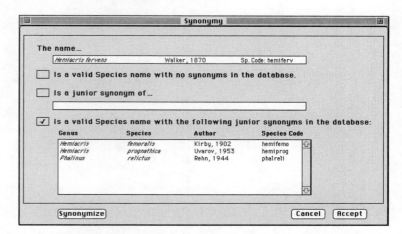

The three alternatives correspond to the three *proper* forms of synonymy status, detailed in the previous section. The next three steps explain what happens if a Species record falls into the improper categories of orphan junior synonym or compound junior synonym, or if it has compound junior synonyms linked to it.

3. **If the Species is an orphan junior synonym** (see the previous section for definition), the junior synonym alternative is checked, but the message indicates orphan status.

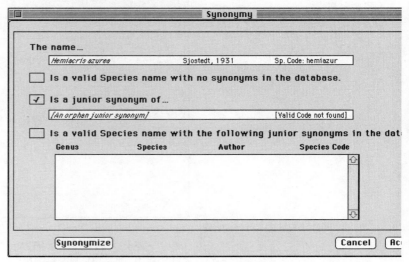

4. **If the Species is an compound junior synonym** (see the previous section for definition), the junior synonym alternative is checked, and a special legend for the third alternative (also checked) appears:

"…which is itself a junior synonym of…" followed by the Valid Species Code for the record of which the current record is a junior synonym.

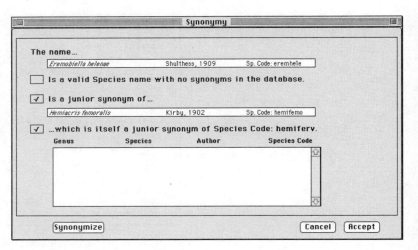

5. **If the Species is itself a proper junior synonym, but has one or more compound junior synonyms linked to it** (see the previous section for definitions), the junior synonym alternative is checked, and a special legend for the third alternative (also checked) appears: "The following is a compound junior synonym of…" followed by a scrollable list of compound junior synonyms.

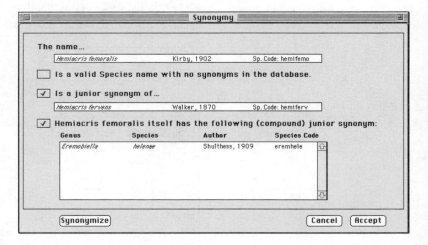

6. **To return to the Species input screen,** click the Cancel button in the Synonymy screen. The current Species record reappears, with a more complete synonymy status display based on the search of the database carried out for the Synonymy screen. See the next section for details on the synonymy status display.

The Synonymy Status Display

The small display area in the Synonymy panel of the Species input screen, the *synonymy status display*, shows the synonymy status for the current Species record. The possible status alternatives are discussed in the section "How Biota Keeps Track of Species Synonymies," earlier in this chapter (pp. 380–381).

The level of detail shown in the synonymy status display depends on whether or not the "Show full Synonymy status in Species input screen" checkbox in the Preferences screen has been checked and on whether or not you have already used the Show Synonymy button for the current Species record. The only advantage of leaving the Full Synonymy checkbox unchecked (disabled—the default setting) is a slightly faster display of Species records in the Species input screen.

To Enable or Disable Full Synonymy Display

1. **From the Special menu, select Preferences.** The Preferences screen appears.

2. **Check or uncheck the box** labeled "Show full Synonymy status in Species input screen." The default setting is unchecked, to allow faster display of Species records in the Species input screen.

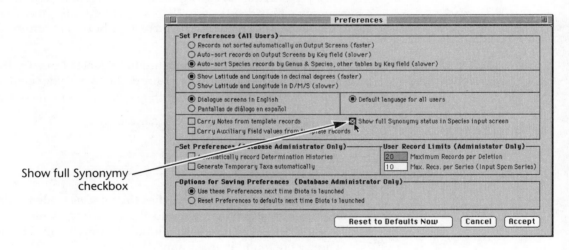

3. **Click the Accept button** in the Preferences screen.

If Full Synonymy Display Is Not Enabled

If you have not enabled the Full Synonymy option (see the previous section), Biota makes a *preliminary* check of synonymy status of each Species record just before it is displayed in the Species input screen and reports accordingly in the Synonymy status display area.

- **If the Valid Species Code matches the Species Code for the record,** the Synonymy status display reads "A valid Species name. Click Show Synonymy to check for synonyms."

- **If the Valid Species Code does not match the Species Code for the record,** the Synonymy status display reads "A junior synonym. Click Show Synonymy for details." The message is in red type, to alert you not to link Specimen records to a junior synonym.

If Full Synonymy Display Is Enabled

If you have enabled the Full Synonymy option (see "To Enable or Disable Full Synonymy Display," earlier in this chapter, p. 385), Biota will investigate the full synonymy status of each Species record just before it is displayed in the Species input screen and report accordingly in the Synonymy status display area. For each record, this takes a second or two longer than displaying the record without performing the Synonymy check (previous section).

- **If the Valid Species Code matches the Species Code for the record and no junior synonyms are found,** the Synonymy status display reads "A valid Species name with no synonyms in the database."

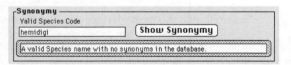

- **If the Valid Species Code matches the Species Code for the record and *n* junior synonyms are found,** the Synonymy status display reads "A valid Species name with *n* junior synonyms in the database." (Three were found in the example below).

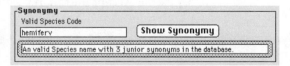

To display the full list of junior synonyms, click the Show Synonymy button. (See "Displaying the Synonymy Status of a Species Record." earlier in this chapter, pp. 381–384.)

- **If the Valid Species Code does not match the Species Code for the record but matches the Species Code of a valid Species,** the Synonymy status display reads "A junior synonym of [full name of the senior synonym]." The message is in red type, to alert you not to link Specimen records to a junior synonym.

To check for any compound junior synonyms linked to record, click the Show Synonymy button. (See "Displaying the Synonymy Status of a Species Record," earlier in this chapter, pp. 381–384.)

- **If the Species is an orphan synonym,** the Synonymy status display reads "An orphan junior synonym." The message is in red type, to alert you not to link Specimen records to an orphan junior synonym and to highlight the need to correct this anomaly.

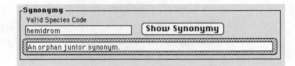

For more information, click the Show Synonymy button. (See "Displaying the Synonymy Status of a Species Record," earlier in this chapter, pp. 381–384.)

- **If the Species is a compound junior synonym,** the Synonymy status display reads "An compound junior synonym." The message is in red type, to alert you not to link Specimen records to a junior synonym and to highlight the need to correct this anomaly.

For more information, click the Show Synonymy button. (See "Displaying the Synonymy Status of a Species Record" earlier in this chapter, pp. 381–384.)

The Synonymy Display after Using the Synonymy Screen

Regardless of whether you have enabled the Full Synonymy option (see "To Enable or Disable Full Synonymy Display," earlier in this chapter, p. 385), Biota investigates the complete synonymy status of a Species record when you click the Show Synonymy button. (See "Displaying the Synonymy Status of a Species Record" earlier in this chapter, pp. 381–384.)

When you return to the Species input screen, the displays in the Synonymy display area match the alternatives in the preceding section.

Declaring a Species a Junior Synonym and Transferring Its Specimens

In a survey or inventory, you may conclude that two groups of specimens linked to different temporary Species (morphospecies) records are, in fact, conspecific. In systematics, you may need to record that two species names are synonymous.

Biota makes it easy to synonymize two Species records. As explained in detail in the first two sections of this chapter, Biota keeps a record of the junior synonym by *synonymizing*—but retaining—its original Species record.

With your confirmation, Biota will automatically switch the links for Specimen records from the junior synonym to the senior synonym. If you have activated the Determination History system (Chapter 19), the switch will be recorded for each Specimen (p. 375).

Here are the steps to take to synonymize one Species record with another.

1. **Display the Species record** to be synonymized (to be declared a junior synonym) in the Species input screen (Chapter 9).

2. **Click the synonymy button in the Species input screen**. The Synonymy screen appears, with a complete report on the synonymy status of the Species record. (See "Displaying the Synonymy Status of a Species Record," pp. 381–384, for details.)

3. **Click the Synonymize button** in the bottom left corner of the Synonymy screen (below).

The standard Look Up Species screen appears. (See pp. 145–146 if you need help with this tool.)

4. **Enter the Genus** (or the first letters of the Genus) of the Species with which the current Species is to be synonymized, then press tab.

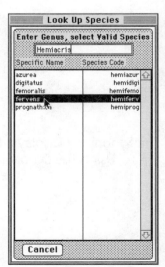

The scrolling list in the Look Up Species tool will display all specific names (Species Names) linked to the Genus.

5. **Click the specific name of the valid Species** (senior synonym) of which the current Species is to be declared a junior synonym.

   ♦ **If the Species you clicked is a valid species, and the Species to be synonymized has no junior synonyms of its own,** you will see the report below.

   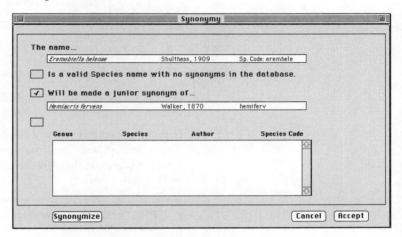

   ♦ **If the Species you clicked is a valid species, and the Species to be synonymized has one or more junior synonyms of its own,** you will see the report below.

   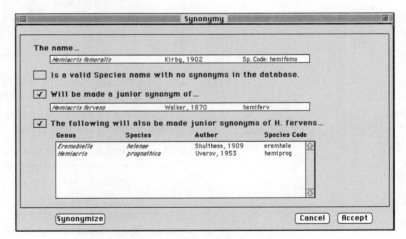

   ♦ **If you attempt to make a Species record a junior synonym of an existing junior synonym** (i.e., if you attempt to create a compound junior synonym), Biota will post the following message.

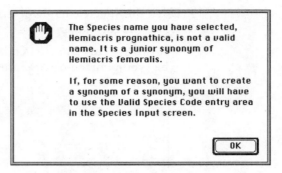

6. **Accept or Cancel the proposed changes** by clicking the appropriate button (Accept or Cancel) in the Synonymy screen.

7. **If you clicked Accept and the Species just designated a junior synonym has linked Specimen records,** Biota will offer to update the Species Code in the Specimen records to the Species Code of the senior synonym.

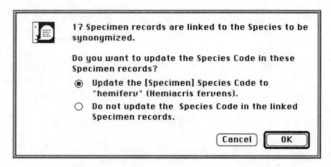

- **If you choose "Update the [Specimen] Species Code...,"** Biota will "move" the Specimens from the junior synonym to the senior synonym, automatically, by changing the [Specimen] Species Code field in the Specimen records.

- **If you choose "Do not update the Species Code...,"** no change will be made in any Specimen record. The Specimens linked to the junior synonym will remain linked to it.

NOTE: If you have enabled the Determination History system, and you choose the first option above, Biota will record the change in Specimen determination for each Specimen record that was "moved" to the senior synonym. See p. 375.

8. **Select an option and click OK.**

Clearing All Synonymies

You can undo any synonymy by setting the Valid Species Code equal to the Species Code for an individual Species record. In addition, Biota offers a tool to do this for every Species record in a Biota Data File, all at once.

1. **From the Special menu, choose Clear All Synonymies.** A warning appears.

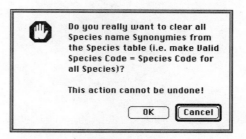

2. *Understanding that this action cannot be undone*, click OK to confirm the proposed action or Cancel to dismiss it.

NOTE: If the user password system has been enabled (pp. 409–412), only a user with Administration access privileges can use this command.

Chapter 21 Host–Guest Relations

In the web of life, biological interactions link one species with another in myriad patterns. A relational database offers a natural way to document these links, enriching the biological information you record about specimens or living individuals.

In addition, organs, tissues, or molecules derived from whole-organism specimens are specimens in their own right and often require their own records, but these records should be linked to the master record for the whole-organism specimen.

How Biota Handles Links between Specimens

Biota uses a *host–guest* metaphor to express linkage between two Specimen records—whether the "guest" is a welcome one or not. The relationship may be an interspecific one, such as between a host and a parasite, pathogen, commensal, or mutualist (guests); or between a food plant (host) and an herbivore (guest). Alternatively, the relationship may be intraspecific, such as between a parent (host) and its several offspring (guests). Using the same metaphor, organ, tissue, or molecular specimens are considered "guests" of the individual organism to which they belonged—the host specimen.

Structurally, Biota handles relationships between Specimen records through a recursive (circular) relationship mediated by the Collection table (see Appendix A). The justification is simple: the collection site for a guest specimen is its host specimen. Thus, Host Specimen Code appears as a field in the Collection table, as illustrated in the Collection input screen below. (The field is abbreviated *Host Spcm Code* in Biota's internal structure. See Appendix A.)

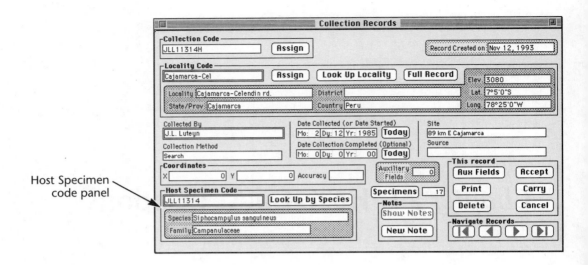

Host Specimen code panel

Host–Guest Relations: An Example

Suppose, for example, that a caterpillar (the guest specimen) has been collected from its food plant (the host specimen), and a Specimen record has been created for each.

To begin with, you link the Specimen record for the plant (the *host Specimen record*) to an ordinary Collection record (the *host Collection record*) through the Collection Code field, defining details of the collecting event (collector, date collected, etc.). Of course, additional Specimen records, such as duplicate plant specimens, may be linked the same Collection record. These are simply the normal procedures for data entry (Chapter 10).

You link the Specimen record for the caterpillar (the *guest Specimen record*), however, to a *separate* Collection record (the *guest Collection record*). The guest Collection record includes the Specimen Code for the host plant in the [Collection] Host Specimen Code field (illustrated above). If Specimen records exist for additional guest specimens collected from the same plant host, you would link them to the same guest Collection record. These links are depicted in the illustration that follows the next paragraph.

Host–guest relations are recursive in Biota. Most often, you will need to use only a single level of host–guest relations, but you can link Specimen records recursively to any level necessary. For each level in a recursive host–guest relationship, a guest Collection record—distinct from the host Collection record—must be created. The illustration below shows the Specimen and Collection records necessary to link the Specimen records for two host–guest levels: (1) a host plant and its caterpillar "guests"; and (2) one of those caterpillars, now as a host, and its parasitic wasp "guests." (Only the linking fields are shown.)

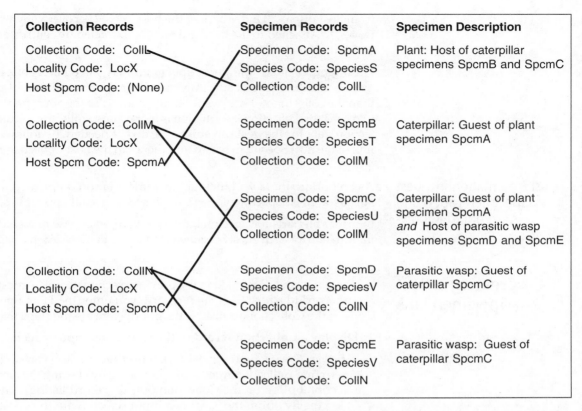

Notice that, at each level in this hypothetical example, guest Specimen records potentially bear a many-to-one relationship with the guest Collection record, and thus also with the host Specimen record. Specifically, two caterpillar specimens (SpcmB and SpcmC) are linked to a single guest Collection record (CollM), and thus to a single plant host Specimen record (Spcm A). Likewise, two wasp specimens (SpcmD and SpcmE) are linked to a single guest Collection record (CollN), and thus to a single caterpillar host Specimen record (SpcmC). Moreover, multiple guests of a single host need not be conspecific. In the illustration, caterpillar SpcmB has been identified as a representative of SpeciesT whereas caterpillar SpcmC belongs to SpeciesU.

Notice, also, that all three Collection records in the illustration are linked to the same Locality record (LocX).

Recording Information in the Guest Collection Record

If the guest specimen was collected by the same person at the same time and place as the host specimen, then most of the fields of the host Collection record can be repeated in the guest Collection record. Biota offers a quick way to transfer this information (see "Creating Guest Collection Records Automatically" later in this chapter, pp. 398–401). Even so, you may wish to add details in the Site or Source fields specifying the precise collection site or collection method for the guest specimen: "Reared from host pupa," "Under bark of the host tree," "Recovered by washing leaves," "Attached to the third gill arch," "Extracted by centrifugation," or "Amplified by PCR."

On the other hand, the guest specimen may have been collected or extracted from the host specimen at a later date by a different person than the collector of the host specimen. In this case the appropriate place to record this collection information is in the guest Collection record. Collection data for the host can still be located, if needed, by following the link to the host Specimen record and thence to the host Collection record.

Host Information in Guest Specimen Labels and Printed Reports

As an option, Biota will include host information on printed locality labels for guest Specimens. See Chapter 13, especially pp. 245–246.

In printed reports for Specimen records, you can choose to include host information for each guest Specimen record. See pp. 266–268 in Chapter 14.

Creating a Host Specimen Link

As explained and illustrated in the previous section, to link a guest Specimen record to a host Specimen record, you enter the Host Specimen Code in the Collection record for the guest. Here are the steps to take.

1. **Display a Collection record in the Collection input screen.**

   - *Either:* **Create a new guest Collection record.** You can do this by selecting Collection from the Input menu, by clicking the New Collection button on the Collection output (record listing) screen, or "on the fly" from the Specimen input screen (with the guest Specimen record displayed). See Chapter 10.

   - *Or:* **Find and display the existing guest Collection record.** You can do this by clicking the Full Record button in the Collection panel of the Specimen input screen with the guest Specimen record displayed (see pp. 124–125), or by using tools from the Find menu (Chapter 11).

2. **Complete or update the fields of the Collection record, except for the Host Specimen Code field,** following the instructions in Chapter 10 (pp. 152–159).

3. **Enter the Host Specimen Code.**

   - *Either:* **Enter the Host Specimen Code manually.** When you press TAB to complete the entry, the Genus, Species Name (specific name), and Family for the host Specimen should appear in the Host Specimen display area, as illustrated below.

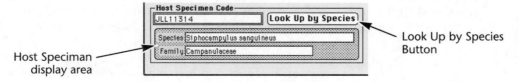

   ◊ **If the information is correct,** click the Accept button in the Collection input screen. You are done; skip the rest of the steps in this section.

◊ **If the host specimen information does not appear or is incorrect,** you have probably entered the wrong Specimen Code. Try the alternative method below.

♦ *Or:* **Click the "Look Up by Species" button** in the Host Specimen Code panel of the Collection input screen (illustrated above). The Look Up Host by Species window appears (below).

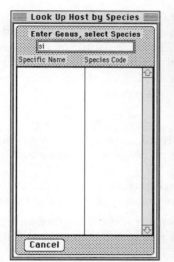 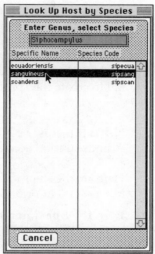 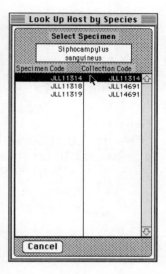

a. **Enter a Genus, or the first letter or letters of a Genus and press TAB** (above left). If a matching Genus record is found, all Species records for the Genus will be appear in the scrollable area of the Lookup window (above center). If more than one match is found, a list of matches will appear from which you can select the correct Genus (see pp. 145–146 for an illustration).

b. **When you find the Species you want, click it** (above center). A list of all Specimen records for that Species will appear (above right).

c. **Click the host Specimen record in the scrollable list** (above right) to enter the Host Specimen Code in the guest Collection record.

WARNING (MACINTOSH ONLY): If the list of Species or Specimens is longer than will fit in the window, the vertical scroll bar will be enabled. To move up and down the list, use only the arrows at the top and bottom of the scroll bar or the "thumb-slider" box inside the scroll bar. Due to a bug in 4D, clicking above or below the slider box in the gray area of the scroll bar displays the first or last screen of values, not the next or previous screen of values.

d. **If the information is correct,** click the Accept button in the Collection input screen.

Creating Guest Collection Records Automatically

Commonly, most fields of a guest Collection record require the same values as the corresponding fields in the linked host Collection record. For example, if a collector removes and preserves ectoparasites from a host when the host is collected, or collects herbivores and food plant vouchers at the same time and place, then the Locality Code, Collected By, Date Collected, Coordinates, and perhaps other fields of the host and guest Collection records will share the same values.

Biota takes advantage of this circumstance to offer a quick way to create guest Collection records, using the corresponding host Collection records as templates. For this reason, you must first have linked each host Specimen record to a host Collection record (through the Collection Code field) before using this tool, although Biota will warn you if you fail to create the appropriate links.

Creating Guest Collection Records Automatically: Step by Step

NOTE: If the user password system has been enabled, only a user with Administration access privileges can use this tool (pp. 409–412).

Here are the steps to take to create guest Collection records automatically, based on information in host Specimen and host Collection records.

1. **In the Specimen output screen, display only the host Specimen records that will be referenced by the guest Collection records that you want to create.** You can use any of the tools of the Find menu to find the host Specimen records. If other Specimen records are also displayed, select the host Specimen records (p. 128) and use the Sub-Selection button (pp. 128–129) to dismiss the other records.

2. **Make a note of how many host Specimen records are displayed.**

3. **Click the Done button and declare the host Specimen records the current Specimen Record Set.** See pp. 13–15 if you need help.

4. **If you wish to make sure that a host Collection record already exists for each host Specimen record,** follow the lettered steps below. (If you are reasonably confident that each host Specimen is linked to a host Collection record, proceed directly to step 5, instead. Biota checks for missing links in any case.)

    a. **From the Find menu, choose Places for Specimens or Species.** The query screen for this tool appears.

    b. **Set up the query "Find all Collections for the Specimen Record Set" and click the Accept button.** (See pp. 208–210 if you need help with this tool.) The Collection records for the host Specimen Record Set will appear.

    c. **Compare the number of host Collection records with the number host Specimen records you noted in step 2.** The two numbers should be the same. If there are fewer Collection Records than Specimen records, this means that one or more Specimen records in the Specimen record set are not linked to any Collection record (through the Collection Code field). In this event, display the Specimen Record Set (choose Specimens from the Display menu) and correct the problem before proceeding.

d. Dismiss the Collection output screen and (if it is open) the Specimen output screen.

5. **From the Special menu, choose Make Guest Collection Records.** An instruction and warning message appears.

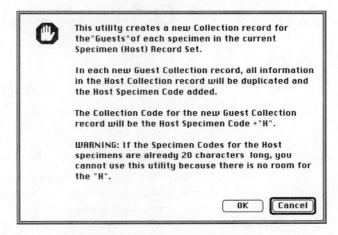

6. **Click the OK button to proceed,** or Cancel to dismiss the procedure. (Notice that Cancel is the default button, for safety.) If you click OK, Biota creates one new *guest Collection* record based on *each host Specimen* record in the current Specimen Record Set, with the following characteristics:

- **The Collection Code of the new guest Collection record** is composed of the *Specimen* Code of the host Specimen record, with the letter *H* appended.

- **The Host Specimen Code field of the guest Collection record** contains the Specimen Code of the host Specimen record.

- **All other fields of the guest Collection record** have the same value as the corresponding fields of the *host Collection* record.

The new guest Collection records are displayed in the Collection output screen.

WARNING: As the instruction screen above reminds you, if a Specimen Code for any host Specimen is already 20 characters in length (the maximum length allowed), you cannot use this tool to create a corresponding guest Specimen record because there is no room to append the *H*. If you try, Biota will display a Duplicate Key Error message, as described in the next section. In this circumstance, you will have to create a guest Collection record manually. You can use the existing host Collection record as a template and the Carry button (pp. 41–42) to speed the process.

Creating Guest Collection Records Automatically: Error Messages

Two error messages may appear during the process of creating new guest Collection records.

- **No host Collection record.** For each Specimen record in the current Specimen record set that is not linked to any Collection record (through the Collection Code field) Biota displays the message below.

Make a note of the Specimen Code. The creation of other new guest Collection records proceeds when you click the OK button.

When the process is complete, create the missing link for each Specimen record flagged for this error, declare them the current Specimen Record Set, then repeat the steps in the previous section to create guest Collection records based on the Specimen records that were flagged.

- **Duplicate Key Error.** If Biota attempts to create a new guest Collection record using a Collection Code that matches the Collection Code for an existing Collection record, you will receive the warning below.

Make a note of the Collection Code. The creation of other new guest Collection records proceeds when you click the OK button.

A Duplicate Key Error can arise in this procedure for one of two reasons. Check the Collection Code in the warning to determine which case applies.

◊ *Either:* **The guest Collection record you want to create already exists.** In this case, just ignore the warning.

◊ *Or:* **The Specimen Code for the host Specimen is already 20 characters in length** (the maximum length allowed). You cannot use this tool to create a corresponding guest Collection record because there is no room to append the *H* to create the guest Collection Code (see p. 399).

When the process is complete, you will have to create a guest Collection record manually for each such case. You can use the existing host Collection record as a template and the Carry button (pp. 41–42) to speed the process.

Finding Host and Guest Records

> Host Specimens & Collections
> Guest Specimens & Collections
> Host Spcms for Guest Spcms
> Guest Spcms for Host Spcms

Although you can find linked host and guest records by following the links shown in the illustration on page 395, one link at a time, the Find menu provides four tools to help trace host–guest links more efficiently.

Finding Host Specimens and Host Collections

1. **If you want to base the search for host Specimen and host Collection records** on a certain group of Specimen records (or even just one host Specimen record), find the Specimen records and declare them the current Specimen Record Set (pp. 13–15).

2. **From the Find menu, choose Host Specimens & Collections.** An option window appears. (See the next step.)

3. **Choose a search option.**

   ♦ **If the active Specimen Record Set is not empty,** the search option window offers two alternatives (below).

   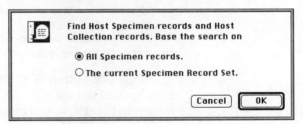

   ◊ **All Specimen records.** If you choose this option (the default) Biota will find and display (1) all Specimen records in the Data File for which the Specimen Code appears in the Host Specimen Code field of any Collection record (all host Specimen records) and (2) the Collection records linked to those Specimen records through the Collection Code field (all host Collection records).

   ◊ **The current Specimen record Set.** If you choose this option Biota will find and display (1) all Specimen records in the current Specimen Record Set (which may contain only a single record, if you wish) for which the Specimen Code appears in the Host Specimen Code field of any Collection record (host Specimen records) and (2) the Collection records linked to those Specimen records through the Collection Code field (host Collection records).

- **If the active Specimen Record Set is empty,** the search option window offers only the first alternative (below). The second alternative is disabled.

4. **Click OK in the search option window to launch the search.** When the search is complete, Biota reports the results and offers three display options.

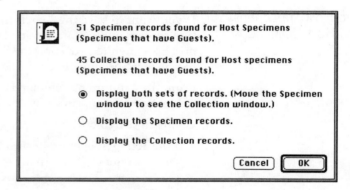

You can choose to display Specimen and Collection records, only the Specimen records, or only the Collection records.

5. **Click the OK button to display the records.** The records appear in the standard output screen for each table. When you click the Done button in the output screen, the usual Record Set options are made available (pp. 13–15).

Finding Guest Specimens and Guest Collections

1. **If you want to base the search for guest Specimen and guest Collection records** on a certain group of Specimen records (or even just one guest Specimen record), find the Specimen records and declare them the current Specimen Record Set (pp. 13–15).

2. **From the Find menu, choose Guest Specimens & Collections.** An option window appears. (See the next step.)

3. **Choose a search option.**

    - **If the active Specimen Record Set is not empty,** the search option window offers two alternatives (below).

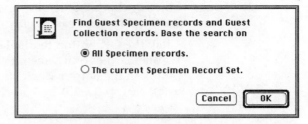

◊ **All Specimen records.** If you choose this option (the default), Biota will find and display (1) all Collection records in the Data File that have a nonblank value in the Host Specimen Code field (guest Collection records) and (2) all Specimen records linked to those Collection records through the Collection Code field (guest Specimen records).

◊ **The current Specimen record Set.** If you choose this option, Biota will find and display (1) all Collection records in the Data File that have a nonblank value in the Host Specimen Code field (guest Collection records) and are linked through the Collection Code field to a record in the current Specimen Record Set, and (2) the Specimen records linked to those Collection records through the Collection Code field (guest Specimen records).

♦ **If the active Specimen Record Set is empty,** the search option window offers only the first alternative (below). The second alternative is disabled.

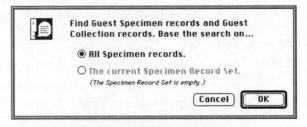

4. **Click OK in the search option window to launch the search.** When the search is complete, Biota reports the results and offers three display options.

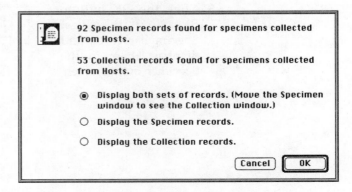

You can choose to display Specimen and Collection records, only the Specimen records, or only the Collection records.

5. **Click the OK button to display the records.** The records appear in the standard output screen for each table. When you click the Done button in the output screen, the usual Record Set options are made available (pp. 13–15).

Finding Host Specimens for Guest Specimens

This tool finds the host Specimen record linked to the Specimen record for a guest or the host Specimen records linked to a group of guest Specimen records.

1. **Create a Specimen Record Set for a guest Specimen record or records** (pp. 13–15). If you wish, you can first ensure that all the putative guests are linked to host records by using the Guest Specimens and Collections tool (pp. 402–404).

2. **From the Find menu, choose "Host Spcms for Guest Spcms."** An explanatory message appears.

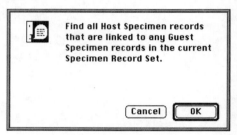

3. **Click the OK button.** Biota finds the Host Specimen records and displays them in the Specimen output screen.

Finding Guest Specimens for Host Specimens

This tool finds the guest Specimen records linked to the Specimen record for a host or to a group of host Specimen records.

1. **Create a Specimen Record Set for a host Specimen record or records** (pp.13–15). If you wish, you can first ensure that all the putative hosts are linked to guest records by using the Host Specimens and Collections tool (pp. 401–402).

2. **From the Find menu, choose "Guest Spcms for Host Spcms."** An explanatory message appears.

3. **Click the OK button.** Biota finds the Guest Specimen records and displays them in the Specimen output screen.

Chapter 22 Temporary Taxa for Approximate Determinations

In virtually any study that involves new collections, many specimens will initially be undetermined to the species level, particularly if you work with highly diverse or poorly known taxa or with groups that are unfamiliar to you. Nonetheless, it is often useful to record approximate identifications of specimens that can later be made more precise as new information or authoritative determinations become available.

A natural (and familiar) way to provide an approximate identification is to identify a specimen accurately at a higher taxonomic rank—"*Drosophila* sp.," "drosophilid fly" (Family Drosophilidae), "fly" (Order Diptera), or simply "insect" (Class Insecta). In fact, *not* recording known, accurate determinations at higher ranks, when specific identities are unknown, means a loss of potentially useful information that is already in hand.

The problem with recording approximate determinations in a relational database, like Biota, is that records for higher taxa must be linked to Specimen records through the appropriate records at all intervening ranks. Otherwise, the Specimen record remains an orphan record (pp. 7–8), invisible to searches based on higher ranks, even though the specimen may have been identified accurately at a higher rank.

The easiest solution to this dilemma is to link any Specimen record that you have identified as representing a higher taxon (a taxon at a rank above species) to the record for that taxon through *temporary taxon* records at all intervening ranks. (The term refers to the temporary nature of the specimen determination. Temporary taxon records themselves need not be temporary.)

Biota's Convention for Temporary Taxon Records

Biota uses a specific convention to indicate a temporary taxon: its name is enclosed in parentheses.

Suppose, for example, that you have identified a particular plant specimen as a composite—a member of the Family Asteraceae—but you do not know its genus and specific name. Using Biota's convention, you would link it to a Species record (with any unique Species Code) for the Species (Asteraceae). The Species record would be linked to a Genus record for the Genus (Asteraceae), which, in turn, would be linked to the Family record for Asteraceae. Subsequently, you would link all Specimen records identified only as members of the Family Asteraceae the same temporary taxon record in the Species table.

With this approach, if you set up the query "Find all Specimens of the Family Asteraceae" in the Lower Taxon for Higher Taxa search tool (pp. 201–205), for example, Biota would find and display not only all Specimens fully determined to Species within Genera of the Family Asteraceae, but also all Specimen records identified only as (Asteraceae).

Automatic Creation of Temporary Taxon Records

If you decide to use Biota's convention for temporary taxa (see the previous section), you need not create the temporary taxon records one by one, although there is nothing wrong with that approach. Biota offers an option for the automatic creation of temporary taxon records.

Enabling or Disabling Automatic Creation of Temporary Taxon Records

NOTE: If the user password system has been enabled, only a user with Administration access privileges (Chapter 23) can enable or disable the automatic creation of temporary taxon records.

1. **From the Special menu, choose Preferences.** The Preferences screen appears.
2. **To enable the automatic creation of temporary taxon records,** click the checkbox labeled "Generate Temporary Taxa automatically" to check it. To disable this option, click the checkbox to uncheck it.

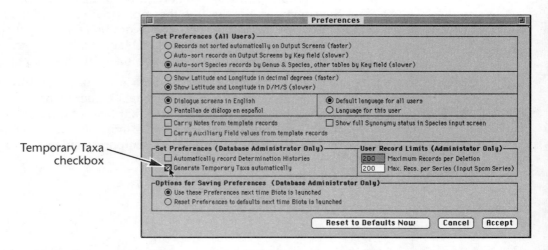

Temporary Taxa checkbox

3. **Choose an option for saving Preferences** (above). You can instruct Biota to use the same Preferences next time it is launched, or use the default settings (automatic creation of temporary taxa disabled) next time.

4. **Click the Accept button** in the Preferences screen.

How Biota Creates Temporary Taxon Records

If you have enabled the automatic creation of temporary taxon records (see the previous section), each time you create a new record for any of Biota's taxon tables above the rank of Species (Kingdom, Phylum, Class, Order, Family, or Genus), a temporary taxon record is created automatically for all tables of lower rank, including Species.

For example, if you create a new record in the Order table for (Asterales), Biota will immediately create each of the following records, automatically:

Table	Key Field	Key Field Value	Linking Field	Linking Field Value
Family	Family	(Asterales)	Order	Asterales
Genus	Genus	(Asterales)	Family	(Asterales)
Species	Species Code	Aster001234	Genus	(Asterales)

Biota assigns a Species Code for the Species record (in the table above) automatically, prefixing the first five characters of the taxon name (*Aster*, in this example) to a six-digit integer based on the Species table sequence number (*0001234* is just an example; see p. 102 for information on sequence numbers). The Species Name field of the automatically created Species record is also assigned the temporary taxon name—*(Asterales)* in this example.

If, instead or in addition, you create a Family record for *Asteraceae*, Biota will automatically create the following records:

Table	Key Field	Key Field Value	Linking Field	Linking Field Value
Genus	Genus	(Asteraceae)	Family	Asteraceae
Species	Species Code	Aster009946	Genus	(Asteraceae)

The Species Name field of the automatically created Species record is also assigned the temporary taxon name—*(Asteraceae)* in this example.

Eliminating Unused Temporary Taxon Records

If you leave the option for the automatic creation of temporary taxon records enabled over a long period, you may discover that some of the temporary taxon records that Biota has created are not needed to link undetermined Specimen records with higher taxa.

You can find all currently unused temporary taxon records quite easily with the Childless Records tool in the Find menu (pp. 215–216). Use it, first, on the Species table. Delete all childless temporary taxon Species records that you do not anticipate using to link records for undetermined specimens. Next, repeat the process with the Genus table, then the Family table, and so on up the taxonomic hierarchy.

Temporary Taxa on Determination Labels

If you follow Biota's parenthesis convention for temporary taxon names (see "Biota's Convention for Temporary Taxon Records," p. 406), temporary taxon names are ignored in preparing determination labels.

Biota looks for an initial left parenthesis in specific names (Species Name), Genus names, and Family names (if you have checked the data option to include Family on the labels, pp. 245–246). The appropriate field on a label is simply left blank if a left parenthesis is encountered. The result is the same whether you create your own temporary taxa using the parenthesis convention or rely on Biota's option for creating them automatically (see "Automatic Creation of Temporary Taxon Records," pp. 406–408).

Chapter 23 Security: Passwords and Access Privileges

Biota offers an optional *user password system* that can be configured to your needs. You can use the Password Editor to set up a system of user passwords for yourself and other users and assign each user individualized access privileges. Each user can change his or her own password, but the Administrator has Super User privileges to change any password or access privileges. (See "Using the Password Editor," pp. 415–424.)

If your database requires extra security, the Administrator can also activate an optional, secure link between your active copy of Biota and one or more Data Files (see "Using the Data File Password Link," pp. 425–431).

For the highest level of security, running Biota4D under 4D Server in a secure room with password-protected access from remote clients is recommended (see "If You Need High Security," p. 432).

Activating and Deactivating the User Password System

When you first launch a new copy of Biota from the distribution disks, the user password system is not activated. Biota opens without displaying the Password screen, shown here. You may choose to leave the user password system deactivated if you are the only user or if you have confidence that no one else with access to your computer will launch Biota and alter a Data File in unintended or undesirable ways.

The cue Biota looks for to activate or deactivate the user password system is the *current password* of the Administrator. If the Administrator's current password is the null string (no characters at all—*not* one or more space characters), the user password system is deactivated. Assigning any other value to the Administrator's password activates the user password system. You can deactivate the system at any time by reassigning the null string as the Administrator's password.

Activating the User Password System

1. **If Biota is not already launched, launch Biota, opening any Data File.** (If Biota is already launched, proceed to step 2 instead.) You have been automatically signed on as "Administrator."

2. **From the Special menu in Biota, choose Change Password.** The "Enter new password" request window appears (illustrated in the next step).

3. **Enter a password** and click the OK button. The password can consist of up to 15 characters, including letters, numbers, and symbols. *This will be the Administrator's password—don't forget it!*

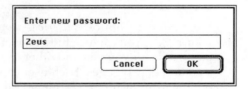

WARNING: *Passwords are case sensitive.* If you create a password that includes uppercase and/or lowercase letters, you must always reproduce the case of each letter precisely.

Biota responds by requesting the new password a second time (illustrated in the next step).

4. **Enter the same password again,** carefully repeating the case of each letter. Click the OK button.

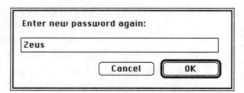

♦ **If the two entries match,** Biota confirms the new password.

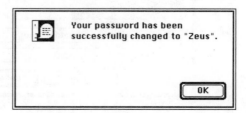

♦ **If the two entries do not match,** Biota posts the message below. Your "old password" in this case is the null string (no characters at all) and the user password system remains inactivated. To activate the user password system, begin again with step 2, above.

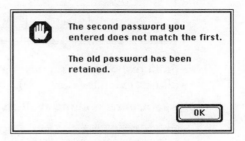

Deactivating the User Password System

1. **If Biota is not already launched, launch Biota with any Data File** (see pp. 8–9). (If Biota is already launched, proceed to step 2 instead.)

    ♦ **When the Password screen appears,** enter the Administrator's user name and the Administrator's current password, then click the Connect button.

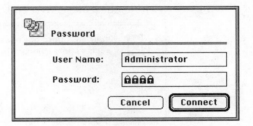

   NOTE: "Administrator" is the user name assigned to the database Administrator when you first activate the user password system. You can change the Administrator's user name (see pp. 418–420).

   WARNING: *Passwords are case sensitive.* If the password includes uppercase and/or lowercase letters, you must reproduce the case of each letter precisely. User names are not case sensitive.

    ♦ **If the Password screen does not appear** during the launch, the user password system has not been activated. Skip the rest of this section.

2. **From the Special menu in Biota, choose Change Password.** The Password screen appears.

3. **Enter the Administrator's user name and the Administrator's *current* password** (as shown in step 1, above), then click the Connect button. If the password is correct, the "Enter new password" request window appears (illustrated in the next step).

4. **Enter nothing at all and click the OK button.**

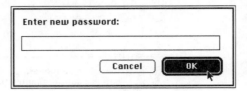

Biota responds by requesting the new "password" a second time (illustrated in the next step).

5. **Again, enter nothing at all and click the OK button.**

- **If the two null entries match,** Biota confirms the "new password" (the null string), indicated by two sets of double quotation marks, with no character between them.

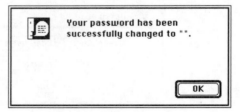

- **If you made a mistake,** Biota posts the message below. The old Administrator's password is still in force and the user password system is still activated. To deactivate the user password system, begin again with step 2, above.

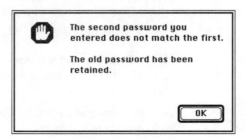

Launching a Password-Protected Copy of Biota

If the user password system has been activated, you must launch Biota by following these steps.

1. **Double-click the Biota icon or application name.** The Password screen appears.

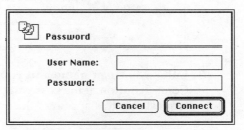

2. **Enter your user name and press TAB.** User names are not case sensitive.

3. **Enter your user password and press TAB.** *Passwords are case sensitive.* If the password includes uppercase and/or lowercase letters, you must reproduce the case of each letter precisely.

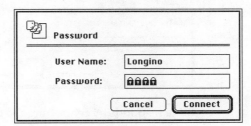

- **If you want to open the same Biota Data File** used the last time this copy of Biota was launched, click the Connect button in the Password window.

- **If you want to open a different Biota Data File** (or verify that the correct file will be opened), hold down the OPTION key (Macintosh) or the ALT key (Windows) while you click the Connect button in the Password window. Use the Find File window to navigate to the correct Data File.

Changing Your User Password

If the user password system has been activated, any user can change his or her own password at any time by following these steps.

1. **If Biota is not already launched, launch Biota with any Data File.** (If Biota is already launched, proceed to step 2 instead.)

2. **From the Special menu in Biota, choose Change Password.** The Password screen appears.

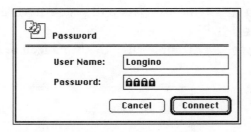

3. **Enter your user name and your *current* password,** then click the Connect button. If the password is correct, the "Enter new password" request window appears (illustrated in the next step).

WARNING: *Passwords are case sensitive.* If the password includes uppercase and/or lowercase letters, you must reproduce the case of each letter precisely. User names are not case sensitive.

4. **Enter a new password** and click the OK button. The password can consist of up to 15 characters, including letters, numbers, and symbols.

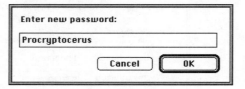

Biota responds by requesting the new password a second time (illustrated in the next step).

5. **Enter the same password again,** carefully repeating the case of each letter. Click the OK button.

- **If the two entries match,** Biota confirms the new password.

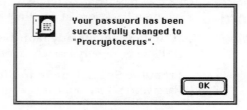

- **If the two entries do not match,** Biota posts the message below. Your old password is still in force. To change it, start again with step 2.

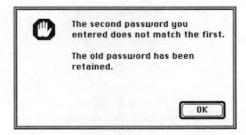

Using the Password Editor

The Password Editor (an off-the-shelf 4th Dimension utility) is easy to use, once mastered, but the interface is not particularly intuitive. You will probably save time by reading or at least skimming this section before using the Password Editor.

NOTE: Once the user password system has been activated, only the database Administrator can access the Password Editor.

Users, User Names, Passwords, and Access Privilege Levels

Biota's implementation of the 4th Dimension user password system is built on several key concepts.

A *user record* is an individual record in the user password system. (The user password system is not an ordinary database table but, rather, an internal one that is not accessible to Biota's import and export tools.)

Each user record has an assigned *user name*, *password*, and *access privilege level* (or *access level*, for short).

- **The user name** can be any name the Administrator assigns to the user record, up to 30 characters, including letters, numbers, and symbols.

- **The password** can consist of up to 15 characters, including letters, numbers, and symbols.

WARNING: *Passwords are case sensitive.* If you create a password that includes uppercase and/or lowercase letters, you must always reproduce the case of each letter precisely.

- **The access level.** Biota offers five access levels, each with its own set of privileges. The privileges and access levels are outlined below and detailed in the table that follows. To assign an access privilege level to a user, you assign the user to an *access group* that has those privileges (see the next section).

    ◊ **Super User.** The database Administrator, *only*, is assigned to the Super User access level, which allows the Administrator to use the Password Editor and enable or disable the Data File password link (pp. 415–431). The Administrator also has all the privileges of the Administration level (below).

◊ **Administration.** Any user assigned to the Administration level can find, display, print, export, create, modify, delete, or import records; change master settings; and use all the tools of the Special menu, with the exception of Edit Password System and Edit Data File Password Link. Only the Administrator can assign passwords and access levels using the Password Editor and enable or disable the Data File password link (pp. 415–431). Other users assigned to the Administration access level do not have these privileges.

◊ **Read Write Export.** A user assigned to the Read Write Export access level can find, display, print, export, create, modify, or delete records, change certain settings in the Preferences screen, and change his or her own password. The Administrator, or another user with Administration privileges, can set limits on the number of records that can deleted (pp. 129–131, Appendix C) or created (pp. 228–234, Appendix C) by users with Read Write Export privileges.

◊ **Read Export.** A user assigned to the Read Export access level can find, display, print, or export records, change certain settings in the Preferences screen, and change his or her own password.

◊ **Read Only.** A user assigned to the Read Only access level can find and display records, change certain settings in the Preferences screen, and change his or her own password.

Privilege	Super User	Administration	Read/Write/Export	Read/Export	Read Only
Find records	✓	✓	✓	✓	✓
Display records	✓	✓	✓	✓	✓
Change password	✓	✓	✓	✓	✓
Print records, reports, and labels	✓	✓	✓	✓	
Use all export tools	✓	✓	✓	✓	
Create records	✓	✓	✓		
Modify records	✓	✓	✓		
Delete records	✓	✓	✓		
Set Preferences for: Determination Histories, Temporary Taxa, Record Deletion and Record Series Creation limits	✓	✓			
All other settings in the Preferences screen	✓	✓	✓	✓	✓
Use Import Editor	✓	✓			
Use Find & Replace tool	✓	✓			
Set Field Value Defaults	✓	✓			

Privilege	Super User	Administration	Read/Write/ Export	Read/Export	Read Only
Enable/Disable Entry Choice Lists	✓	✓			
Set Core Field Aliases	✓	✓			
Set default Record Code Prefixes	✓	✓			
Create Guest Collection Records automatically	✓	✓			
Clear All Synonymies	✓	✓			
Find, modify, or delete Determination Histories using tools of the Special menu	✓	✓			
Use the Password Editor; Set the Data File Link	✓				

Opening and Closing the Password Editor

1. **Launch Biota and sign on as Administrator.** If you are already signed on as a different user, you need not quit Biota. Choose Change Password from the Special menu and sign on as Administrator without shutting down Biota. *Be sure to reenter the Administrator's password correctly if you use this trick.*

   NOTE: "Administrator" is the user name assigned to the database Administrator when you first activate the user password system. You can change the Administrator's user name (see pp. 418–420).

   WARNING: *Passwords are case sensitive.* If a password includes uppercase and/or lowercase letters, you must reproduce the case of each letter precisely. User names are not case sensitive.

2. **From the Special menu, choose Edit Password System.** The Password Editor appears.

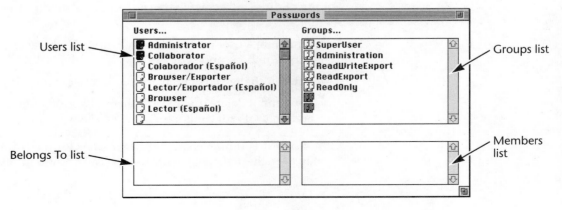

The Password Editor consists of four scollable lists: the Users list, the Belongs To list, the Groups list, and the Members list. The Password Editor has its own menu bar with a Passwords menu.

When you first activate the user password system (see "Activating the User Password System," pp. 410–411), Biota offers a set of generic user names (e.g. Collaborator, Browser, etc.), in addition to Administrator, as shown above. These default user records have different access privileges and startup Language settings that you can use as models. You can use the Password Editor to change these entries to suit your own needs, or leave them as they stand.

3. **To close the Password Editor and register any changes,** click the close box in the upper-left corner (Macintosh) or the close button in the upper-right corner (Windows) of the Password Editor window. (There is no Close command in the menus.)

Editing a User's Password and Profile

1. **Open the Password Editor** (see the previous section).
2. **Double-click a user name in the Users list** or select a user name and choose Edit User from the Passwords menu. The "Edit user" screen appears.

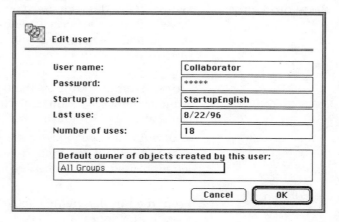

In the "Edit user" screen, you can assign or modify the following settings:

- **User name.** A user name can include up to 30 characters, including letters, numbers, and symbols.
    ◊ **You can change** any of the default user names to suit your needs.
    ◊ **You cannot delete** user records themselves, although you can add additional user records (see "Adding a New User Record," p. 420).

    WARNING: You may change the *user name* for the Administrator. However, the *first record* in the user password system, and *only* that record, is permanently assigned the privilege of access to the Password Editor and Data File password link. Furthermore, the first record *must* be assigned to the Super User group to activate these privileges (p. 415).

- **Password.** A password can include up to 15 characters, including letters, numbers, and symbols.
  - ◊ **When you first activate the user password system in Biota,** each of the default users has a nonsense password assigned, for safety's sake. You will need to replace the password for any of the default user records you want to use.
  - ◊ **Once you assign a password,** it will appear as a string of asterisks in the "Edit user" screen, as shown above. If you are not sure about an existing password (the user may have changed it), delete the characters and reenter the correct password.
- **Startup procedure.** Biota has two alternative startup procedures, StartupEnglish (no spaces) and StartupSpanish (no spaces).
  - ◊ **If you assign StartupEnglish** (no spaces) to a particular user, Biota will display the English language version of all dialogue screens (instructions, options, warnings, errors, and result reports) for that user, regardless of the saved Language setting in the Preferences screen (Special menu).
  - ◊ **If you assign StartupSpanish** (no spaces) to a particular user, Biota will display a Spanish language version of all dialogue screens (instructions, options, warnings, errors, and result reports) that are accessible to users with Read Write Export, Read Export, or Read Only privileges, regardless of the saved Language setting in the Preferences screen (Special menu). Dialogue screens accessible only to users with Administration access privileges are displayed in English for all users.
  - ◊ **If you leave the "Startup procedure" blank,** Biota will display dialogue screens in English or Spanish, depending on the saved Language setting in the Preferences screen. If no user is likely to change the Language setting in the Preferences screen, you can leave the Startup procedure setting blank for all users.
- **"Last use" and "Number of uses."** These areas display usage statistics for the user. You can delete the displayed value to reset the counters. (Oddly, the counters are also editable, if you can think of some legitimate reason to edit them.)
- **"Default owner of objects created by this user."** This panel has no function in a compiled application such as Biota. Ignore it.

3. **When the user profile is complete, click the OK button** in the "Edit user" screen to record it.

4. **To close the Password Editor and register any changes,** click the close box in the upper-left corner (Macintosh) or the close button in the upper-right corner (Windows) of the Password Editor window. (There is no Close command in the menus.)

Adding a New User Record

You can add as many new user records as necessary to the user password system. However, *once a user record is added, it can be edited but cannot be deleted.* (This puzzling limitation is built into 4th Dimension.) Although user records require very little additional memory, you may wish to "recycle" the existing user records by editing them, before adding any new ones.

1. **Open the Password Editor** (see "Opening and Closing the Password Editor," pp. 417–418).
2. **From the Password menu, choose New User.** The "Edit user" screen appears with a temporary name for the new user.

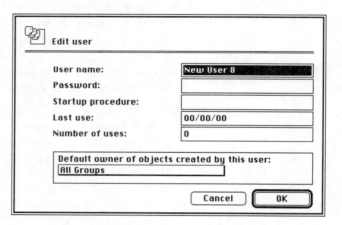

3. **Follow the guidelines in the previous section** ("Editing a User's Password and Profile," pp. 418–420) to complete the new user record and close the Password Editor.

Assigning Users to and Removing Users from Access Groups

Access privilege levels (see "Users, User Names, Passwords, and Access Privilege Levels," pp. 415–417) are determined for individual users in the Users list by assigning users to *access groups*. Biota has five preset access groups, each named for the access level assigned to the group. These access groups appear in the Groups list when you open the Password Editor.

1. **Open the Password Editor** (see "Opening and Closing the Password Editor," pp. 417–418).
2. **Select a user name in the Users list** by clicking it once. The Belongs To list shows the access group (or groups) to which the user currently belongs. The Belongs To list will be blank if the user has not been assigned to any access group.

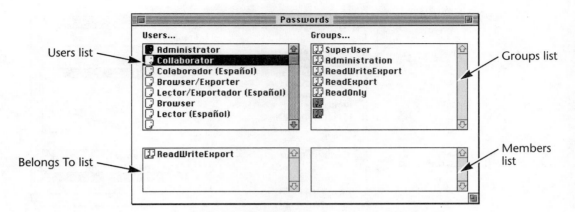

Users list → (Users list)
Belongs To list → (Belongs To list)
Groups list ← (Groups list)
Members list ← (Members list)

In the example above, the user named Collaborator belongs to the Read Write Export access group.

3. **To remove a user from an access group,** follow these steps.

   a. **In the Users list, select (single-click) the user** to verify which group or groups the user belongs to (see step 2, above).

   b. **In the Groups list, select (single-click) the access group** from which you wish to *remove* the user. The user's name should appear in the Members list in the lower right panel.

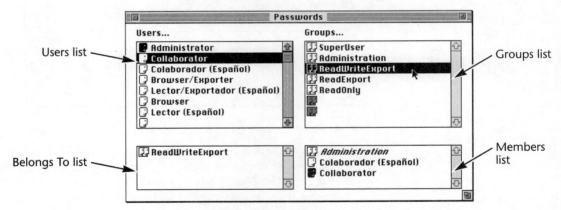

Users list → (Users list)
Belongs To list → (Belongs To list)
Groups list ← (Groups list)
Members list ← (Members list)

Notice that Collaborator is included in the Members list for the Read Write Export group along with Colaborador and *Administration*—the Administration group. (Groups that are members of other groups are shown in italics with a "double-head" icon; see p. 424.)

   c. **Using the mouse, select the name of the user in the Members list and drag it anywhere outside the Members list panel.** In the illustration below, the user called Collaborator is being removed from the Read Write Export group.

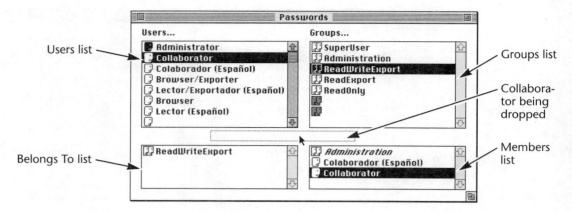

When you release the mouse button, the user's name disappears from the Members list, and the access group name disappears from the Belongs To list.

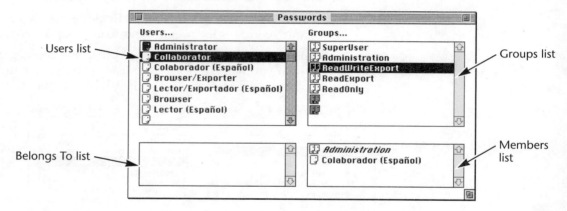

4. **To add a user to an access group,** follow these steps.

a. **In the Users list, select (single-click) the user** to verify to which access group or groups the user currently belongs (see step 2, above). If the user belongs to the wrong group, remove the user from the group using the technique in step 3, above, before proceeding.

NOTE: In Biota, since access privileges are hierarchical, there is no point in a user belonging to more than one access group (although the Password Editor permits it).

b. **Drag the user's name from the Users list over the name of the group (in the Groups list)** to which you want to assign the user. In the illustration below, the user called Collaborator is being dragged towards the Groups list.

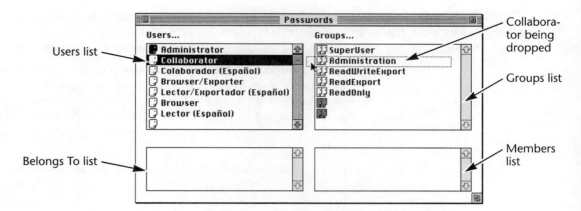

Users list
Belongs To list
Collaborator being dropped
Groups list
Members list

c. **When the correct access group name is highlighted, release the mouse button** ("drag and drop"). In the illustration below, the user called Collaborator is being "dropped" onto the Administration group name.

As shown below, the user's name now appears in the Members list, along with all other members of the group. The name of the access group now appears in the Belongs To list.

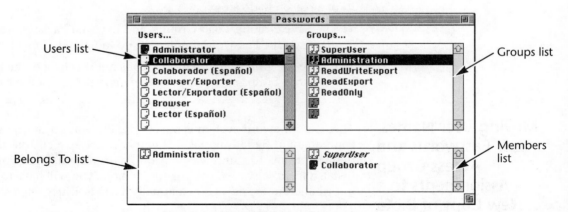

Users list
Belongs To list
Groups list
Members list

d. **To close the Password Editor and register any changes,** click the close box in the upper-left corner (Macintosh) or the close button in the upper-right corner (Windows) of the Password Editor window. (There is no Close command in the menus.)

Pitfalls to Beware and Features to Ignore in the Password Editor

The Password Editor and its Password menu include a number of features that have no application in Biota. This section of the chapter is intended to warn against certain changes and satisfy the curious.

- **If you double-click the name of a group or select Edit Group from the Password menu,** an "Edit group" window appears, in which you can change the name of the group. *Do not change the name of Biota's preset access groups, or the access restrictions will not work.* Biota looks for these particular group names in order to limit access appropriately.

- **Do not change the hierarchical (italicized) group memberships** for higher access groups (although the Password Editor permits it). Specifically, the Super User group must remain a member of the Administration group, the Administration group must remain a member of the Read Write Export group, and the Read Write Export group must remain a member of the Read Export group for the hierarchical privilege system to work.

  If you accidentally change these settings, drag the appropriate group name in the Groups list and drop it on the name of the group (also in the Groups list) of which it should be a member, to recreate the hierarchy—or install a fresh copy of Biota from the distribution disks.

- **If you select the New Group command from the Password menu,** a new group icon will appear in the Groups list. *The Password Editor does not allow you to remove groups,* once created. (In fact, there are some functionless vestigial group icons in the Groups list for this reason.) You can use this command to create your own groups, if you can think of a reason to do so.

- **The Save Groups, Load Groups, and External Package Access** commands in the Password menu have no function in Biota.

- **Ignore the *gray* user and group icons** at the *bottom* of the lists in the Password Editor. You cannot change or make use of them.

Moving User Names, User Passwords and Access Group Assignments to a New Copy of Biota

The user password system is part of the Biota application itself, not a component of Biota Data Files. If you have set up a custom list of user names, passwords, and access group assignments, and you need to install a fresh copy or a new version of Biota, you will have to move the information to the new copy manually. Here is a suggested strategy.

1. **Launch the old copy or old version of Biota** (sign on as the Administrator).

2. **Open the Password Editor** (pp. 417–418).

3. **Copy down the list of user names** (or, better, use a screen capture utility, if you have one, to copy it). You can expand the Password Editor window as large as your monitor screen permits.

4. **Select each user, one at a time, and record his or her group membership** from the Belongs To list.

5. **Double-click each user, one at a time, and record his or her Startup procedure,** if any.

6. **Quit the old copy or version of Biota and launch the new one.**

7. **Recreate the password and access scheme** in the new version, based on your notes or screen captures.

Using the Data File Password Link

The user password system protects your copy of the Biota application from unauthorized use (pp. 409–424). By assigning access privileges to each user, authorized users can be restricted to activities appropriate to their training and responsibilities (pp. 415–417). The user password and access privilege system, however, does not in itself protect Biota *Data Files* from unintended or unauthorized access, which could occur by means of an unprotected copy of Biota.

You can create a secure link between a particular copy of Biota and a particular Data File (or Files) using a Data File password link, as detailed below. You need not activate the user password system in order to activate the Data File password link, although it usually makes sense to activate the user password system if the Data File link is used. (The two systems are technically independent.)

Here's how the Data File password functions.

- **New Biota Data Files** do not have the password link activated.

- **Any Biota Data File that has not had the password link activated** can be opened by any copy of Biota. (If the user password system has been activated, however, the user must have a valid user name and password to launch Biota. Such a user is referred to, below, as an *authorized user*.)

- **Once you activate the password link** for a particular Data File (pp. 426–428), the copy of Biota that you used to activate the link opens that Data File without requiring authorized users to enter the Data File password.

- **If you use a fresh copy or a new version of Biota** to open a protected Data File, Biota will request the Data File password at startup (pp. 430–431).

  ◊ **If the correct Data File password is entered,** that copy of Biota will open the file thereafter without requiring authorized users to enter the Data File password.

  ◊ **If the correct password is not entered** (in up to three attempts), Biota quits without opening the Data File.

- **If you assign the same Data File password to more than one Data File,** by using the same copy of Biota, all the files will open for any authorized user, without requesting the Data File password.

Activating the Data File Password Link

When you create an empty Biota Data File (pp. 9–10), the Data File password in the Data File is blank (a null string); the file can be opened with any copy of Biota. (If the user password system has been activated, you must have a valid user name and password to launch Biota, however.) To activate the Data File link, take these steps.

1. Launch Biota.

- **If the user password system has been activated,** sign on as Administrator.

- **If the Password screen does not appear** when you launch Biota, the user password system has not been activated. You are automatically signed on as Administrator.

- **If you are already signed on as a different user,** you need not quit Biota. Choose Change Password from the Special menu and sign on as Administrator without shutting down Biota. *Be sure to reenter the Administrator's password correctly if you use this trick.*

NOTES:

a. "Administrator" is the user name assigned to the database Administrator when you first activate the user password system. You can change the Administrator's user name (see pp. 418–420).

b. You need not activate the user password system in order to activate the Data File password link. The two systems are functionally independent.

2. From the Special menu, choose Edit Data File Password Link. If the Data File password link has not been activated, the message at the right appears:

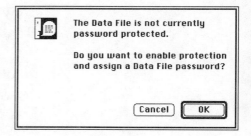

NOTE: If the screen at the right appears instead, go to step 4 in the next section "Changing the Data File Password," pp. 428–429).

3. Click the OK button in the message window. A password request window appears (illustrated in the next step).

4. **Enter the new Data File password** (below) and click the OK button. The password can consist of up to 15 characters, including letters, numbers, and symbols.

   You will need this password to open the Data File if you install a fresh copy or a new version of Biota, or if you need to change the Data File Password—don't forget it!

   WARNING: *Data File passwords are case sensitive.* If you create a password that includes uppercase and/or lowercase letters, you must always reproduce the case of each letter precisely.

   Biota responds by requesting the new password a second time (illustrated in the next step).

5. **Enter the same password again,** carefully repeating the case of each letter. Click the OK button.

   - **If the two entries match,** Biota confirms the new password and warns you not to forget it.

   - **If the two entries do not match,** Biota posts the message below. The Data File password link remains inactivated. To activate it, begin again with step 2, above.

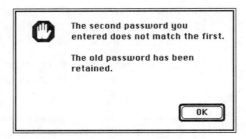

6. **Click the OK button** in the message window.

Changing the Data File Password

To change the password for the Data File link, take these steps.

1. **Launch Biota,** signing on as Administrator if the user password system has been activated. See step 1 in the previous section "Activating the Data File Password Link," p. 426, for details.

2. **From the Special menu, choose Edit Data File Password Link.** If the Data File password link has been activated, the following message appears:

Note: If the screen below appears instead, the Data File password link has not been activated. Go to step 4 in the previous section "Activating the Data File Password Link," p. 427).

3. **Click the OK button** in the message window. A password window appears, requesting the *current* Data File password.

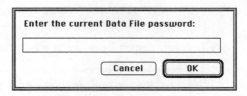

4. **Enter the current Data File password** and click the OK button.

   WARNING: *Data File passwords are case sensitive.* If the password includes uppercase and/or lowercase letters, you must reproduce the case of each letter precisely.

   - **If you enter the correct password,** Biota requests the new Data File password.

   See step 4 in the section "Activating the Data File Password Link," p. 427, for instructions on completing the process of entering the new password.

   - **If you do not enter the correct password,** Biota posts this message.

   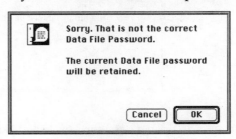

Deactivating the Data File Password Link

To deactivate the Data File link, take these steps.

1. **Follow steps 1 through 4 in the previous section** "Changing the Data File Password," pp. 428–429.
2. **When Biota requests the new Data File password, enter nothing at all and click the OK button.**

Biota responds by requesting the new "password" a second time (illustrated in the next step).

3. **Again, enter nothing at all and click the OK button.**

- **If the two null entries match,** Biota confirms the deactivation of the Data File password link.

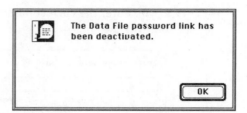

- **If you made a mistake,** Biota posts the message below. The old Data File password is still in force and the Data File password link is still activated. To deactivate the link, begin again with step 1, above.

Opening a Password-Protected Data File with a New Copy or New Version of Biota

To open a password-protected Data File with a new copy or a new version of Biota, take these steps.

1. **Launch the new copy or new version of Biota,** finding and opening the existing Data File as detailed in Chapter 2, pp. 8–9. If the Data File is password-protected, the following request will appear near the end of the startup process:

2. **Enter the Data File password** and click the OK button.

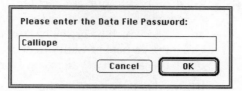

WARNING: *Data File passwords are case sensitive.* If the password includes uppercase and/or lowercase letters, you must reproduce the case of each letter precisely.

You have three chances to enter the correct Data File password.

- **If you enter the correct password,** Biota completes the startup process. *This copy of Biota is now authorized to open this particular Data File in the future, without requesting the Data File password.*

- **If you do not enter the correct password** in three attempts, Biota posts the message below.

Using the Data File Link with Backup Files

One motivation for implementing a Data File link is to guard against well-intended but mistaken use of the wrong Data File. Backup copies of Data Files are particularly easy to confuse with the active copy of the same file.

Of course you should always make regular backups of any actively used Biota Data File, using a clear naming convention for the backup files (Appendix F). If you decide to use the Data File password link, consider changing the Data File password (pp. 428–429) in each backup, using the same naming convention for the password that you use for the file name, so you can open the backups if necessary. With this strategy, the Data File password for the *active* Data File remains constant and each backup has a distinct Data File password that you can remember easily, should it ever be needed.

If You Need High Security

The user password system and Data File password link features of Biota are designed to prevent accidental damage to your database through carelessness, ignorance, or foolishness. A clever hacker can break into any file to which he or she has direct access, including Biota Data Files. Of course, a hacker with direct access to the machine that holds your database can find a way to copy any file in its entirety anyway.

If your database requires the highest level of security (if it includes collection or incidence localities for endangered species of commercial value, for example), you should not rely on software tools, alone, for security. Instead, run Biota4D under 4D Server (Appendix E) in a physically secure room, and allow access to your database only through 4D Client on remote machines, with the user password system enabled. All authentication of passwords is done by Biota4D on the Server—and, of course, the Data File itself resides only on the Server. There is no data caching on Client machines between Biota sessions. This is not an inexpensive solution, but good security is rarely cheap.

PART 6

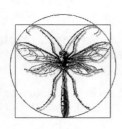

Exporting and Importing Data

Chapter 24 Exporting Data

Any database that cannot easily export any required set of fields and records in digital form is likely to become a La Brea tarpit of extinct information. Biota has been designed for maximum longevity for your data. There is nothing you can put into a Biota Data File that you cannot later export to files that can be opened, read, and edited by text processors, spreadsheet applications, and database management applications.

Biota's Export Editor (pp. 434–443) makes it easy to control the scope and format of any text data you need to export from individual Biota tables to plain text flatfiles. (A flatfile is a row-by-column table of records and fields.) Images can also be exported, but require different tools (pp. 449–451).

The Quick Report Editor (a 4th Dimension utility) offers an alternative way to create your own, custom export format, by using fields from several tables at once (pp. 457–462). With the Quick Report Editor, you can save the format, as well as exported text files, to disk.

In addition to these general-purpose data export utilities, Biota offers a variety of tools designed for exporting specific kinds of biodiversity data.

- Tools for exporting Notes, Auxiliary Fields, and Images (pp. 443–451).
- A tool for exporting Specimen-based flatfiles, including virtually any field in any Biota table (pp. 454–457).
- A tool for exporting Taxonomic Flatfiles that show the taxonomic classification for any selection of taxa (pp. 451–453).
- A tool for exporting formatted text for "Specimens Examined" sections of taxonomic monographs (pp. 463–469).

- A tool for exporting character matrices (created in Biota Auxiliary Fields) in NEXUS format for input to MacClade[1] or PAUP[2]. (pp. 310–311).

- A tool for exporting Collections-by-Species incidence or abundance matrices for input to ordination and other statistical procedures (pp. 469–472).

- A tool for exporting hyperlinked Web pages for any selection of records and taxonomic scope, including Images and host-guest links (Chapter 25).

NOTE: Exporting data from a Biota Data File has no effect on the records in the file. A *copy* of the data is exported to a text file, leaving the records themselves unaltered.

Using the Export Editor

Biota's Export Editor (Export by Tables and Fields, from the Im/Export menu), can export data to disk files from any Biota table, one table at a time (Appendices A and B). If you need to combine fields from several Biota tables in an export file, use a different tool (see the later sections of this chapter, pp. 443–472).

The Export Editor creates free-format, delimited text (ASCII) files, such as the column-by-row text files you can read and create with a spreadsheet application (e.g., Microsoft Excel), or import as text flatfiles to database management applications (e.g., 4th Dimension, FoxPro, dBase, Paradox, Access, or FileMaker) or Geographic Information Systems (GIS).

Text files created by the Export Editor can be read in unaltered form by Biota's Import Editor (Chapter 26), making it easy to move records from one Biota Data File to another.

The Export Editor is designed to make exporting data flexible and easy. You choose a table to export from, then choose which of its fields you want to export. You have complete control over the number and order of fields in text files you create by exporting data, which need not match the number or order of fields in the Biota source table. You can specify your choice of field and record delimiters for the text file. If you wish, Biota can create column headings for the text file, based on Biota field names.

NOTE: Although you may assign any character you wish as a record delimiter (p. 441), this section assumes that you are using a line end character or characters (ASCII Character 13 for Macintosh, ASCI 13 + 10 for Windows), so that Biota records become table rows, and Biota fields become table columns in the exported text file.

---

[1] Maddison, W. P., and D. R. Maddison. 1992. *MacClade: Analysis of phylogeny and character evolution.* Sinauer Associates, Sunderland, MA.

[2] Swofford, D. L. 1993. *Phylogenetic analysis using parsimony (PAUP), version 3.1.1.* Smithsonian Institution, Laboratory of Molecular Systematics, Washington, D. C.

Key Fields

Most of the tables in Biota's structure (Appendix A) have a Key field (Record Code, taxon name, etc.; pp. 6–7) that contains a unique value for each record. The exceptions are the Notes tables and Det History table, which do not require a unique Key (the Record Code fields in these child tables are "foreign keys" that need not be unique either), and the Auxiliary Field Value tables, which have a two-field (composite) Key.

Generally speaking, you will need to include the Key field, if there is one, among the fields to be exported, so you can tell records apart in the text file that may be otherwise identical. The Key field *must* be included (for any table that has one) if you intend later to use the Import Editor to import text files into another Biota Data File (see Chapter 26).

Field Types and Field Lengths

Appendices A and B show the field type and field length (maximum number of characters) for each field in the Biota data structure. When you export data from a Biota field, the exported data will reflect the type (and length) of the Biota field it represents, as detailed below.

- **Alphanumeric fields.** A value exported from an alphanumeric field may include any ASCII characters. Biota exports only what it finds in an alphanumeric field without adding trailing space characters. (In other words, Biota exports free-format fields, not fixed-length fields.) For a particular alphanumeric field, the number of characters exported will vary from none (if there was no entry in that field for a particular record) to the maximum indicated for the field in Appendix A. If a field to be exported is blank in a record, Biota nonetheless exports the field delimiter character, so that columns in the exported file will be correctly aligned.

- **Date fields.** Before exporting dates, please read or review the sections on Dates (pp. 111–114) in Chapter 8. The format for all exported date fields, using the Export Editor, is MM/DD/YYYY. For example, April 9, 1993 will be exported as 04/09/1993, not 09/04/1993.

If you want exported dates in some other format, you must either convert them after export or else use a different export tool in Biota, such as Export Specimen Flatfile (pp. 454–457). Spreadsheet applications (e.g., Microsoft Excel) can easily convert columns of dates from the Biota date output format to any other format, by applying format commands for date columns in the spreadsheet.

- **Partial Dates: Date Flag fields.** As explained in full in Chapter 8 (pp. 112–113), for some date fields, Biota can display and print partial dates—Month-Year dates and Year-only dates. These fields are [Collection] Date Collected, [Collection] Date Coll End, [Specimen] Date Prepared, and [Specimen] Date Determined. Each of these *Date fields* is paired with an integer *Date Flag field* (Appendix A).

If you are exporting complete dates only, you need not export any values for Date Flag fields, which have the value 0 (zero) for complete dates.

If you will be exporting partial dates, you should include the corresponding Date Flag field along with the Date field in the list of fields to be exported. In the text file produced by the Export Editor, you can then decode partial dates, as follows:

◊ **Month-year dates.** For a month-year date (e.g., May 1975), the Date field itself will be exported as a complete date. The corresponding Date Flag field will contain a 1—the code for a month-year date. The value of the Day element in the Date field of month-year dates is arbitrary. If the records were created in Biota, however, Biota's own convention will have been applied: the first day of the month (e.g., May 1, 1975 for May 1975).

◊ **Year-only dates.** For year-only dates (e.g., 1975), the Date field itself will be exported as a complete date. The corresponding Date Flag field will contain a 2—the code for a year-only date. The values of the Day and Month elements in the Date field of year-only dates are arbitrary. If Biota created the records, however, Biota's own convention will have been applied: the last day of the year (December 31, 1975 for 1975).

- **Latitude and Longitude fields.** These two fields in the Locality table are of type Real. Regardless of the coordinate display setting in the Preferences screen (pp. 116–117), Latitude and Longitude are exported by the Export Editor in decimal degrees, the standard for GIS, using the GIS conventions for hemispheres:

  ◊ **North and East** are positive values.

  ◊ **South and West** are negative values.

  After export, you can convert decimal degrees to traditional Degree-Minute-Second coordinates or to Degrees and Decimal Minutes (used by some GPS devices) by the formulas:

  $$AbsDD = Abs \text{ (Decimal Degrees)}$$

  $$Degrees = Int \text{ (AbsDD)}$$

  $$Decimal\ Minutes = (AbsDD - Degrees) * 60$$

  $$Minutes = Int \text{ (Decimal Minutes)}$$

  $$Decimal\ Seconds = (Decimal\ Minutes - Minutes) * 60$$

  $$Seconds = Round \text{ (Decimal Seconds)}$$

  where *Abs* means absolute value, *Int* means the integer part of a real number, and *Round* means rounded to the nearest integer. The variables AbsDD and Decimal Seconds are computational intermediates that simplify conversion. These conversions can be made quite easily in a spreadsheet by inserting columns with the appropriate formulas.

- **Text fields.** Biota uses text fields for Note Text fields of Notes tables, for the [Personnel] Notes field, and the [Loans] Description field. A text field may contain up to 32,000 characters.

- **Boolean fields.** Boolean fields have one of two values: True or False. In an exported text file, a Boolean field will have either a zero (for False) or a 1 (for True). There is only one Boolean field in Biota, [Personnel] Group, which, when True, identifies a Personnel record as a Group Name record, rather than an individual person's record. The Group field has the value False for an individual (see pp. 174–177).

- **Picture fields.** You cannot export Picture (Image) fields using the Export Editor. Images can be imported to disk files using the Export Images tool (pp. 449–451).

Exporting by Tables and Fields: Step by Step

Follow these steps to export records from a Biota Data File using the Export Editor. This tool exports the fields and records you choose, for one Biota table at a time.

1. **Choose Export by Tables and Fields** from the Im/Export menu. The Export Editor appears.

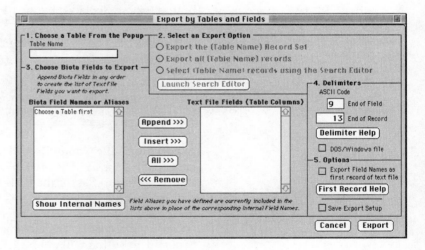

2. **Select a table.** From the Table Name popup list at the upper left, select the name of the Biota table from which you want to export records (see Appendices A and B).

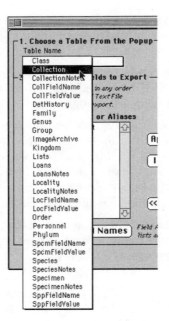

Choosing an export table activates the options available for selecting records (in the panel at the upper right, Select an Export Option) and displays the fields for the table you selected (in the panel at the lower left, Biota Field Names or Aliases).

3. **Select the records for export.** The three option buttons in the upper-right panel define your options for selecting records for export. Once you choose a table from the popup list, the words *Table Name* in the button text are replaced by the name of the table you selected.

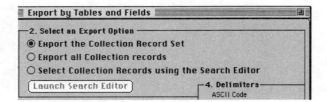

- **Export the (Table Name) Record Set.** If the table you choose is a Core table (p. 5) or the Det History table (Chapter 19), and you have defined a current Record Set for the table (pp. 12–16), the first option button will be enabled. If you click this button when it is enabled, values for the fields you choose (in a later step in this section) will be exported for each record in the current Record Set when you launch the export. The button will not be activated under two circumstances:

    ◊ If you select a Peripheral table other than Det History.

    ◊ If you select a Core table or the Det History table, but the current Record Set for the table is empty. In this case, Biota posts the message illustrated here.

- **Export all (Table Name) records.** This option is available for all tables. If you click this button, values for the fields you choose (in a later step in this section) will be exported for *all* records in the table when you launch the export.

- **Select (Table Name) records using the Search Editor.** This option is available for all tables except for the Lists table, from which all records must be exported (pp. 319–322). If you choose this button, the Launch Search Editor button is activated.

When you click the Launch Search Editor button (above), the Search Editor appears, set to search among all records for the table you selected. (See pp. 192–200 for help using the Search Editor.) If you use this method to select records, Biota will inform you how many records it found, before you continue with the rest of the Export procedure. When you launch the export, the fields you select will be exported for the records you found using the Search Editor, as detailed below.

4. **Using the Field Name/Alias button.** The button beneath the Biota Field Names or Aliases panel controls whether Biota lists strictly Internal Field Names (as given in Appendix A) or substitutes a Field Alias for the corresponding Internal Field Name for each field you have given an Alias. (See pp. 323–326 to learn about Field Aliases.)

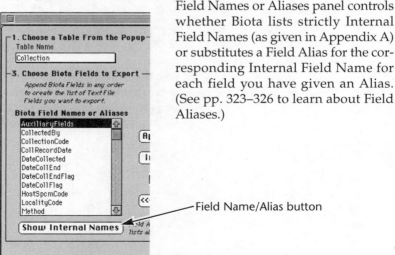

Field Name/Alias button

- **If no Aliases have been defined,** the button is disabled and reads: Show Field Aliases. The caption to the right of the button reads: "Internal Field Names appear in the lists above. No Field Aliases have been defined."

- **If any Aliases have been defined,** when you first open the Export Editor the button is enabled and reads: Show Internal Names. The caption to the right of the button reads: "Field Aliases you have defined are currently included in the lists above in place of the corresponding Internal Field Names."

- **If you click the Show Internal Names button,** the button text changes to Show Field Aliases and the caption reads: "Internal Field Names appear in the lists above. Click the button to the left to include the Field Aliases you have defined."

5. **Choose the fields to export.** The Biota Field Names or Aliases panel is simply a source of field names. The panel labeled Text File Fields (Table Columns), at the right, is the list Biota will consult in exporting data from your Biota Data File to a text file.

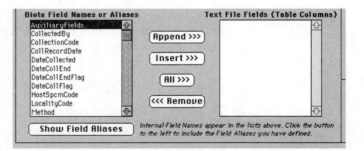

6. **Use the buttons between the two lower panels** in the Export Editor to create the list of fields in the Text File Fields panel. The list of fields you create can include any or all of the field names or aliases from the list available, in any order. You can even repeat fields if you want.

- **Append button.** Click a field in the Biota Field Name panel to select it, then click the Append button. The field name is added to the end of the list in the Text File Fields panel.

- **Insert button.** Click a field already listed in the Text File Fields panel to select it. Then click the field in the Biota Field Name panel that you want to insert. When you click the Insert button, the new field is inserted just *above* the selected field in the Text File Fields panel.

- **All button.** The All Button is a toggle. When the button text reads "All>>>," it means "Append All." If you click the button, all fields in the Biota Field Name panel are entered in the Text File Fields panel (in the order listed in the Field Name Panel), *replacing* any fields already listed in the Text File Fields panel. The button text then changes to read "<<<All," meaning "Remove All." If you click the button again, the Text File Fields panel is cleared and the button text changes back to "All>>>."

- **Remove button.** To remove a single field already entered in the Text File Fields panel, select the field by clicking it in the list, then click the Remove button.

7. **Set the Field and Record Delimiters.** In the Delimiters panel of the Export Editor, you specify the characters Biota will insert between fields (columns) and at the end of records (rows) in the exported text file. You may use any characters in the standard ASCII set (Characters 33–126), but, of course, they must make sense in the context of other applications you intend to use to work with the text file. The Delimiter Help button presents a list of common options.

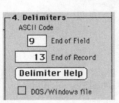

- *Macintosh:* The default delimiters are TAB (ASCII Character 9) for the field delimiter and RETURN (ASCII Character 13) for the record delimiter—the characters that Macintosh spreadsheet applications use for "TAB-separated values" text files. If you are exporting data to a PC (DOS or Windows) text file that uses the standard PC line termination (RETURN + LINE FEED, ASCII Characters 13 + 10), click the DOS/Windows File checkbox. The End of Record box will then read "1310."

- *Windows:* The default delimiters are TAB (ASCII Character 9) for the field delimiter and RETURN + LINE FEED (ASCII Characters 13 + 10), for the record delimiter—the characters that Windows spreadsheet applications use for "TAB-separated values" text files. If you are exporting data to a Macintosh text file, click the DOS/Windows File checkbox to *uncheck* it. The End of Record box will then read "13."

8. **Set the Options.** There are two checkboxes in the Options panel. Both are unchecked, by default.

- **Column Headings Checkbox.** If the want the first row (record) of the exported text file to consist of column headings, you should check this option. Biota will use the Internal Field Names to create column headings. If you have included any Aliases in the list of fields to be exported, however, Biota will instead create headings from Aliases, where they appear.
- **Save Export Setup Checkbox.** If you check this option, Biota will remember the way you have set up the Export Editor, even if you Cancel without attempting to export, for the duration of the current Biota session.

9. **Launch the export.** When the list of Text File Fields and the settings are ready, click the Export button to launch the export process.

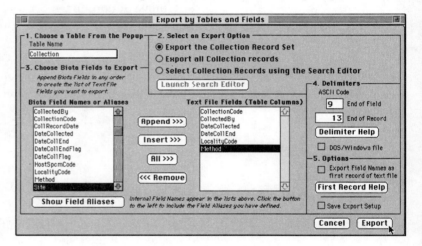

The standard Save File window for your operating system appears.

10. **Name the text file.** Find the folder or directory where you want to create the text file, name the file, and click the Save button (Macintosh) or the OK button (Windows). The progress indicator appears.

When the export is complete, a message informs you.

11. **To halt the export,** click the Cancel button in the progress window. Biota posts a warning that the text file may be incomplete.

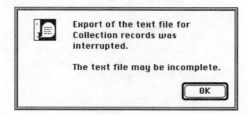

NOTE: You can carry out other tasks with Biota or other applications while the export is in progress.

Exporting Notes

For five Biota Core tables—Specimen, Species, Collection, Locality, and Loans—you can create as many Notes as you want for each Core table record (p. 5). Each Note is actually a separate (child) record in a Notes table, linked to a parent record in the Core table (Appendix A).

Biota provides two tools for exporting Notes records, the Export Editor and the Export Notes tool from the Im/Export menu. The Export Editor produces text files containing fields you select from Notes tables *only*. In contrast, the Export Notes tool exports certain fields from each parent record, along with full records for each selected Note.

Exporting Notes with the Export Editor

The Export Editor is discussed in detail earlier in this chapter (pp. 434–443). Using the Export Editor, you can export any of the following:

- All Notes records linked to records in the current parent-table Record Set (see the next section).
- All Notes records for a particular Notes table (p. 444).
- Notes records you find using the Search Editor (from the Export Editor itself), based on content of the Notes records themselves, or on the content of parent records (pp. 445–446).

Exporting Notes Records Linked to Records in the Current Parent-Table Record Set

To illustrate, suppose you want to export all *Collection* Notes for a particular group of Collection records. Although Collection Notes will be used as an example, *the technique is the same for any Core table with Notes* (Species, Specimen, Locality, and Loans, as well as Collection).

1. **Display the group of Collection records for which you want to export Notes,** using any technique from the Find menu (Chapter 11).
2. **Designate these records as the current Collection Record Set** (pp. 13–15).
3. **From the Im/Export menu, select Export by Tables and Fields** to open the Export Editor.

4. **Select the Collection Notes table** from the Table Name popup list.

5. **Click the first option button,** which will be activated and will read: Export CollectionNotes records for the Collection Rec Set.

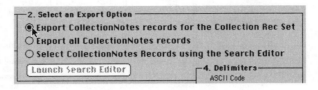

If the Collection Record Set is currently empty, this option is disabled (dimmed), and you will see this message:

6. **Complete the export Editor setup** and launch the export by following the instructions on pp. 439–443, steps 4–11.

Exporting All Notes Records for a Notes Table

To illustrate, suppose you want to export all *Locality* Notes from a Biota Data File. Although Locality Notes will be used as an example, *the technique is the same for any Core table with Notes* (Species, Specimen, Collection, and Loans, as well as Locality).

1. **From the Im/Export menu, select Export by Tables and Fields** to open the Export Editor.

2. **Select the Locality Notes table from the Table Name popup list.**

3. **Click the second option button, labeled Export All LocalityNotes records.** (If the Locality Record Set is empty, this will be the default option and the first option will be disabled—dimmed, as illustrated below.)

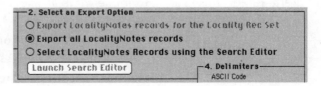

4. **Complete the export Editor setup** and launch the export by following the instructions on pp. 439–443, steps 4–11.

Exporting Notes Records Based on Their Own Content

To illustrate, suppose you want to export all *Specimen* Notes written by "A. Packer." Although Specimen Notes will be used as an example, *the technique is the same for any Core table with Notes* (Species, Collection, Locality, and Loans, as well as Specimen).

1. **From the Im/Export menu, select Export by Tables and Fields to open the Export Editor.**
2. **Select the Specimen Notes table** from the Table Name popup list.
3. **Click the third option button,** labeled Select Specimen Notes Records using the Search Editor.

4. **Click the Launch Search Editor button.** The Search Editor appears.
5. **In the Search Editor, set up the search criterion** "Note By *is equal to* A. Packer" and launch the search. (See pp. 192–200 for detailed instructions on using the Search Editor.)

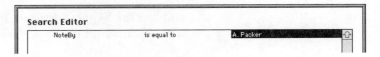

NOTE: The Search Editor looks for precise matches when you use the "is equal to" operator. For this example it might be wiser to use the "contains" operator instead, so that notes by "A. Packer" as well as "Adolph Packer," "A. L. Packer," "A Packer," and "APacker" would all be found.

6. **Follow the directions on pp. 439–443, steps 4–11** to complete the Export Editor setup and launch the export.

Exporting Notes Records Based on the Content of Parent Records

To illustrate, suppose you want to export to a text file all *Collection* Notes for *collections made by* (not for all notes written by) "A. Packer." Although Collection Notes will be used as an example, *the technique is the same for any Core table with Notes* (Species, Specimen, Locality, and Loans, as well as Collection).

1. **From the Im/Export menu, select Export Table and Fields** to open the Export Editor.
2. **Select the Collection Notes table** from the Table Name popup list.
3. **Click the third option button,** labeled "Select Collection Notes Records using the Search Editor" (see the previous section for an illustration).

4. **Click the Launch Search Editor button.** The Search Editor appears.
5. **In the Search Editor, display the fields for the Collection table** (not the Collection Notes table) in the field list panel. Set up the search criterion "Collected By *is equal to* A. Packer" and launch the search. (See the note at step 5 of the previous section (p. 445) for a suggestion regarding variations in personal names. See pp. 192–200 for detailed instructions on using the Search Editor.)

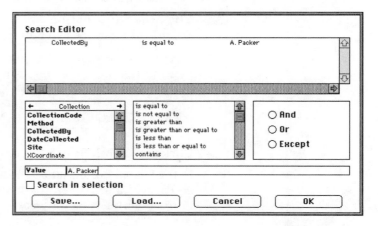

6. **Complete the Export Editor setup and launch the export** by following the instructions on pp. 439–443, steps 4–11.

NOTE: In addition to using parent record content to select Notes records for export, you can also use the content of *any related table farther up the same hierarchy* to select Notes records, using the Search Editor directly from the Export Editor, illustrated above. For example, you could select and export all Specimen Notes for a Family, or all Collection Notes for Localities in Peru.

Exporting Notes Records Using the Export Notes Tool

The Export Editor (above) produces text files containing fields only from Notes tables themselves. In contrast, the Export Notes tool exports not only the full record for each attached Note, but also certain fields from each parent record in a Record Set, as well as important fields from other linked Core tables, where appropriate.

For example, if you print Specimen Notes using the Export Notes tool, Biota exports, for each Specimen record in the current Specimen Record Set: Specimen Code, Specimen Record Date, Genus, Specific Name, Collected By, Date Collected, Locality Name, and all fields (Note Date, Note By, and Note Text) for each Note associated with the Specimen Record (see the example below).

Here is the procedure for using the Export Notes tool.

1. **Find the Species, Specimen, Collection, Locality, or Loans records whose Notes you want to export** (or the Specimen or Species records for which you want to export linked Collection Notes).
2. **Establish the records as the current Record Set for the table.** (See pp. 13–15 for help if necessary.)
3. **From the Im/Export menu, select Export Notes.** The Export Notes option screen appears.

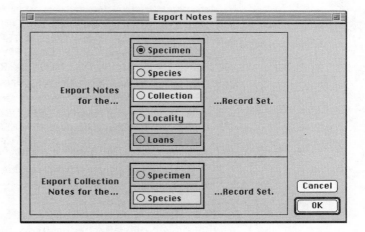

Notice that the option screen is divided into two panels.

- **The upper panel** offers export of Notes records and information from their parent records (and other related records, as appropriate) for each of the five Core tables that support Notes.
- **The lower panel** permits you to export Collection Notes based on the current Specimen or Species Record Set. Each Collection Note is printed only once, even if several Specimen or Species records are linked to the Note's parent Collection Record.

4. **Click the option you want, then click the OK button** in the Export Notes screen The standard Save File window for your operating system appears.
5. **Name the text file.** Find the folder or directory where you want to create the text file, name it, and click the Save button (Macintosh) or the OK button (Windows). The progress indicator appears. You can click the Cancel button in the progress indicator window at any time to halt the export.

Sample output from the Export Notes tool is illustrated below, for the Specimen table. Notice that the first Specimen record has two notes. Biota exports all notes for each parent record in the parent table Record Set. Notice, also, that information from related Core tables (Species and Collection) is included for each Specimen record.

```
*******************************************************
SPECIMEN CODE: HMSB001; RECORD DATE: May 21, 1996; GENUS:
Geospiza; SPECIES: fortis; COLLECTED BY: C. Darwin; DATE
COLLECTED: Oct 9, 1832; LOCALITY NAME: San Cristóbal Island
NOTE DATE: Oct 9, 1832; NOTE BY: C. Darwin
   NOTE TEXT: Time of collection 7:20 a.m.
NOTE DATE: Jun 2, 1939; NOTE BY: D. Lack
   NOTE TEXT: Bill exceptionally wide for this species.
*******************************************************
SPECIMEN CODE: HMSB002; RECORD DATE: May 14, 1996; GENUS:
Geospiza; SPECIES: fortis; COLLECTED BY: R. Fitzroy; DATE
COLLECTED: Sep 18, 1832; LOCALITY NAME: San Cristóbal Island
NOTE DATE: Sep 14, 1832; NOTE BY: R. Fitzroy
   NOTE TEXT: Collected 350 yds. inland from the landing.
*******************************************************
SPECIMEN CODE: HMSB003; RECORD DATE: May 14, 1996; GENUS:
Geospiza; SPECIES: fuliginosa; COLLECTED BY: R. Fitzroy; DATE
COLLECTED: Sep 18, 1832; LOCALITY NAME: San Cristóbal Island
NOTE DATE: Jun 21, 1955; NOTE BY: D. Lack
   NOTE TEXT: Specimen typical of this species for this population.
```

NOTE: If you want to create a Record Set for *all* parent (Core table) records that have Notes attached to them, for a particular Core table, you can use the following trick. From the Find menu, select By Using Search Editor. From the popup list of tables, select the *parent* table for the Notes (e.g., the Specimen table), then click the Search button. In the Search Editor, select the related *Notes* table (e.g., the Specimen Notes table) in the Fields panel (see pp. 197–200 for help switching to the Notes table in the Fields panel). Set up a query of this form: *SpecimenCode is not equal to !@#$%^&**. In this query, *be sure to use the Record Code field from the Notes table, not the parent table.* For the text string to search for, you can use any garbage you like instead of *!@#$%^&**, just as long as it cannot possibly be a real Record Code in your Data File. When the Core table records (Specimen records, in this example) are found, click Done and create a Record Set for them. Then use the Export Notes tool (explained in this section) to export the related Notes.

Exporting Auxiliary Fields

Auxiliary Fields are special fields that you define and name yourself (Chapter 15). Like data in any other table in Biota, the data in the four Auxiliary Field Value tables (Spcm Field Value, Spp Field Value, Coll Field Value and Loc Field Value) can be exported directly to a text file using the Export Editor (see pp. 434–443, earlier in this chapter). In fact, if you are moving Auxiliary Fields data between Biota Data Files, you must use this method to create a text file in a format suitable for importing into another Biota Data File.

The information in Auxiliary Fields appears internally in triplet format (see "How Auxiliary Fields Work, pp. 282–283, and Appendix A), rather than in columns and rows. For this reason, the Export Auxiliary Fields command from the Im/Export menu offers special tools for exporting Auxiliary Field Values in matrix format, just as they appear in the Auxiliary Field display screens (pp. 308–310) or in NEXUS format (pp. 310–311).

These matrix export tools, available through the Export Auxiliary Fields command, are described in detail in Chapter 15, pp. 308–311.

Exporting Images

Biota records Images in a table called Image Archive, which is linked to the Species table (Appendix A). Each Species record (parent record) can have many Image Archive records (child records) linked to it. An Image Archive record has four fields: Species Code (the link to the Species record), Image Number (used internally to order Image records within Species), Image Name, and the Image itself.

Options for Exporting Images

Biota offers three ways to export Images to disk files—in each case, one Image per disk file.

- **The Export Images tool** in the Export menu lets you export many Images, each to its own file, with a single command. This tool is the subject of the remainder of this section.

- **The Image input screen** offers a Save to Disk File button, which lets you export the Image displayed to a file that you name, one Image at a time (pp. 355–357).

- **The Create Web Pages tool** has a checkbox option, Include Images, for exporting one Image per Species to a disk file. The Image files are 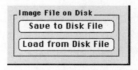 linked automatically to Species records on the Species Web pages that Biota creates from your Data File (Chapter 25, pp. 480–481).

Using the Export Images Tool: Step by Step

WARNING: Be sure you have sufficient disk space available. Images—especially color images—require substantial disk space.

1. **Find the Species records for which you want to export Images, and make them the current Species Record Set** (pp. 13–15). You can use any of the tools of the Find or Tree menus to gather the records you want (Chapter 11).

2. **From the Im/Export menu, select Export Images.** An option window appears.

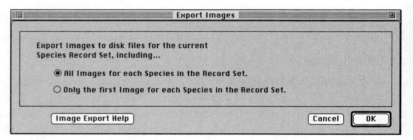

3. **Choose an option for Image export.**
   - **If you choose the first option** (the default) Biota exports all Images linked to each of the Species records in the current Species Record Set. Each Image is exported to its own disk file.
   - **If you choose the second option,** only the first Image, according to the Image order you have determined in the Species input screen, is exported to a disk file, for each Species in the current Species Record Set. (To make a different Image the first one, see "Changing the Order of Image Records for a Species," pp. 359–363.)

4. **Click the OK button in the Export Images option screen to launch the export.** The progress indicator appears during the export.

5. **To halt the export,** click the Cancel button in the progress indicator window. Any files already created when the export was stopped will remain on your disk.

NOTE: You can carry out other tasks with Biota or other applications while the export is in progress.

How Biota Assigns Disk File Names to Exported Image Files

The Image files that Biota creates when you use the Export Images Tool (see the previous section) are placed in the same folder or directory as Biota.

File names for the exported Image files are assigned automatically, according to the following rules, which apply equally to Macintosh and Windows 95:

- **The file name assigned will be of the form "SpeciesCode_ImageName,"** with additional underscore characters substituted for any internal space characters in the Species Code or Image Name field value.

- **Each occurrence of the characters \ / : * ? " < > | is replaced with an underscore** in the assigned file name, to comply with Windows file name limitations. (Macintosh excludes only the colon.) These substitutions are made on both platforms, for compatibility.

- **If the file name would be longer than 31 characters, Biota truncates the name to 31 characters,** cutting off the final part of the Image Name string as necessary. Because Species Codes may be up to 20 characters, and Image Names can be as long as 30 characters, the file name could potentially be as long as 51 characters, counting the underscore character between the Species Code and the Image Name. (Although Windows 95 allows file names up to 255 characters in length, Biota uses the Macintosh limit of 31 on both platforms for compatibility.)

If a file name, created by these rules, duplicates an existing file name, the new file will replace the old one.

Duplicate file names can arise in two ways:

- **You have previously exported an Image with the same Image Name for the same Species record.**

- **You have very long Species Codes and Image Names,** and the truncation rule (above) creates identical file names by truncating the distinct part of two otherwise-identical Image names (beyond the 10th character of the Image Name) for the same Species record.

Exporting Taxonomic Flatfiles

A common way of organizing lists of taxa is a *taxonomic flatfile*, with taxa of a lower taxonomic level (rank) as rows and information on the higher classification of those taxa in columns. In the example below, the rows are plant species, and the columns show the specific name, genus, and family for each of these plant species.

Family	Genus	Specific Name
Campanulaceae	*Centropogon*	*caoutchouc*
Campanulaceae	*Centropogon*	*erianthus*
Campanulaceae	*Lobelia*	*laxiflora*
Campanulaceae	*Lobelia*	*salicifolia*
Campanulaceae	*Siphocampylus*	*ecuadoriensis*
Campanulaceae	*Siphocampylus*	*sanguineus*
Campanulaceae	*Siphocampylus*	*scandens*
Ericaceae	*Anthopterus*	*verticillatus*
Ericaceae	*Cavendishia*	*forreroi*
Ericaceae	*Cavendishia*	*gilgiana*
Ericaceae	*Cavendishia*	*leucantha*
Ericaceae	*Cavendishia*	*lindauiana*
Ericaceae	*Cavendishia*	*tenella*
Ericaceae	*Ceratostema*	*nodosum*
Ericaceae	*Ceratostema*	*peruvianum*
Ericaceae	*Ceratostema*	*reginaldi*
Ericaceae	*Macleania*	*bullata*
Ericaceae	*Macleania*	cf. *ericae*
Ericaceae	*Macleania*	*coccoloboides*
Ericaceae	*Macleania*	*glabra*
Ericaceae	*Macleania*	*loeserneriana*

The Export Taxonomic Flatfile tool creates text files organized in this way, based on whatever range of taxonomic levels (ranks) that you specify, building from the current Record Set for the lowest level in the specified range. (You have to add your own borders with a text processor or spreadsheet application.)

NOTE: The text files created by this tool use the TAB character as a field delimiter.

Using the Export Taxonomic Flatfile Tool: Step by Step

1. **Decide which taxonomic level will form the *base* of the flatfile to be exported.** In the illustration above, the rank of Species is the base level. Records of the base level become the rows of the flatfile.

2. **Find the records, at the base level, for which you want to export a taxonomic flatfile.** You can use any of the tools of the Find, Series, or Tree menus to gather the records you want (Chapters 11 and 12).

3. **Establish these records as the current Record Set for the base level table.** (See pp. 13–15 for help if necessary.)

4. **From the Im/Export menu, select Export Taxonomic Flatfile.** The Export Taxonomic Flatfile setup screen appears.

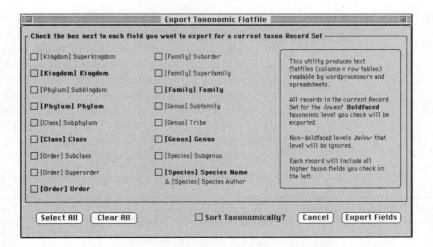

5. **Click the checkbox for the base level,** which will form the rows of the flatfile to be exported. *The base level must be one of the obligatory taxonomic levels* (in **boldface** type). If the current Record Set for the base level is empty, Biota posts an error message.

6. **Click the checkboxes for the columns you want in the flatfile.** *The base level must be the lowest rank checked.*

   NOTE: The Species Author field is exported automatically if Species is checked. You can delete the Author column from the flatfile later if you don't want it.

7. **To order the base level taxa in the exported file according to the higher taxonomic levels checked,** click the "Sort taxonomically?" checkbox at the bottom of the screen. Within taxa, subtaxa will be ordered alphabetically, as in the table of plant species at the beginning of this section.

8. **When the setup is ready, click the Export Fields button in the setup screen.** The standard Save File window for your operating system appears.

9. **Name the text file.** Find the folder or directory where you want to create the text file, name the file, and click the Save button (Macintosh) or the OK button (Windows). The progress indicator appears.

10. **To halt the export, click the Cancel button** in the progress indicator window. If you stop the export before it is done, the exported file may be incomplete.

    NOTE: You can carry out other tasks with Biota or other applications while the export is in progress.

Exporting Specimen Flatfiles

Using the Export Specimen Flatfile tool, you can export any or all of 72 fields from nine Biota tables, for each record in the current Specimen Record Set. Because the Specimen table lies at the base of both the place (Collection/Locality) and taxonomic table hierarchies (pp. 5–8 and Appendix A), a Specimen-based flatfile may logically include any field from any table in either hierarchy. (In technical terms, the Specimen file has many-to-one relations, directly or through intermediate tables, with Collection, Locality, and all Taxon tables.) The order of columns in exported Specimen flatfiles corresponds to the order of fields in the two Export Specimen Flatfile screens (see steps 2 and 4 below).

Here is an example of a file exported using this tool.

Specimen Code	Date(s) Collected	State/ Prov	Country	Latitude	Longitude	Genus	Species Name	Family
JLL5650	3 Apr 1978	Pichincha	Ecuador	0°0'0"N	0°0'0"E	Anthopterus	verticillatus	Ericaceae
JLL14691	15 Oct 1992	Napo	Ecuador	0°37'0"S	77°51'0"W	Ceratostema	nodosum	Ericaceae
JLL5657	5 Apr 1978	Napo-Pastaza	Ecuador	0°0'0"N	0°0'0"E	Ceratostema	peruvianum	Ericaceae
JLL14757	6 Nov 1992	Carchi	Ecuador	0°50'0"N	78°2'0"W	Macleania	loeserneriana	Ericaceae
JLL14789	17 Nov 1992	Pichincha	Ecuador	0°6'0"N	78°30'0"W	Macleania	cf. ericae	Ericaceae
JLL13382	22 Nov 1989	Pichincha	Ecuador	0°35'0"S	78°25'0"W	Macleania	loeserneriana	Ericaceae
JLL13330	7 Nov 1989	Pichincha	Ecuador	0°4'0"S	78°44'0"W	Macleania	bullata	Ericaceae
JLL8474	20 May 1982	Pichincha	Ecuador	0°15'0"S	78°45'0"W	Macleania	coccoloboides	Ericaceae
JLL7269	19 Apr 1979	Chocó	Colombia	4°40'0"N	76°25'0"W	Cavendishia	tenella	Ericaceae
JLL5642	2 Apr 1978	Pichincha	Ecuador	0°0'0"N	0°0'0"E	Cavendishia	gilgiana	Ericaceae
JLL11106	19 Jan 1985	Cañar	Ecuador	2°35'0"S	78°49'0"W	Siphocampylus	scandens	Campanulaceae
JLL10953	10 Jan 1985	Carchi	Ecuador	0°45'0"N	77°40'0"W	Siphocampylus	ecuadoriensis	Campanulaceae
JLL11314	12 Feb 1985	Cajamarca	Peru	7°5'0"S	78°25'0"W	Siphocampylus	sanguineus	Campanulaceae
JLL11280	28 Jan 1985	Loja	Ecuador	3°58'0"S	79°20'0"W	Centropogon	erianthus	Campanulaceae
JLL11176	21 Jan 1985	Azuay	Ecuador	3°20'0"S	78°53'0"W	Centropogon	caoutchouc	Campanulaceae
JLL10573	15 May 1984	Chocó	Colombia	4°40'0"N	76°25'0"W	Cavendishia	forreroi	Ericaceae

Using the Export Specimen Flatfile Tool: Step by Step

NOTE: The text files created by this tool use the TAB character as a field delimiter.

1. **Find the Specimen records for which you want to export the Specimen Flatfile, and make them the current Specimen Record Set** (pp. 13–15). You can use any of the tools of the Find, Series, or Tree menus to gather the records you want (Chapters 11 and 12).

2. **From the Im/Export menu, select Export Specimen Flatfile.** The first field export options screen of the two-screen Export Specimen Flatfile tool appears. This first screen has fields from the Specimen, Collection, and Locality tables. Notice that the [Specimen] Specimen Code field is already checked. This field is the only obligatory one.

3. **In the first screen, click the checkbox for each field that you want to export.**

NOTES:

a. **Collection dates:** In the first screen of options, notice that there is a single checkbox for collection dates, labeled [Collection] Date(s) Collected. If you check this box, Biota combines information from the [Collection] Date Collected and [Collection] Date Coll End fields, formatted as described on pp. 113–114.

b. Latitude and Longitude: If you check [Locality] Latitude and/or [Locality] Longitude in the first screen of options, the format of the values exported will be controlled by the current lat/long display setting in the Preferences screen (Special menu). You can choose either decimal degrees or Degrees-Minutes-Seconds in the Preferences screen. See pp. 116–117.

4. **Click the Second Screen of Fields button to access the remaining options** (the button is being clicked, above). The second screen of field export options appears. This screen shows fields for the Taxon tables—Species, Genus, etc.

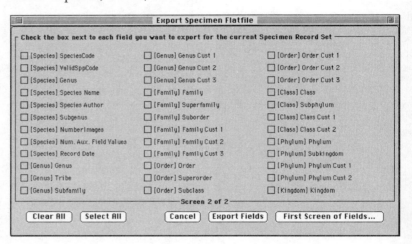

5. **In the second screen, click the checkbox for each field that you want to export.**

6. **If you want to return to the first screen for any reason,** click the First Screen of Fields button on the second screen. (You can toggle back and forth as often as you wish without launching the export or losing the settings.)

7. **To select all fields on *both* screens,** click the Select All button on *either* screen.

8. **To clear the checkboxes for all fields on *both* screens,** click the Clear All button on *either* screen.

9. **To launch the export, click the Export Fields button** on *either* screen. The standard Save File window for your operating system appears.

10. **Name the text file.** Find the folder or directory where you want to create the text file, name the file, and click the Save button (Macintosh) or the OK button (Windows). The progress indicator appears.

11. **To halt the export,** click the Cancel button in the progress indicator window. If you stop the export before it is done, the exported file may be incomplete.

NOTE: You can carry out other tasks with Biota or other applications while the export is in progress.

Exporting Custom Flatfiles

Using the Export Custom Flatfile tool from the Im/Ex menu, you can use the Quick Report Editor to create, use, save, and reload your own custom formats for exporting Biota records and fields to disk files. A Custom Flatfile export must be based on a Record Set from any of Biota's Core tables (p. 5, Appendix A). The flatfile exported automatically includes column headers based on Biota field names, and you may choose to create an "indented table" as an option (see the example on p. 462, lower table).

When to Use the Custom Flatfile Tool

In general, it is easier and more efficient to use other tools from the Im/Export menu when they are appropriate for your needs. But there are some things the Custom Flatfile tool can do that the others cannot.

- **The Custom Flatfile tool differs from the Export Editor** (Export by Tables and Fields, pp. 434–443) in that the Custom Flatfile tool allows you to include Core fields from related parent tables (up to any level of "grandparent" records), whereas the Export Editor exports fields from one table at a time. For example, using the Custom Flatfile tool, you can design an export format based on Species records that includes the fields [Species] Genus, [Species] Species Name, [Species] Author, [Family] Family, and [Genus] Custom1. (You could *not* include any fields from Specimen, Collection, or Locality tables, or from Species Notes or Auxiliary Fields tables, since these are a not parent tables to the Species table.)

- **The Export Taxonomic Flatfile tool** (pp. 451–453) can produce a file with the fields [Species] Genus, [Species] SpeciesName, [Species] Author, and [Family] Family, based on Species records, but has no provision for including any of the "Custom" fields in the higher taxon tables, such as [Genus] Custom 1 (see Appendix A)—unlike the Custom Flatfile tool.

- **The Export Specimen Flatfile tool** can export virtually any field from any Core table, but is based strictly on Specimen records. The Custom Flatfile tool can base the export on records from any Core table.

You cannot use the Custom Flatfile tool to export data in any useful way from any Peripheral table, including Notes, Images, and Auxiliary Fields. Use the Export Editor or the appropriate special tools discussed elsewhere in this chapter.

Using the Export Custom Flatfile Tool: Step by Step

1. **Find the Core Table records on which to base the export, and make them the current Record Set for the table** (pp. 13–15). For example, designate a selection of Species records as the Species Record Set. You can use any of the tools of the Find, Series, or Tree menus to gather the records you want (Chapters 11 and 12).

2. **From the Im/Export menu, select Export Custom Flatfile.** The Custom Flatfile option window appears.

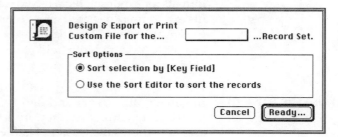

3. **Select a Core table from the table name popup list in the option window.**

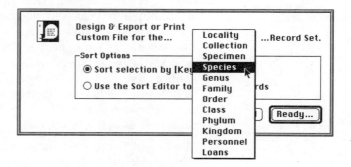

4. **Choose a Sort Option in the option window:**

   ♦ **To sort the records by Key field** (Record Code for Locality, Collection, Specimen, Species, or Loans; taxon for Genus or higher taxon tables), make sure the first option (the default) is selected.

   ♦ **To use the Sort Editor to sort the records,** click the second option. See pp. 131–139 for help using the Sort Editor.

   NOTE: You can use the Quick Report Editor's own sorting capabilities (pp. 273, 460) if you prefer, which will override either option in the Custom Flatfile option window, to produce "indented tables." See step 8, below, and the examples on p. 462 (lower table).

5. **Click the Ready button in the options window.** An instruction screen appears.

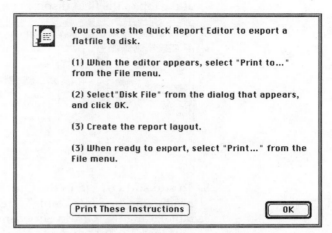

Biota provides these onscreen instructions (which you can print if you want, using the button on the instruction screen) because of the complex interface that 4D provides for exporting files using the Quick Report Editor.

6. **Click the OK button in the instruction screen to proceed.** The Quick Report Editor appears.

7. **Set up the columns for the fields you want to export.** For detailed help with the Quick Report Editor, see pp. 269–278. Briefly, drag field names from the upper-left panel to the empty form on the right, as shown here. (Use commands from the Edit menu to delete columns or insert intervening ones.)

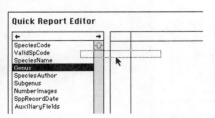 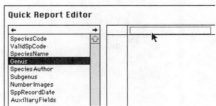 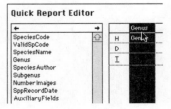

You can access fields from other tables by selecting the table name from the popup list that appears when you click and hold in the box at the top of the field display panel.

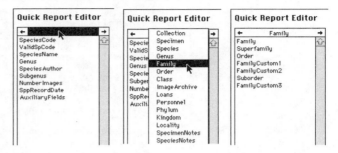

8. **To add sort criteria in the Quick Report Editor itself, which will produce an "indented table" effect** (see the example on p. 462, lower table), drag the <<Add Sort>> marker from the lower left panel of the Quick Report window to the column you want to sort. The name of the column will be added to the Sort list in the panel. (Select Delete Last Sort from the Special menu to remove Sort levels.)

9. **To specify export to a text file,** instead of printing a report, select "Print to..." from the File menu while the Quick Report Editor is open.

The "Print to" option window appears.

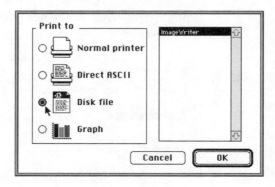

10. **Click the Disk File button, then click the OK button** in the "Print to" option window (above).

11. **From the File menu, select Print,** while the Quick Report Editor is open.

A Save File window appears, entitled "Export data" (Macintosh) or "Save file into" (Windows).

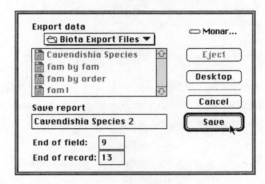

12. **Set up the "Export data" window.** Find the folder or directory you want, name the text file, and confirm or change the "End of field" and "End of record" delimiters.

NOTE: The Macintosh version is illustrated above. The "End of field" and "End of record" delimiter options are not available in the Windows version, which uses TAB (ASCII 9) as the end-of-field delimiter and the standard Windows end-of-line sequence (ASCII 13 + ASCII10) as the end-of-record delimiter.

---

13. **To launch the export,** click the Save button in the Export Data window.

14. **To save the Custom Flatfile format** you have created, select Save As from the File menu while the Quick Report Editor is still open.

15. **To load a Custom Flatfile format** that you have previously saved, use the Open command from the File menu while the Quick Report Editor is open.

16. **To dismiss the Quick Report Editor,** click the OK button at bottom of the Quick Report window or select Close from the File menu while the Quick Report Editor is open.

Examples of Custom Flatfiles Exported by Biota

Here are two examples of flatfiles exported using the Quick Report Editor, accessed through the Export Custom Flatfile tool. The same Species records formed the basis for both.

Notice, in the second example, the "indented table" effect produced by adding sorts to the Quick Report Editor setup (no repeated values for Family and Genus).

Sorted on Family, Genus, and Species Name using the Sort Editor (step 4, above); no sorts set in the Quick Report Editor:

Family	Genus	Species Name	Species Author	Species Code
Campanulaceae	Centropogon	caoutchouc	(H.B.K) Gleason	cencaou
Campanulaceae	Centropogon	erianthus	(Benth.) Benth. & Hook. f.	ceneria
Campanulaceae	Siphocampylus	ecuadoriensis	Wimmer	sipecua
Campanulaceae	Siphocampylus	sanguineus	A. Zahlbr.	sipsang
Campanulaceae	Siphocampylus	scandens	(HBK) G. Don	sipscan
Ericaceae	Cavendishia	forreroi	Luteyn	cavforr
Ericaceae	Cavendishia	lindauiana	Hoer.	cavlind
Ericaceae	Cavendishia	palustris	A.C. Smith	cavpalu
Ericaceae	Cavendishia	pseudospicata	Sleumer	cavpseud
Ericaceae	Cavendishia	tenella	A.C. Smith	cavtene

Sorted on Family, Genus, and Species Name within the Quick Report Editor (step 8, above); Sort Editor not used (step 4):

Family	Genus	Species Name	Species Author	Species Code
Campanulaceae	Centropogon	caoutchouc	(H.B.K) Gleason	cencaou
		erianthus	(Benth.) Benth. & Hook. f.	ceneria
	Siphocampylus	ecuadoriensis	Wimmer	sipecua
		sanguineus	A. Zahlbr.	sipsang
		scandens	(HBK) G. Don	sipscan
Ericaceae	Cavendishia	forreroi	Luteyn	cavforr
		lindauiana	Hoer.	cavlind
		palustris	A.C. Smith	cavpalu
		pseudospicata	Sleumer	cavpseud
		tenella	A.C. Smith	cavtene

Exporting Specimens Examined Lists for Publications

One of the most tedious tasks in preparing taxonomic or floristic/faunistic manuscripts is the compilation and formatting of lists of specimens examined for each species treated. Biota produces virtually journal-ready "Specimens Examined lists" for any Record Set of species you designate, exported to a text file in an easily edited format.

What the Specimens Examined Tool Exports

Specimens Examined sections of published works in systematics follow different traditional formats for different taxa—and vary among journals and among different authors within disciplines. Although Biota's Specimens Examined tool provides some key format options, you will need to add the finishing touches using a text editor (e.g., Microsoft Word, WordPerfect) to reflect specific styles for your field and to suit your own taste. Changes needed will almost certainly include punctuation and fonts, but perhaps also the ordering and repetition rules for certain fields. Even so, Biota does the hard part: collating and condensing the Collection and Locality data for the specimens of the species you designate.

The exported text file has the following format (see pp. 468–469, later in this chapter, for actual examples):

1. **To mark the beginning of the text section for each species, a line of asterisks is exported.** You will want to remove this marker line in a text editor when you edit the file for format and details.

2. **The phrase "Specimens examined for Genus species" follows the marker line** (where "Genus species" is the name of a species in the Species Record Set). You will need to underline or italicize Latin binomials using a text editor.

3. **The following Biota fields are exported—if neither repeated nor blank—**for each Specimen record you choose to include, in this order (see Appendix A):

   - **[Locality] Country.** Set in uppercase characters. Not repeated if Country is the same as for the previous specimen.
   - **[Locality] State/Province.** Not repeated if State/Province is the same as for the previous specimen.
   - **[Locality] District.** Not repeated if District is the same as for the previous specimen.
   - **[Locality] Locality Name.** Not repeated if this Specimen record is linked to the same Locality record as the previous one.
   - **[Locality] Latitude** and **[Locality] Longitude.** Exported in Degree-Minute-Second format, *regardless* of the Lat/Long format setting in the Preferences screen (Special menu). Not repeated if this Specimen record is linked to the same Locality record as the previous one.

- **[Locality] Elevation.** Not repeated if this Specimen record is linked to the same Locality record as the previous one.
- **[Collection] Site.** Not repeated if this Site is the same as for the previous specimen.
- **Host Specimen information.** If the [Collection] Host Spcm Code field is not blank in the Collection record (and is not the same for the previous Specimen), Biota exports the phrase "Ex Genus species" based on the [Species] Genus and [Species] Species Name fields for the Host Specimen that is linked to the Collection record (see pp. 493–496).
- **Collection date(s)** The collection date entry combines information from the [Collection] Date Collected and [Collection] Date Coll End fields, formatted using the [Collection] Date Coll Flag and [Collection] Date Coll End Flag fields, as described on pp. 111–114, "Dates." Not repeated if this Specimen record has the same collection date(s) as the previous one.
- **[Collection] Collected By.** Not repeated if this Specimen record is linked to the same Collection record as the previous one.
- **[Specimen] Specimen Code (optional).** Including specimen codes (collector's numbers or museum or herbarium accession numbers) in Specimens Examined lists is standard in some fields (e.g., botany), and with the advent of barcodes and specimen-level databases, may well become standard in others (e.g., entomology).

NOTE:

In systematic botany, published "Specimens Examined" lists are traditionally based on collector's numbers, further identified by appending the standard abbreviation for the herbarium holding individual specimens examined. For example, "Smith 6557 (US, NY)" means that two specimens (herbarium sheets) from Smith's collection number 6557 were examined, one held by the Smithsonian and one by the New York Botanical Garden.

To produce such lists with Biota, one strategy is to use the collector's number followed by the herbarium abbreviation as the Specimen Code for individual sheets examined. If you use this format, the exported Specimens Examined list will need little modification in a text editor. (Lists of specimens with the same collector's number will need to be condensed and details of the collection locality deleted, if you follow the usual tradition.)

- **[Specimen] Stage/Sex (optional).** Not repeated if this Specimen record has the same value for Stage/Sex and is linked to the same Collection record as the previous one.

NOTE: In an attempt to deal with plurals, if the number of specimens for a Stage/Sex category is greater than 1, an "s" is appended to the value for Stage/Sex: e.g., *female* becomes *females*. You will have to correct most irregular plurals in the text file with a global search and replace. The only exception is that Biota, revealing its entomological heritage, looks specifically for *larva* and *pupa*, and correctly renders the Latin plurals *larvae* and *pupae*.

- **If individual Specimen Codes are not listed, the number of specimens examined for each Collection or for each Stage/Sex in each Collection.** The number of specimens examined is computed in one of two ways, depending on the settings you establish for the export (see next section in this chapter):
  ◊ **Option 1:** The number of specimens is simply the number of Specimen records (among those you choose to include in the export) that share the same value for Collection Code—or for Collection Code and Stage/Sex.
  ◊ **Option 2:** The number of specimens is computed by accumulating the sum of values for the [Specimen] Abundance field, for Specimen records (among those you choose to include in the export) that share the same value for Collection Code—or for Collection Code and Stage/Sex.

Exporting a Specimens Examined List: Step by Step

1. **Find the *Species* records for which you want to export a "Specimens Examined" list, and make them the current Species Record Set** (pp. 13–15). You can use any of the tools of the Find or Tree menus to gather the records you want (Chapters 11 and 12).

2. **Determine the scope for Specimen records:**
   - **If you want to include *all* Specimen records in the Data File** that are linked to records in the Species Record set, proceed directly to step 3.
   - **If you want to limit the export to certain Specimen records** for the species in the Species Record Set, you can do so by designating the Specimen records you want as the current Specimen Record Set.

   NOTE: Here is one way to find the Specimen records you want to include. From the Find menu, select Lower Taxa for Higher Taxa. Set up the query "Find all Specimen Records for the Species Record Set" and click the Accept button in the Lower Taxa for Higher Taxa screen. All Specimen records for the Species in the Species Record Set will be displayed. Find the ones you want and make them a Subselection, then click Done and make them the Specimen Record Set.

3. **From the Im/Export menu, select Export Specimens Examined List.** An option window appears.

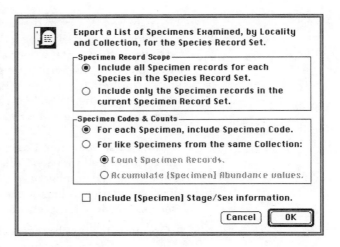

4. **Choose an option from the Specimen Record Scope panel.**

    - **To include *all* Specimen records in the Data File** that are linked to records in the Species Record set (the default option), make sure the first option is selected in the Specimen Record Scope panel.

    - **To limit the export to the Specimen records** in the Specimen Record Set, choose the second option in the Specimen Record Scope panel.

5. **Choose options from the Specimen Codes and Counts panel.**

    - **To include the Specimen Code for each Specimen record** in the exported Specimens Examined list (the default option), make sure the first option is selected in the Specimen Codes & Counts panel. If you choose this option, the lower pair of buttons remain disabled, since they concern counts of specimens.

    - **To report counts of specimens instead of Specimen Codes**, choose the second option in the Specimen Codes & Counts panel. If you choose this option, the lower pair of buttons in the Specimen Codes & Counts panel are enabled and Biota offers another pair of options:

        ◊ **To report simple counts of specimens for each Collection**, make sure the Count Specimen Records option (the default) is selected from the lower pair of buttons in the Specimen Codes & Counts panel.

        ◊ **To use the [Specimen] Abundance field to add up numbers of specimens examined**, choose the "Accumulate [Specimen] Abundance value" option from the lower pair of buttons in the Specimen Codes & Counts panel. This is the option to choose if you have used each Specimen record for a "lot" (group) of conspecific specimens. For botanical specimens, you could select this option for duplicates that you have examined, if you choose not to list them individually (see the note on p. 464).

6. **To include information from the [Specimen] Stage/Sex field,** click the checkbox at the bottom of the options screen.

   • **If you chose to include Specimen Codes in step 5**, the Stage/Sex value is reported in parentheses after each Specimen Code—e.g., "USNMH362145 (female)." See the example below (p. 469, Example 2).

   • **If you chose to report counts in step 5,** counts of specimens for each value of Stage/Sex within each Collection are reported with the name of the stage or sex—e.g., "5 females." See the example below (p. 469, Example 4).

7. **Click the OK button in the options window to launch the export.**

   If you have included more than 20 species in the Species Record Set, a confirmation message appears. You can cancel the export now, if you wish, by clicking the Cancel button in the confirmation window.

   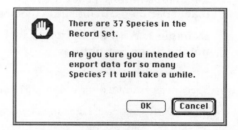

   If you click the OK button in the confirmation window, or there are fewer than 20 species in the Species Record Set, the Save File window for your operating system appears.

8. **Name the text file.** Find the folder or directory where you want to create the text file, name the file, and click the Save button (Macintosh) or the OK button (Windows). The progress indicator appears.

   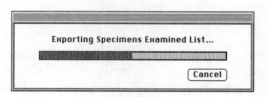

   WARNING: The progress indicator advances each time a Species is completed. If you include species with numerous Specimen records, the indicator will not change quickly. Be patient! Biota is working through hundreds of conditionals to save you from making all the same decisions yourself. You can do other tasks with Biota or with other applications while the export continues.

9. **Error messages during export.** Biota assumes that you do not want to list either a species with no specimens, or a specimen for which either Collection or Locality table data (see Appendix A) are missing.

If Biota finds incomplete information during the export, the export is aborted and you will see one of the following messages, with the genus and specific name substituted for *Genus species* in the messages:

- "No Specimen records found for *Genus species*."
- "No Collection records found for *Genus species*."
- "No Locality records found for *Genus species*."

10. **To halt the export for any reason,** click the Cancel button in the progress indicator window. If you stop the export before it is done, the exported file may be incomplete.

Examples of Specimens Examined Lists Exported by Biota

Here are some examples of Specimens Examined lists, just as Biota exported them, to illustrate the effects of different sets of Specimen Codes & Counts and Stage/Sex options. The same specimens were used for all illustrations. (They are hummingbird flower mites collected from plant specimens, to illustrate the host specimen option.)

Example 1. With the option to include Specimen Codes selected, but not the Stage/Sex option:

****
Specimens examined for Rhinoseius luteyni:

ECUADOR: Cotopaxi, Latacunga-Quevedo rd., 0°58'0"S, 78°56'0"W, 274-3350m, 3-14km E Pilaló, ex Sphyrospermum buxifolium, 4 Apr 1992 (coll. J.L. Luteyn) — JLL14400:01, JLL14400:02, JLL14400:03, JLL14400:04, JLL14400:05, JLL14400:06, JLL14400:07, JLL14400:08, JLL14400:09, JLL14400:10.

****
Specimens examined for Rhinoseius steini:

ECUADOR: Cañar, 7-9km NE Pindilíg tow. Rivera, 2°35'0"S, 78°49'0"W, 3000-3200m, None, ex Siphocampylus scandens, 19 Jan 1985 (coll. J.L. Luteyn) — JLL11106:01, JLL11106:02, JLL11106:04, JLL11106:05, JLL11106:06, JLL11106:07, JLL11106:08, JLL11106:09, JLL11106:10, JLL11106:13, JLL11106:14;

Carchi, Tulcán-El Caramelo rd., 0°45'0"N, 77°40'0"W, 3300m, 27 km E Panam Hwy., ex Siphocampylus ecuadoriensis, 10 Jan 1985 (coll. J.L. Luteyn) — JLL10953:01, JLL10953:02, JLL10953:03, JLL10953:05, JLL10953:06, JLL10953:07, JLL10953:08, JLL10953:09;

PERU: Amazonas, 19km SW Leimebamba, 6°45'0"S, 77°48'0"W, 3020m, along rd. to Balsas, ex Vaccinium mathewsii, 13 Feb 1985 (coll. J.L. Luteyn) — JLL11382:01, JLL11382:02, JLL11382:03, JLL11382:04, JLL11382:05, JLL11382:06, JLL11382:07, JLL11382:10.

Notice that each collection begins on a new line, to help you work with the text file in a text editor when you add the finishing touches.

In the remaining examples, only the first species (*Rhinoseius luteyni*), will be illustrated.

Example 2. With Specimen Code and Stage/Sex options selected:

****

Specimens examined for Rhinoseius luteyni:

ECUADOR: Cotopaxi, Latacunga-Quevedo rd., 0°58'0"S, 78°56'0"W, 274-3350m, 3-14km E Pilaló, ex Sphyrospermum buxifolium, 4 Apr 1992 (coll. J.L. Luteyn) — JLL14400:09 (Adult female), JLL14400:05 (Adult male), JLL14400:03 (Deutonymph), JLL14400:01 (Larva), JLL14400:04 (Larva), JLL14400:06 (Larva), JLL14400:08 (Larva), JLL14400:10 (Larva), JLL14400:02 (Protonymph), JLL14400:07 (Protonymph).

Example 3. With the Count Specimen Records option selected, but not the Stage/Sex option:

****

Specimens examined for Rhinoseius luteyni:

ECUADOR: Cotopaxi, Latacunga-Quevedo rd., 0°58'0"S, 78°56'0"W, 274-3350m, 3-14km E Pilaló, ex Sphyrospermum buxifolium, 4 Apr 1992 (coll. J.L. Luteyn) — 10 specimens,

Example 4. With the Count Specimen Records and Stage/Sex options selected:

****

Specimens examined for Rhinoseius luteyni:

ECUADOR: Cotopaxi, Latacunga-Quevedo rd., 0°58'0"S, 78°56'0"W, 274-3350m, 3-14km E Pilaló, ex Sphyrospermum buxifolium, 4 Apr 1992 (coll. J.L. Luteyn) — 1 adult female, 1 adult male, 1 deutonymph, 5 larvae, 2 protonymphs.

Exporting Collections-by-Species Incidence or Abundance Tables

Biogeographers and conservation biologists often need to generate lists of species for particular sets of sites, and, conversely, lists of sites from which a species or group of species has been recorded. In ecology, quantitative biodiversity data is of interest in its own right. In biodiversity inventories, data on frequencies of species in samples or collections are essential for statistical estimates of inventory progress or approach to completion.

Biota can export Collections-by-Species tables of either incidence data (presence/absence) or abundance data (number of individuals) based on:

- **A Species Record Set** and the Specimen and Collection records linked to it. Use this option if you want an incidence or abundance table of the Collections in which a Species or group of Species has been recorded.

- **A Specimen Record Set** and the Species and Collection records linked to it. Use this option if you need to produce a Collections-by-Species table for a subset of the Specimen records linked to a particular set of Species and Collection records. For example, in a vegetation study, this option was used to produce separate tables for seedlings, saplings, and mature trees, all representing the same set of species from the same set of quadrats.

- **A Collection Record Set** and the Specimen and Species records linked to it. Use this option if you want an incidence or abundance table of the Species recorded for a particular set of Collection records.

In general, these three options produce different results, since the Collection records for a particular group of Species may well be linked to other Species as well.

NOTE: The text files created by this tool use the TAB character as a field delimiter.

Exporting Collections-by-Species Tables: Step by Step

1. **Find the Species, Specimen, or Collection records on which you want to base a Collections-by-Species table and make them the active Record Set for the table** (pp. 13–15). You can use any of the tools of the Find or Tree menus to gather the records you want (Chapter 11).

2. **Select Export Colls x Species Table from the Im/Export menu.** An option window appears.

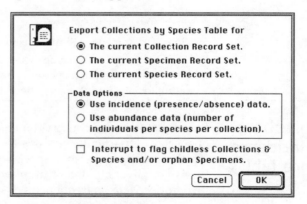

3. **From the upper set of three option buttons in the window (above), choose the Record Set** on which you will base the analysis. This is the Record Set you defined in step 1. See the introduction to this section (pp. 469–470) for help if you need it.

4. **Choose a Data Option.**

    - **If you choose the incidence (presence/absence) option** (the default), Biota records a 1 for each Species that occurs in a particular Collection—regardless the number of Specimens—and a zero otherwise.

    - **If you choose the abundance option** (the lower button in the Data

Options panel), Biota records *the sum of the [Specimen] Abundance field values* for the Specimen records for each Species in each Collection in the analysis. (If the Abundance value is 1 for each Specimen record—entered as the default when you create a Specimen record, the sum exported will be the number of Specimen records.)

5. **To interrupt the export for problem data warnings,** click the checkbox at the bottom on the options window.

   Three kinds of errors are flagged:

   ♦ **Collection records not linked to any Specimen record** (childless or "empty" Collections).

   ♦ **Specimen records not linked to any Species record** (orphan specimens).

   ♦ **Specimen records with zero in the [Specimen] Abundance field,** which are flagged if you chose the Abundance option in step 4. (If you checked the Incidence option, Biota allows zeros in the Abundance field so that Specimen records can be used to link Species and Collection when no specimen exists; see p. 65.)

   If Biota find one of these errors during the export, a warning message like the one below is displayed. (Messages for other errors types are analogous). If you click the OK button in the warning message, the export will proceed by excluding the problem record. If you click the Cancel button in the warning message window, the export is aborted.

   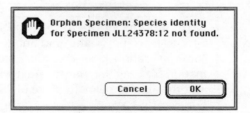

   NOTE: If you want to check for childless (empty) Collections beforehand, select Childless Records from the Find menu then select Collections from the popup list. To find orphan Specimens, select Orphan Records from the Find menu and then select Specimen → Species from the popup list. See pp. 215–216 for help.

6. **Click the OK button in the options window to launch the export.**
   The standard Save File window for your operating system appears.

7. **Name the text file.** Find the folder or directory where you want to create the text file, name the file, and click the Save button (Macintosh) or the OK button (Windows). The progress indicator appears.

8. **To halt the export for any reason,** click the Cancel button in the progress indicator window. If you stop the export before it is done, the exported file may be incomplete.

NOTE: You can carry out other tasks with Biota or other applications while the export is in progress.

An Example of a Collections-by-Species Table Exported by Biota

The table below shows the abundance of several species of hummingbird flower mites (the columns) in a set of host plant collections (the rows). The corresponding incidence table would have a "1" substituted for each non-zero cell value.

SP. CODE: GENUS: SPECIES: SUBGENUS:	antio Rhinoseius antioquiensis Gr. richardsoni	lute Rhinoseius luteyni Gr. rafinskii	rhinhaplo Rhinoseius haplophaedia Gr. richardsoni	rhinrich Rhinoseius richardsoni Gr. richardsoni	stein Tropicoseius steini Gr. chiriquensis
JLL14587H	1	0	0	0	0
JLL11382H	0	0	0	0	8
JLL14400H	0	9	0	0	0
JLL12327H	0	0	0	1	0
JLL12329H	0	0	0	4	0
JLL12414H	0	0	0	7	0
JLL13339H	0	0	3	0	0
JLL11314H	0	0	0	2	27

NOTE: Biota includes 4 rows of Species identifiers: Species Code, Genus, Species (specific) Name, and Subgenus. If you need to code species for some special purpose, consider using the Subgenus field for this information. You can re-name the field with an alias if you want (pp. 323–326) and the alias appears in place of "SUBGENUS" in the exported table. For example, a plant ecologist using Biota might re-name the Subgenus field "Life Form" and use the information to sort columns of the text file into herbs, shrubs, and trees for data analysis.

Chapter 25 Biota and the Internet

Two distinct paradigms have emerged for providing World Wide Web access to biodiversity databases. In the simpler of these, the database software itself produces *static* hypertext (HTML) pages automatically, for whatever selection of records you wish, ready for posting on at your own Web site. Biota fully supports this strategy (see the next section), although you must provide your own Web server or have access to a Web server. To update the information your site offers, when necessary, you ask Biota to generate a new set of pages, then you post them on the server.

In the second paradigm, an Internet visitor to your Web site fills in query forms to request data *dynamically* from your active Data Files. Your Web server interfaces with the Data File by means of intermediary software, which generates pages "on the fly" and passes them to the server. Biota does not yet offer this form of access, but this chapter points you towards tools currently available (others are still on the horizon) that can be used to create dynamic access to Biota Data Files (see p. 8–11).

As a third option, because 4D Server uses TCP/IP as its communications protocol, you can use 4D Server to provide direct Internet access to your active Data File with the full Biota user interface. With this option, the remote user queries your Data File using 4D Client, rather than a Web browser (see p. 484).

Exporting Web Pages Automatically from Biota Data Files

Biota's Create Web Pages tool guides you through the setup process for the completely automated (procedural) export of hyperlinked Web pages, ready to post on a Web server. (The only required further preparation affects exported images, which currently require a format conversion before posting.)

You can export hierarchically-linked pages for any range of taxonomic levels (including intermediate levels such as Subfamily), for Record Sets that you define with either a top-down or bottom-up search in the taxonomic hierarchy.

If you include Specimen records, you can select any set of fields in the Specimen, Collection, or Locality tables for inclusion as Specimen data. You can include an image for each Species record. Other options include control over page length, custom page footnotes, hyperlinking of host and "guest" Specimen records, and the translation of non-ASCII characters to HTML character codes.

Choosing a Strategy for Creating Web Pages

Before you begin, you must decide whether to base the scope of the taxonomic records to be exported to Web pages on a *top-down* or a *bottom-up* or search of tables in the taxonomic hierarchy. If you choose a *bottom-up* search, you may also choose to restrict the export by Collection or Locality. In any case, if you include the Specimen level, you can export data for each selected Specimen record based on related information in the Collection and Locality tables, as well data from the Specimen table itself.

- **The top-down strategy** is based on the Lower Taxa for Higher Taxa tool of the Find menu (pp. 201–205), which the Create Web Pages procedure presents if you choose this strategy. Some typical queries you could construct using this tool are " Find all Species for the Order Orthoptera," "Find all Specimens for the Family Fringillidae," or "Find all Specimens for the Genus Record Set" (where the Genus Record Set contains all genera of Galápagos finches, for example).

    With this strategy, Biota will ask you to choose a *criterion* table at a higher taxonomic rank than the *target* table (pp. 201–205). The search can be based on a single higher taxon record (e.g., an Order record) or a higher taxon Record Set (e.g., the current Order Record Set) in the criterion table. Biota finds all linked records down to the level of the target table (e.g., Species or Specimen).

    The exported Web pages, organized by the taxonomic hierarchy, will include the criterion and target levels you choose, all obligatory intervening ranks, and any optional ranks (e.g., Subfamily) that you indicate later in the setup.

- **The bottom-up strategy** is based on the Higher Taxa for Lower Taxa tool of the Find menu (pp. 205–208), which the Create Web Pages procedure presents (p. 476) if you choose this strategy. Some typical queries you could construct using this tool are "Find all Genera for the Specimen Record Set" or "Find all Orders for the Species Record Set."

    With this strategy, Biota will ask you to choose a criterion table at a lower taxonomic rank than the target table (pp. 205–208). The search is based on the current Record Set at the criterion level. Biota finds all records linked to the criterion Record Set, at ranks up to and including the target level.

The exported Web pages, organized by the taxonomic hierarchy, will include the criterion and target levels you choose, all obligatory intervening ranks, and any optional ranks (e.g., Subfamily) that you indicate later in the setup process.

♦ **To create Web pages based on a particular set of Locality or Collection records,** you must use a two stage strategy. First, create a Locality or Collection Record Set for the records of interest. Then use the Specimens or Species for Places tool (pp. 210–213) to create a Specimen or Species Record Set based on links to the Locality or Collection Record Set. Finally, choose the bottom-up option (above), based on the Specimen or Species Record Set.

Using this strategy, you can create Web pages (for example) for all Specimens in the Collections from a particular site, or all Species from a particular country or province. In either example, you could also include linked records up to the level of, say, Family or Order.

Creating Web Pages: Step by Step

WARNINGS:

a. Before beginning, be certain that you have enough space available on your mass storage device.

b. Biota places the Web page files in the same folder or directory as your active Data File. (You can move them, once they are created.) Move or delete any old Biota-created Web page files from your Data File folder or directory *now*, since the new files will have the same names as the old ones (e.g., "Species1.html").

Here are the steps to follow to create Web pages, once you have settled on a strategy (see the previous section regarding strategies).

1. **Define the appropriate criterion Record Set,** using tools from the Find or Tree menus (Chapter 11). See the guidelines in the previous section. (If you intend to use the *top-down* strategy based on a single higher taxon record, there is no need to create a Record Set. You may proceed directly to the next step.)

2. **From the Im/Export menu, choose Create Web Pages.** The Strategy option screen appears.

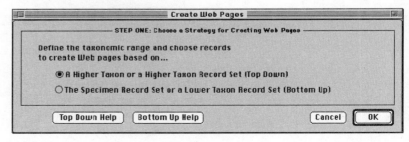

3. **Choose either the top-down strategy (the default) or the bottom-up strategy** and click the OK button.

   ♦ **If you choose the top-down strategy,** a special version of the Lower Taxa for Higher Taxa tool appears. If you are not familiar with this tool, see "Finding Records for Lower Taxa Based on Higher Taxa" in Chapter 11, pp. 201–205.

   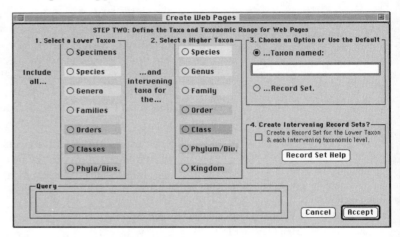

   ♦ **If you choose the bottom-up strategy,** a special version of the Higher Taxa for Lower Taxa tool appears. If you are not familiar with this tool, see "Finding Records for Higher Taxa Based on Lower Taxa" in Chapter 11, pp. 205–208.

   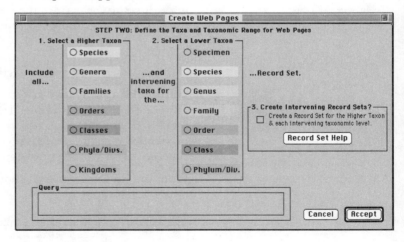

4. **Set up the appropriate query** in one of the screens above.

   > **NOTE:** It is not necessary to check the "Create Intervening Record Sets?" box to create the Web pages, but it might help you check the pages against records in the Data File. Use the Display menu to see the Record Sets created.

5. **Click the Accept button in the search screen.** Biota performs the search, which may take a few moments; then the Intermediate Taxon option screen appears.

6. **Click the check box for each intermediate taxonomic rank that you want Biota to recognize** in creating the Web page hierarchy.

   Any ranks above or below the criterion and target ranks (tables) are dimmed (disabled). (In the example above, the criterion table was Family and the target table was Specimen.)

   All *obligatory* ranks within the taxonomic scope you chose in the previous step are automatically enabled and may not be manually disabled (it would make no sense to disable them).

   You can use the Select All Optional Levels and Clear All Optional Levels to speed up the selection process, if you wish.

7. **Click the Accept button in the Intermediate Taxon option screen** (above). The Web Page options screen appears (below).

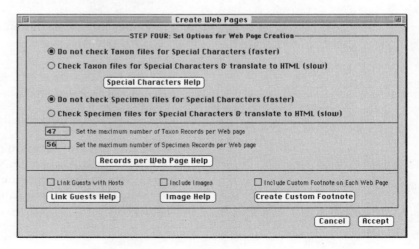

If you included the Specimen table in taxonomic scope you specified in step 4, above, the options screen will appear as illustrated above. If the Specimen table was not included, all options that apply to Specimen records will be dimmed (disabled).

The next five steps treat each of the options in the Web Page options screen, one at a time.

8. **Set options for HTML translation of special characters** (top panel of the Web Page options screen, above).

    Hypertext Markup Language (HTML), which Web browsers read, does not recognize characters with diacritical marks and a few other special characters.

    If the records you are exporting contain HTML special-meaning characters (& < >), characters with diacritical marks (á é í ó ú à è ì ò ù â ê î ô û ä ë ï ö ü ã å õ ç ñ and uppercase equivalents), inverted punctuation marks (¿ or ¡), or German ß, these characters will not display correctly with a Web browser unless you select "Check...files for Special Characters and translate to HTML."

    The two sets of option buttons in the top panel of the Web Page options screen (above) allow you instruct Biota to ignore, or to check and translate special characters for taxon files (Species, Genus, Family,..., Kingdom)—the upper pair of buttons—or data associated with Specimen records, including associated Collection and Locality data— the lower pair of buttons.

    NOTE: If you do *not* have any special characters in your records, or you are doing a test run, the export will run *much* faster if you do *not* ask Biota to check for special characters. Biota is programmed so that the degree and second symbols (characters ° and ") in latitudes and longitudes will be translated correctly with *either* setting, so it is not necessary to ask for character translation if these are the only special characters in your data.
    ---

9. **Set guidelines for the maximum number of taxon records or Specimen records per Web page**.

    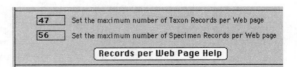

    These two numeric entry areas allow you to specify *guidelines* for the maximum number of taxon (Species, Genus, Family,...,Kingdom) records or Specimen records (including associated Collection and Locality data) per Web page. You can change the numbers to any value you wish.

Here are the rules Biota follows:

- **When the Web Page options screen first appears,** these values (above) are set to the highest *total* number of taxa for any selected rank (47 records, in the example above), and to the *total* number of Specimen records to be included on the Web pages (56 records in the example above), respectively.

- **If you leave these values as they stand,** all taxa within a taxonomic rank will be displayed on a single Web page, and all Specimen records will be displayed on a single Web page. The longest of the taxon pages will have the number of records indicated in the upper box (usually, the page for the lowest rank in the taxonomic scope you defined). The single Specimen page will include the number of records indicated in the lower box.

- **All records for a *given taxon* always appear on the same Web page,** regardless the maximum records setting, and regardless of how many records there may be for the taxon. For example, if you have set a maximum of 50 records per taxon page, and you ask Biota to create a Web page that includes a genus with 75 Species, all 75 will nonetheless appear on a single Web page—but Biota will begin a new page with for the next genus.

- **Likewise, all Specimen records linked to a given Species record will appear on the same Web page,** regardless of the numerical limit you set for Specimen records.

- **If the numerical limit you set has not been reached** at the end of a set of taxon records or the end of a conspecific group of Specimen records, Biota includes the next taxon of the same rank, or the next group of conspecific Specimen records, on the same Web page as the previous taxon or group.

- **If you want to specify only one taxon per Web page,** regardless of how many or how few records it contains, set the numerical limit to 1. The second rule (above) guarantees the desired result.

- **If you want to specify only one group of conspecific Specimen records per Specimen Web page**, regardless of how many or how few Specimen records are linked to each Species, set the Specimen record numerical limit to 1. The second rule (above) guarantees the desired result.

10. **Specify whether to link guest Specimen records with host Specimen records.**

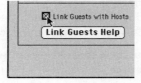

Hypertext links are a natural way to navigate from "guest" Specimen records to the associated host Specimen record at a Web site (see Chapter 21 for a full discussion of host–guest relations).

If you create a Specimen Record Set that includes not only guest Specimen records but also the host Specimen records with which they are linked (see pp. 401–404), Biota will automatically create the hypertext links between guests and their hosts on the appropriate Specimen Web pages, as illustrated below.

> Species *Rhinoseius richardsoni* Hunter 1972
>
> Specimen Code: JLL12329:02; Collection Code: JLL12329H; Collected By: J.L. Luteyn; Host Spcm Code: JLL12329 (Ex Psammisia sp. nov., Ericaceae); Locality Name: Mpio.ElCairo,Correg. Boquer—n; State/ Prov: Valle; Country: Colombia; Loc Latitude: 4°45'0"N; Loc Longitude: 76°20'0"W
>
> Specimen Code: JLL12329:01; Collection Code: JLL12329H; Collected By: J.L. Luteyn; Host Spcm Code: JLL12329 (Ex Psammisia sp. nov., Ericaceae); Locality Name: Mpio.ElCairo,Correg. Boquer—n; State/ Prov: Valle; Country: Colombia; Loc Latitude: 4°45'0"N; Loc Longitude: 76°20'0"W

NOTES:

a. Be certain to include [Collection] Host Specimen Code among the fields you select for export with each Specimen record, in step 14, below.

b. Biota does not check to see if you have included the proper records for the links to work. If you fail to include the host Specimens, the links from the guest Specimens are created properly, but there are no target host records.

c. This option is disabled if you did not include the Specimen table in the taxonomic scope of the search.

11. **Specify whether to include an image with each Species record that has associated images.**

If this option is checked, Biota will export one Image for each Species record in the selection, if that Species has any associated Image Archive records. (See Chapter 18 for full information on the Image Archive and its relation to Species records). Each image is exported to its own file, which is automatically linked and displayed with the appropriate Species record on a Species Web page. This option is disabled if you did not include the Species table in the taxonomic scope of the information to be exported.

- **The Image exported**—if more than one Image is linked to a particular Species record—is the *first* Image for the Species, according to the Image order you have determined in the Species input screen. (To make a different Image the first one, see "Changing the Order of Image Records for a Species," pp. 359–363.)

- **Web file names for Image files, when exported,** consist of the Species Code with the extension *.pict*. These files are in the PICT format (pp. 348–349).

- **You must convert the format of each image to the GIF format** (pp. 348–349), using another application, for them to be displayed in the Web pages Biota that exports. After conversion, *the Web file names for the Image files must consist of the Species Code with the extension .gif* to match the HTML image file references that Biota places automatically in the exported Species pages.

> NOTES: Several freeware or shareware image file converters are available over the Internet. For Macintosh, the freeware application *clip2gif* is an excellent choice. You can simply drag all the .pict files that Biota exports for your Web pages, as a group, and drop them on the clip2gif icon for batch conversion to GIF files. The suffix is automatically changed from .pict to .gif. The *clip2gif* application is available from several sites: to get a current list of them, search on *clip2gif* at http://pubweb.nexor.co.uk/public/mac/archive/doc/search.

12. **Specify a custom footnote to be included on each Web page**, if you wish.

    a. **Click the Create Custom Footnote button** in the Web Page options screen.

    The Custom Footnote editing window appears. (It will be blank unless you have previously entered text in the footnote window during this Biota session.)

    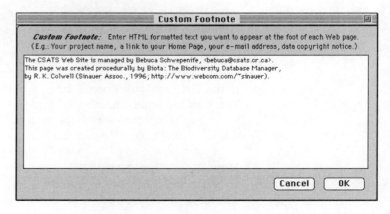

b. Enter whatever you wish to include in the footnote (e.g., your project name, Web site name, email address, data copyright notice, disclaimers, etc.). You can enter text only, or HTML coded objects such as hypertext links (e.g., a link to your home page) or image references for project or institutional logos.

   c. Click the OK button in the Custom Footnote window to accept the entry.

   The footnote button now reads Edit Custom Footnote (instead of Create Custom Footnote), to indicate that a footnote entry now exists.

   d. Click the checkbox labeled Include Custom Footnote on Each Web Page (see below).

13. **Click the Accept button in the Web Page options window** to accept the settings from steps 8–12, above.

- **If you did not include the Specimen table** in the taxonomic scope for the Web pages, the creation of the Web pages begins now. Skip to step 16, below.

- **If you included the Specimen table** in the taxonomic scope for the Web pages, the Field options screen, below, appears.

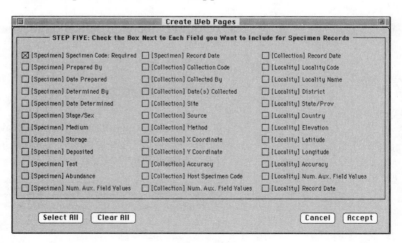

14. **In the Field options screen (above), check each field for which you want Biota to export information** for each Specimen record in the selection. The Specimen Code field is obligatory. You can use the Select All and Clear All buttons to speed the selection process if you wish.

   NOTE: If you checked the Link Guests with Hosts box in the Web page options screen in step 10, above, be certain to include [Collection] Host Specimen Code among the fields you select for export with each Specimen record.

15. **Click the Accept button in the Field options screen** (above) to launch the Web page export. Biota posts a series of progress indicators as the pages are made for each table. You can carry out other activities with Biota (or other applications) while the pages are being created in the background, as long as you do not open any records in input screens that might be involved in the export.

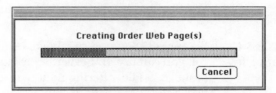

16. **To abort the creation of Web pages,** click the Cancel button in any progress window (example above).

17. **Check the Web pages using a Web browser,** when the process is complete. Open the page for the highest taxonomic rank, using the Open File command from the File menu of Netscape or the equivalent command in another Web browser. Use the hypertext links to move up and down the taxonomic hierarchy.

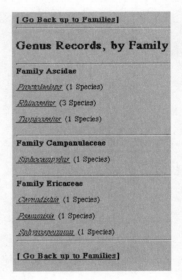

18. **Move the Web pages to a Web server** and add additional material to the pages (species descriptions, backgrounds, headers, etc.), if you wish. You will need to create a link or links in your home page to direct visitors to the Biota pages.

Other Internet Options

In addition to the Web page export option (above), two other strategies for Internet access to Biota Data Files should be mentioned.

Dynamic Access to Biota Data Files

Biota produces standard 4th Dimension Data Files. Because the table and field structures of Biota's files are fully disclosed in Appendices A and B, you can use commercially available applications to design a link between a Biota Data File and a Web server.

At present, the best product available for building such a server link appears to be NetLink/4D (from Foresight Technology, www.fsti.com). NetLink/4D is a set of 4th Dimension "externals" that you can use to access 4D Data Files from WebStar, a capable Web server for Macintosh (from StarNine/Quarterdeck, at www.starnine.com).

For this solution, you need to own 4th Dimension, NetLink/4D, and WebStar and have some programming experience in 4th Dimension.

4D Client-Server Access to Biota Data Files

Because 4D Server is capable of using TCP/IP as its data transfer protocol, you can use 4D Server to provide direct Internet access to an active Biota Data File, offering remote users the full Biota user interface. The user needs 4D Client to connect, but you can legally provide 4D Client to your colleagues, gratis.

For this approach you need to own 4D Server. See Appendix E.

Limited experience with this strategy suggests that it is most effective on local networks. Over most Internet connections, the transfer rate from 4D Server is too slow (as of this writing) to make it a competitive alternative to Web server/browser options. You can see an example of this strategy at http://herbaria.harvard.edu/treebase.

If the current promotional material from ACI (the makers of 4th Dimension) is to be believed, 4D Server itself will eventually become a Web server, capable of serving Web clients (browsers) directly from 4th Dimension Data Files.

Chapter 26 Importing Data

Biota's Import Editor (Import by Tables and Fields, in the Im/Export menu), can import data into any Biota table from free-format, delimited text (ASCII) files, such as the column-by-row text files you can create with a spreadsheet application (e.g., Microsoft Excel), or export as text flatfiles from database management applications (e.g., 4th Dimension, FoxPro, Access, dBase, Paradox, or FileMaker).

Biota's Import Editor has two related functions. You can use it either to import new records or to update fields in existing records in a Biota Data File. Most of what you need to know about importing data applies to both functions, but there are some important distinctions, which are described later in this chapter (pp. 487–488, 495–499).

The Import Editor is designed to make importing data as pain-free as possible. The number and order of fields (columns) in the text file need not match the number and order of fields in the Biota table receiving the data. Text fields can be skipped. The text file can have column headings, or not. Any field and record delimiters can be used. The Import Editor does extensive error checking, and you can undo imports that run into trouble. When Biota finds an error in the text file, it displays the offending record and tells you what went wrong.

But importing data into any relational structure, however powerful the import tools, takes planning and effort. Expect some frustration.

Import Step 1: Match Fields

The first step is to compare the fields in your data files with the fields in Biota's data structure (Appendix A). If you are importing data from another relational data structure, you will need to consider tables and relational links as well.

Almost inevitably, there will be fields that don't match. Remember that:

- Many fields in Biota (all those in italics in Appendix A) can be renamed (pp. 323–326).
- If you need more fields than Biota offers in its Core tables, you can create your own Auxiliary Fields to accommodate them (Chapter 15).
- If you need longer fields than Biota offers, some kinds of data may work well as Notes (pp. 179–183).

If there are too many fields in your data file that have no natural or adoptive home in the Biota structure, however, you are probably better off not trying to use Biota for those data.

Import Step 2: Prepare the Text Files

You will need to prepare a separate text file for each Biota table that will receive data.

The rows in the text file will become Biota records, if you are importing new records, or will match Biota records, if you are updating existing records. The only potential exception is the first row of the text file, which you can use for column headings if you wish; see the next section, "The Column Heading Row Option."

The columns in the text file correspond to fields in a Biota table. The columns can be in any order within a text file, and the file can include columns (even blank ones) that you do not want to import.

If you intend to import records from an existing text flatfile (a columns-by-rows table) that includes columns destined for fields in several different, linked Biota tables (see Appendix A for tables and fields), you may wish to follow the suggestions in "A Strategy for Preparing Text Files to Import into Hierarchically Linked Tables" (pp. 507–520), after reading the sections below on "The Column Heading Row Option," "Key Fields," and "Field Types and Field Lengths."

The Column Heading Row Option

If you are working from a table in an existing text file from which you intend to import, chances are that the first row of the table consists of descriptive column headings, rather than data. Likewise, for a newly created text file, you will probably find it much easier to keep track of what you are doing if you use column headings.

Biota does not use the column headings from your text file to match up text file columns with Biota fields; the Import Editor has a much more flexible way to do that. But you can tell Biota's Import Editor to ignore the first row of the text table so that you can leave your existing headings in the text file, or create new ones.

Because column headings are ignored during import (if you tell Biota to ignore them), it does not matter whether they are exactly the same as the names of the Biota fields they are destined to fill, or not. On the other hand, you may find it easier to keep track of what you are doing if column headings in the text file match or at least approximate Biota field names.

Automatic Stripping of Initial and Terminal Space Characters

In preparing text files for import, it is easy to start a field entry with a space character accidentally, or inadvertently append one or more spaces at the end of a field entry. These initial and terminal space characters wreak havoc in relating fields and make sorts and searches behave unintelligibly in any field (until you finally figure out what is wrong!).

To eliminate this problem, the Import Editor automatically strips off all initial and terminal space characters (ASCII Character 32). No error comments are issued.

"Internal" spaces—those with some other character both before and after—are not removed and are imported as text as part of the character string.

NOTE: Biota automatically strips terminal space characters and flags initial space characters in all Key and relating fields during input using Biota's record input screens.

Key Fields

Most of the tables in Biota's structure (Appendix A) have a Key field (Record Code, taxon name, etc.; pp. 6–7) that requires a unique value for each record in a Biota Data File. (The exceptions are the Notes tables and Det History tables, each of which has a linking field that accepts non-unique values, and the Auxiliary Field Value tables, which have a two-field (composite) Key. These exceptions are discussed later, pp. 502–507.)

Biota checks Key fields carefully during the import or update process. However, the way Key fields are handled depends entirely on whether you have indicated that you are importing new records or updating existing ones.

If you are importing new records, and a unique value in the Key field is required, Biota will halt the import and post an error message if it encounters a record with any of the following:

- A blank Key field.
- A Key field value that duplicates a value already imported.
- A Key field value that duplicates a value that was already in the Data File when you started the import.

On the other hand, if you are updating existing records, and a unique value in the Key field is required, Biota will halt the import and post an error message if it encounters a record with either of the following:

- A blank Key field.
- A Key field value that does *not* match any value already in the Data File—Biota cannot tell what record to update if there is no match.

Import error messages are discussed in detail later in this chapter (pp. 495–502).

If you are importing new records, it is likely that you will need to create Record Codes for records in one or more of the Specimen, Species, Collection, and Locality tables. Read or review the Chapter 7, "Record Codes" before assigning codes. Once you decide on a system, you can use spreadsheet tools (e.g., the Series command in Microsoft Excel) to assign Record Codes in a column of your text file, by using an alphanumeric prefix with a consecutive integer counter, e.g., ABC001, ABC002, ABC003, etc. (see "A Strategy for Preparing Text Files to Import into Hierarchically Linked Tables," pp. 507–520).

Field Types and Field Lengths

Appendix A shows the field type and field length (maximum number of characters) for each field in the Biota data structure. The fields in the text file from which you import data should match the type and length of the Biota fields they will fill, although Biota allows as much flexibility as possible, as explained for each field type, below.

- **Alphanumeric fields.** An alphanumeric field accepts all characters. You can import a number or a date into an alphanumeric field, but the characters are treated as text, not numbers or dates. The number of characters that can be imported ranges from zero (a blank value) to the maximum indicated in Appendix A for each field. If you try to import a longer value, Biota will issue a warning, giving you the option to accept a truncated value or abort the import (see "Displays and Error Messages During Record Importing or Updating," pp. 495–502).

- **Date fields.** Before importing dates, please read or review the sections on Dates (Chapter 8, pp. 111–114). The required text file format for dates is MM/DD/YYYY. For example, April 9, 1993 must be written 04/09/1993 to be imported correctly. (If you import 09/04/1993 instead, Biota will interpret the date as September 4, 1993.)

  If you have dates in some other format, you must first convert them before importing. Spreadsheet applications (e.g., Microsoft Excel) can easily convert columns of dates to the required format from any other standard format.

  Biota checks all imported dates for Gregorian calendar validity as well as format. Month cannot exceed 12 and Day cannot exceed the maximum number of days in the month, given the year. (February 29 is allowed only in years divisible by 4 and only in centennial years divisible by 400.) If a format error is encountered in a date, or an invalid date is found, Biota posts an error message (see "Displays and Error Messages During Record Importing or Updating," pp. 495–502).

- **Partial Dates: Date Flag fields.** As explained in full in the Chapter 8, for some date fields, Biota can display and print partial dates—Month-Year dates and Year-only dates. These fields are [Collection] Date Collected, [Collection] Date Coll End, [Specimen] Date Prepared, and [Specimen] Date Determined. Each of these fields is paired with an integer field for a Date Flag (Appendix A).

If you are importing complete dates only, you need not import any values for Date Flag fields, which are initialized by default to zero, the code for complete dates.

If you are importing complete dates mixed with partial dates, you can either enter a zero in the Date Flag field for individual records that have complete dates, or leave the Date Flag field blank for such records.

When you import partial dates, follow these rules:

◊ **Month-year dates.** For a month-year date (e.g., May 1975), the Date field itself must be imported as a complete date. You must also import a 1 in the corresponding Date Flag field—the code for a month-year date. You can use any convention you like for the Day element of month-year dates in the Date field. If you want to be consistent, however, use Biota's own convention: the first day of the month (May 1, 1975 for May 1975).

◊ **Year-only dates.** For year-only dates (e.g., 1975), the Date field itself must be imported as a complete date. You must also import a 2 in the corresponding Date Flag field—the code for a year-only date. You can use any convention you like for the Day and Month components of year-only dates in the Date field. Biota's own convention is the last day of the year (December 31, 1975 for 1975).

♦ **Number Fields (Long Integer and Real).** Biota checks values imported into Long Integer fields for nonnumeric characters (besides the minus sign). If you import a real number into a long integer field, it is truncated to its integer part.

Values imported into Real number field are checked for nonnumeric characters (besides the minus sign and decimal point).

If an invalid character is found in a number field, Biota posts an error message ("Displays and Error Messages During Record Importing or Updating," pp. 495–502).

♦ **Latitude and Longitude fields.** These two fields in the Locality table are of type Real. Latitude and longitude must be imported in decimal degrees, using the standard conventions for hemispheres: North and East are positive values, South and West are negative values. You can convert degree-minute-second coordinates to decimal degrees (in a spreadsheet, for example) by the formula:

Decimal Degrees = Degrees + (Minutes/60) + (Seconds/3600)

♦ **Text fields.** Biota uses Text fields for Note Text fields in the four Notes tables, for the [Personnel] Notes field, and for the [Loans] Description field. A text field may contain up to 32,000 characters (but they take up only as much memory in the Data File as required).

- **Boolean fields.** Boolean fields have one of two values: True or False. In a text file to be imported, a Boolean field must have either a zero (for False) or a 1 (for True). There is only one Boolean field in Biota, [Personnel] Group, which, when True, identifies a Personnel record as a Group Name record, rather than an individual person's or institution's record. The Group field has the value False for an individual (see pp. 174–177).
- **Picture (Image) fields.** You cannot import or update Picture (Image) fields using the Import Editor. Images can be imported from disk files to the Image input screen, or pasted from the Clipboard (Chapter 18).

Import Step 3: Set Up and Launch the Import or the Update

When you have a text file ready to import, follow these steps to launch the Import Editor.

1. **Choose Import by Tables and Fields from the Im/Export menu.** An option screen appears, offering to import new records or update existing records.

2. **Select the function you want and click the OK button.** The Import Editor appears.

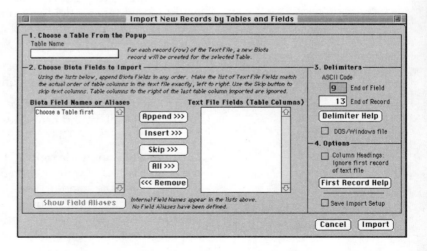

- **If you choose the new record option,** the Import Editor appears in a window with the title "Import New Records by Tables and Fields."

- **If you choose the update option,** the Import Editor appears in a window with the title "Import Fields to Update Existing Records."

> NOTE: In the illustrations in this section, the window title will specify the new record option, but all aspects of setting up the Import Editor are the same for either option.

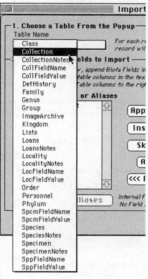

3. **From the Table Name popup list at the upper left, select the name of the Biota table** for which you want to import or update records (see Appendix A).

In the "Biota Field Names or Aliases" panel at the lower left, Biota displays all fields for the table you selected, listed alphabetically.

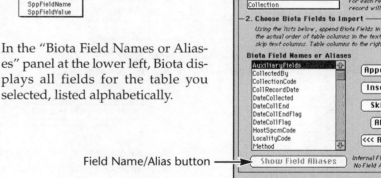

Field Name/Alias button

4. **The Field Name/Alias button.** The button below the "Biota Field Names or Aliases" panel controls whether Biota lists strictly Internal Field Names (as given in Appendix A), or substitutes a Field Alias for the corresponding Internal Field Name for each field you have given an Alias. (See pp. 323–326 to learn about Field Aliases.)

- **If no Aliases have been defined,** the button is disabled and reads "Show Field Aliases." The caption to the right of the button reads "Internal Field Names appear in the lists above. No Field Aliases have been defined."

- **If any Aliases have been defined,** when you first open the Import Editor the button is enabled and reads "Show Internal Names." The caption to the right of the button reads "Field Aliases you have defined are currently included in the lists above in place of the corresponding Internal Field Names."

- **If you click the Show Internal Names button,** the button text changes to "Show Field Aliases" and the caption reads "Internal Field Names appear in the lists above. Click the button to the left to include the Field Aliases you have defined."

5. **List the Biota fields that will receive data from the text file.** The "Biota Field Names or Aliases" panel is simply a source of field names. The panel labeled Text File Fields (Table Columns), at the right, is the list Biota will consult in placing data from your text file into fields in your Biota Data File.

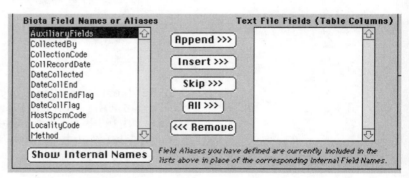

As a rule, the number and order of field names (Internal Field Names or Aliases) in the list you create in the Text File Fields panel must match the number and order of columns in your text file. There are two exceptions to this rule:

- **If you want to skip a column in the text file,** click the Skip button to enter "[Skip a Text File Column]" in the Text File Fields panel (see step 6, below).

- **If there are columns in your text file to the right of the last column you list to import,** Biota will ignore them. There is no need to list "[Skip a Text File Column]" entries for such columns (although it would do no harm).

NOTE: If you have included column headings in your text file, the headings need not match the field names you list in the Text File Fields panel, since headings are ignored (see "The Column Heading Row Option," pp. 486–487, earlier in this chapter, and "Column Headings checkbox," step 8 below).

6. **Use the buttons between the two lower panels in the Import Editor to create the list of fields in the Text File Fields panel.**

- **Append button.** Click a field in the Biota Field Name panel to the select it, then click the Append button. The field name is added to the end of the list in the Text File Fields panel.
- **Insert button.** Click a field already listed in the Text File Fields panel to select it. Then click the field in the Biota Field Name panel that you want to insert. When you click the Insert button, the new field is inserted just *above* the selected field in the Text File Fields panel.
- **Skip button.** If there is a column in your text file that you want Biota to ignore, click the Skip button to append or insert the phrase "[Skip a Text File Column]" in the Text File Fields panel. You must do this for each intervening column to be skipped. (Columns to the right of the last one listed are skipped automatically, however.)
- **All button.** The All Button is a toggle. When the button text reads "All>>>," it means "Append All." If you click the button, all fields in the Biota Field Name panel are entered in the Text File Fields panel (in the order listed in the Field Name Panel), *replacing* any fields already listed in the Text File Fields panel. The button text then changes to read "<<<All," meaning "Remove All." If you click the button again, the Text File Fields panel is cleared and the button text changes back to "All>>>."
- **Remove button.** To remove a single field already entered in the Text File Fields panel, select the field by clicking it in the list, then click the Remove button.

7. **Set or confirm the Field and Record Delimiters.** In the Delimiters panel of the Import Editor, you enter the information that Biota needs in order to tell where one column (field) of your text file stops and the next begins, and where one record stops and the next begins. You may use any characters in the standard ASCII set (characters 33–126), but of course they must match what is used in your text file. The Delimiter Help button presents a list of common options.

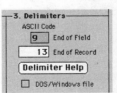

- *Macintosh:* The default delimiters are TAB (ASCII Character 9) for the field delimiter and RETURN (ASCII Character 13) for the record delimiter—the characters that Macintosh spreadsheet applications use for "TAB-separated values" text files. If you are importing data from a PC (DOS or Windows) text file that uses the standard PC line termination (RETURN + LINE FEED, ASCII Characters 13 + 10), click the DOS/Windows File checkbox. The End of Record box will then read "1310."

- *Windows:* The default delimiters are TAB (ASCII Character 9) for the field delimiter and RETURN + LINE FEED (ASCII Characters 13 + 10), for the record delimiter—the characters that Windows spreadsheet applications use for "TAB-separated values" text files. If you are importing data from a Macintosh text file, click the DOS/Windows File checkbox to *uncheck* it. The End of Record box will then read "13."

8. **Set Options.** There are two checkboxes in the Options panel.

- **Column Headings checkbox.** If the first row (record) of your text file consists of column headings, you must check this option so that Biota will ignore the first row. Otherwise, Biota will import, or attempt to import, the headings as field values for a record. (See "The Column Heading Row Option," pp. 486–487, earlier in this chapter.)

- **Save Import Setup checkbox.** If you check this option, Biota will remember the way you have set up the Import Editor, even if you Cancel without attempting to import, for the duration of the current Biota session. If you have a problem while importing data, this option can save you time when you are ready to try again after you have made corrections in the text file.

9. **Click the Import button to launch the import process,** when the list of Text File Fields and the settings are ready.

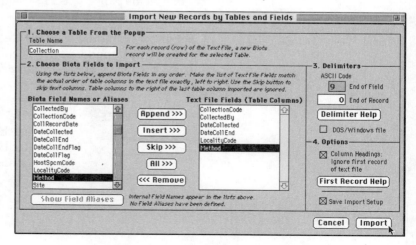

Biota presents the standard Open File window for your operating system. Find the text file you have set up to import, then click the Open button.

Displays and Error Messages during Record Importing or Updating

Once the import or update begins, Biota displays a special progress indicator that shows the first few dozen characters of each record, as it is being imported from the text file. You can click the Cancel button in the progress indicator window at any time to halt the import.

The Import Progress Indicator

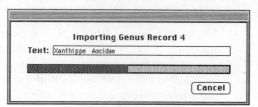

If Biota encounters a problem, it posts an explanatory error message (discussed individually later in this section) and freezes the progress indicator with the first 50 characters or so of the offending record still displayed,

The progress indicator and all import and update error message windows (except for the Missing Key Error, pp. 497–498) include the *Record Number* of the offending record ("Genus Record 4," in the example above). The Record Number is *simply the position of the bad record in the text file*—not counting the column heading record, if any. (The Record Number specified in import error messages is *not* the Record Code or Key value for the bad record.)

Importing New Records: Undoing an Aborted Import

If Biota encounters an error while importing new records from a text file, an error message is displayed. (Errors are discussed individually later in this chapter.)

When you click the OK button on the error screen (or Cancel for Field Length Errors, p. 500), Biota presents the option screen below, to ask you how to proceed.

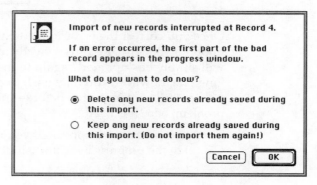

- **If you choose the "Delete any new records..." option** (the default), Biota will undo the entire import, deleting all records already imported in the current episode. While it may seem a pity to delete correctly imported records, this alternative is quite often the better one—unless the error occurs well into a very long import.

- **If, instead, you choose the "Keep any new records already saved..." option,** you will not have to import those records again, but *you must delete them from your text file* before you can import the remaining records (after correcting the bad record). Otherwise, you will get Duplicate Key errors.

The option screen above also appears if you interrupt the import by clicking Cancel in the progress indicator window. In this case, you will almost certainly want to delete all records imported before the interrupt, by selecting the first option in the screen above. You can use the Record Number to find out how far the import progressed in your text file before you canceled it, however, should you decide to keep the records already imported.

Updating Existing Records: Redoing an Aborted Import

If Biota encounters an error while updating existing records from a text file, an error message is displayed. (Errors are discussed individually later in this chapter.)

When you click the OK button on the error screen (or Cancel for Field Length Errors, p. 500), Biota presents the message below to inform you that any records correctly updated before the problem occurred remain in the Biota Data File in updated form.

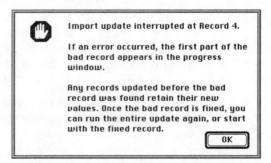

Once you correct the error in the text file, you can simply repeat the import if you wish. Records correctly updated before the error occurred will simply be "updated" again to the same values. If the import is a long one, however, you may wish to create a new text file containing only records not imported before—including the one that caused the problem, once it is fixed.

The screen above also appears if you interrupt the update by clicking Cancel in the progress indicator window.

Too Many Fields Error

In each record (row) in your text file, Biota looks for a field delimiter (pp. 493–494) to match each field you have listed in the Text File Fields panel of the Import Editor. If Biota reaches the end of a record before all fields in the list have been matched, you will see the message below.

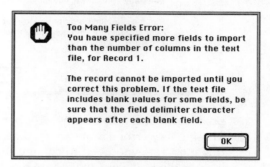

If this message appears in response to the *first* record in your text file (as illustrated above), chances are you truly listed more fields in the Import Editor than the number of columns in the text file, or you specified incorrect field or record delimiters. Open the text file (or better, a copy of it, so you can leave it open) in a spreadsheet application and check the number and order of columns against the number and order of fields in the Text File Fields panel of the Import Editor. There can be fewer fields (plus Skips) listed for import than the number of columns in the text file, but not more.

If the Too Many Fields Error screen appears after one or more records have been successfully imported or updated (the Record number cited in the error message is greater than 1), there is probably a missing field delimiter, or a missing delimiter and a missing field value in the offending record. As the message reminds you, if you have a blank field, you nevertheless must include the delimiter that separates the blank from the next field.

If you can't find the problem using a spreadsheet application, as a last resort open the text file in a word processor, set to display TABS and RETURNS (paragraph marks), if those are the delimiters you have used. *If you make changes, be sure to save the file as Text.*

If you get the Too Many Fields error message on the *last* record of your text file, make sure the last line (record) in your text file ends with the record delimiter you have specified.

Missing Key Error: Importing New Records

When you are importing new records into a Biota table that has a Key field (pp. 6–7), Biota issues a Missing Key Error message if you failed to include the Key field among those listed for import in the Text File Fields panel of the Import Editor.

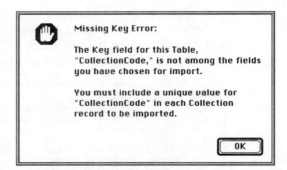

Because Biota checks for this error before starting to import any data, a Missing Key error simply aborts the import before it begins. If you get this message, you need to change the setup in the Import Editor to include the Key Field, make sure Key Field data exist in the correct column in the text file, then launch the import again.

Missing Key Error: Updating Existing Records

If you are updating existing records in a Biota table that has a Key field (pp. 6–7), Biota issues a Missing Key Error message if you fail to include the Key field among those listed for import in the Text File Fields panel of the Import Editor.

Because Biota checks for this error before starting to import any data, a Missing Key Error simply aborts the import before it begins. If you get this message, you need to change the setup in the Import Editor to include the Key Field, make sure Key Field data exist in the correct column in the text file, then launch the import again.

Duplicate Key Error: Importing New Records

If you are importing new records into a Biota table that has a Key field (pp. 6–7) that requires unique values, Biota posts a Duplicate Key Error message whenever it encounters a value for the Key field that is already in the Biota Data File.

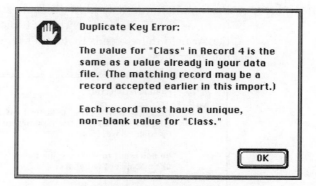

Because each record is saved as it is imported, the duplication may be either between two records in the text file you are importing or between a record in the text file and a record that was in the Biota Data File before the import began. You may have to check both possibilities—or you could simply change the Key field in the offending record.

Update Key Match Error: Updating Existing Records

If you are updating existing records and Biota cannot match a Key field value with any record already in your Biota Data File, Biota cannot tell which record to update and posts the error message at the right.

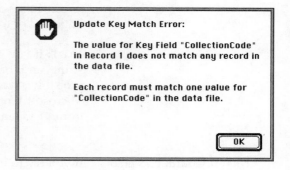

Blank Key Error

If you are importing new records or updating records in a Biota table that has a Key field (pp. 6–7), Biota posts a Blank Key Error message whenever it encounters a blank value (no characters at all). A value entered as all space characters gives the same error, since initial spaces are automatically deleted (see the note on p. 487).

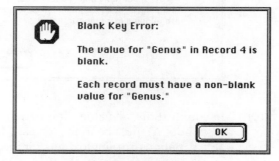

Field Length Error
Each alphanumeric field in the Biota structure has a maximum length (p. 6), as specified in Appendix A. If Biota encounters a value for a particular field that exceeds that field's length limit, Biota posts the following error message.

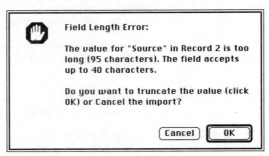

Unlike all other import error messages, the Field Length Error message allows you to continue with the import or the update if you choose—by accepting the problem value, truncated to fit the field. If you click the Cancel button, Biota posts the standard import or update error screen. In the case of new records, you can elect to delete records already imported or to keep them (pp. 495–496).

NOTE: If you get a Field Length Error message claiming a very long string, and/or the import progress indicator shows garbage, you have probably saved the text file in some format other than plain text (e.g., Microsoft Word document format, or Excel spreadsheet or workbook format). Open the file in the application and save it again as Text.

Mixed Characters Error
If you try to import alphabetic characters into a numeric field, Biota posts a Mixed Characters Error message. Fields of type Real allow digits, the decimal point, and the minus sign only. Integer and Long Integer fields allow only digits and the minus sign.

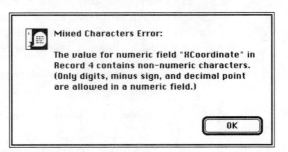

Date Format Error
In your text file, all values destined for a Biota field in Date format (see Appendix A) must be in the format MM/DD/YYYY. The only acceptable variations are single digits for month or day, or YY for the year, in which case Biota imports the year as 19YY. (Be sure to use the full YYYY format after 1999.)

NOTE: See pp. 111–114 for full details on how Biota handles dates, including partial dates and date ranges. For information on *importing* partial dates, see p. 488, earlier in this chapter.

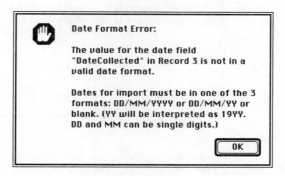

Invalid Date Error

If you attempt to import a date that is in the correct format, but violates the Gregorian calendar, Biota posts one of the two messages below. See p. 488, earlier in this chapter, for the conditions tested.

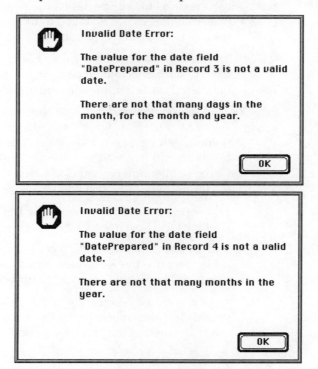

Boolean Field Error

In your text file, a Boolean field variable must take one of two values: a 1 if the variable is True and a 0 if the variable is False. See p. 490, earlier in this chapter, for more details.

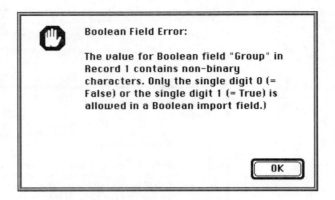

Importing Data to Notes Tables or to the Determination History Table

Five Core tables in Biota's structure—Species, Specimen, Collection, Locality, and Loans—are linked to Notes tables, with corresponding names (Appendix A). In each case, the Core table is the parent table and the Notes table the child table, so that many Notes records may be saved for each Core Table record. For example, for each Species record, you can create as many Species Notes as you wish (see pp. 179–183). Each Species Notes record is linked to the parent Species record through the Species Code field (Appendix A).

The Determination History (Det History) table works in a similar manner. For each Specimen record, any number of Det History records can be created, as determinations are updated over time.

Thus, you can import multiple Notes records with the same value in the Record Code field (which may differ in author, date, and contents), without getting an error message. (The Record Code fields in the Notes tables are : [Species Notes] Species Code, [Specimen Notes] Specimen Code, [Collection Notes] Collection Code, [Locality Notes] Locality Code, and [Loans Notes] Loan Code.)

Likewise, you can import multiple Det History records with the same [Det History] Specimen Code value.

You cannot update existing records in a Notes table or the Det History table. Because the Key values for these tables are not unique, Biota would not know which record to update if there were any ambiguity.

Here are some special error comments you may receive in importing records to Notes tables or the Det History table, in addition to general errors detailed in the previous sections (pp. 495–502).

No Unique Key Update Error

Because the Key value for Notes tables and the Det History table are not unique, it is not possible to update existing records in a Notes table or the Det History table. If you try to do so, Biota issues an error message.

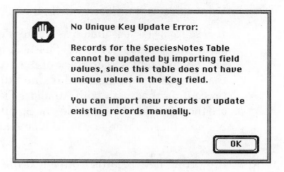

Unknown Key Error

For Biota to find display a Notes or Det History record, it must be properly linked to a record in the corresponding parent table (pp. 7–8). If the Record Code value in a new Notes record or Det History record you want to import does not match any record in the corresponding Core (parent) table, Biota issues an error message.

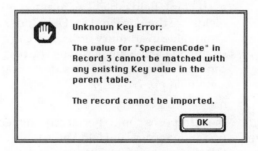

Importing New Records and Updating Existing Records in Auxiliary Fields Tables

Before importing any data to Auxiliary Fields tables, it might be a good idea to read or review Chapter 15, "Auxiliary Fields," although a brief outline follows here.

Biota supports user-defined Auxiliary Fields for four parent (Core) tables: Species, Specimen, Collection, and Locality. Linked to each parent table is a child table for Auxiliary Field Values (Appendix A). In the example below, a child table called Coll Field Value is linked to the Collection (parent) table.

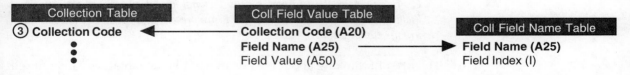

As shown for the Coll Field Value table above, each *Field Value* table has just three fields: Record Code, Field Name, and Field Value (where "Record Code" is Species Code in the Spp Field Value table, Specimen Code in the Spcm Field Value table, Collection Code in the Coll Field

Value table, or Locality Code in the Loc Field Value table).

A *Field Name* table is linked as a parent to each Field Value table, through Field Name fields. (The Coll Field Name table is illustrated above.) The Field Name tables keep track of the ordering of Auxiliary Field Names that you assign in the Field Name Editor (pp. 283–295) and speed up procedures that work with Auxiliary Fields. (The four Field Name tables are the Spp Field Name table, the Spcm Field Name table, the Coll Field Name table, and the Loc Field Name table.

Importing or Updating Records in Field Name Tables

Importing new records from a text file to a Field Name table or updating existing Field Name records is straightforward. The Field Name field, the Key, requires unique values for importing new records, or values that match existing ones for updating. (The previous section outlines the role of Field Name tables in managing Auxiliary Field data.)

The second and only other field, Field Index, is most logically assigned unique, consecutive integers, but in fact Biota will accept any integers or even blank fields (blanks are assigned a zero in the Field Index field). Fields are then ordered by Field Index value, with fields that share the same value in arbitrary order. You can easily correct such anomalies later using the Field Name Editor. (The Sort button in the Field Name Editor can renumber field names automatically, or you can do it manually, as shown on pp. 293–295).

Importing or Updating Records in Field Value Tables

Biota imposes some special requirements for importing new records or updating records in Field Value tables. (The introductory paragraphs of this section, pp. 503–504, outline the role of Field Value tables in managing Auxiliary Field data.)

Field Value tables have a two-field (composite) Key: each record must have a *unique combination* of values in the Record Code and Field Name fields (where "Record Code" is Species Code, Specimen Code, Collection Code, or Locality Code—depending on the table).

Furthermore, there is no point in importing or creating a record in a Field Value table with a blank in the Field Value *field.* (In fact, Biota does not even save such a record if you try to create one within Biota by using an Auxiliary Field input screen.)

For these reasons, the Import Editor does not accept any record for input to a Field Value table unless all three fields are present and the Key fields contain a unique combination of values.

NOTE: In thinking about records in Field Value tables, you may find it helpful to visualize the Record Code value as a row index and the Field Name value as a column index for a columns-by-rows table with Field Values in the cells. Thus each combination of Record Code and Field Name points to a unique cell of the table. You import data only for the *nonempty* cells.

Auxiliary Field Setup Error   If you have set up the Import Editor to import data into a Field Value table (see. pp. 503–504) but you fail to include all three fields (Record Code, Field Name, and Field Value) in the list of fields destined to receive data from the text file, Biota does not allow you to proceed until you correct this problem.

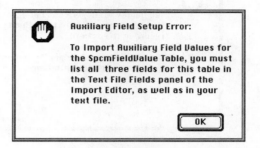

Auxiliary Field Blank Value Error   All three fields must be present in each record you import into a Field Value table (see above, pp. 503–504, for the reasons). If one or more of the three values is blank, Biota posts the following error notice. (This is the message for an update error. For the same error while importing new records, the last sentence of the text begins: "To Import Auxiliary Field Values for....")

Duplicate Auxiliary Field Value Key Error: Importing New Records   If you attempt to import a Field Value record that duplicates an existing one in both the Record Code and Field Name fields, Biota aborts the import and posts the following message.

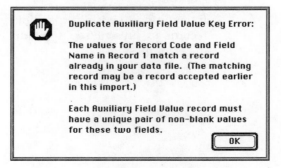

Auxiliary Field Value Key Match Error: Updating Existing Records

If you attempt to update a Field Value record that Biota cannot find a match for in both the Record Code and Field Name fields, Biota aborts the import and posts the following message.

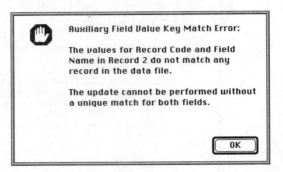

Importing Records for the Lists Table

The Lists table in Biota has two special purposes. It keeps track of Core Field Aliases you define (pp. 323–326) and it manages the Field Value Choice Lists you create (pp. 315–322).

For most users, the only reason you will ever want to import records into the Lists table is to transfer Lists records from one Biota Data File to another (to a new, blank one, for example). The procedure for exporting Lists records from one Biota Data File and importing them into another is detailed in Chapter 16, pp. 319–322.

Appendix B details the fields, default values, and required records for the Lists table. You should study Appendix B is you intend to create the required records for import into the Lists table using another application (e.g., a spreadsheet application) or if you want to modify exported Lists records before importing them again.

The process of importing Lists records has three important special characteristics.

- When you import new Lists records, Biota *first deletes all existing records* in the Lists table. The new records you import replace all existing records.
- For this reason, *you cannot use the Import Editor to update existing Lists records.*
- If you import Lists records, Biota will check that all three required records are present (see Appendix B). If not, you will see the following error message.

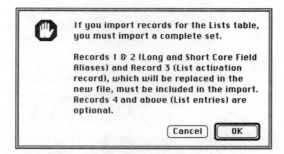

When Biota has completed the successful importation of a set of Lists records, the message below appears.

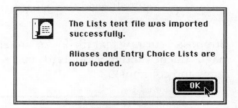

A Strategy for Preparing Text Files to Import into Hierarchically Linked Tables

If your raw text file is a flatfile (a columns-by-rows table) that includes columns destined for several hierarchically linked Biota tables, you may find it helpful to refer to this section, which outlines a strategy for creating the corresponding text files for import into Biota.

For example, you may have data in a taxonomic flatfile (pp. 508–512), with rows for Species (or some other taxonomic level) and columns for the higher taxonomic levels that each belongs to (genus, family, etc.). Or, your text file may be a specimen flatfile (pp. 512–520): one row per specimen, with fields for several levels in both the taxonomic and place hierarchies.

Strategy: Preparing Text Files for Import from an Existing Flatfile

The two sections that follow this one describe, step by step, an efficient strategy for using your existing text file to create new text files in the format Biota requires for import. Don't be intimidated by the number of steps. In fact, they simply fill in the details for a repeated cycle of a few steps, for a pair of linked tables (child and parent).

Here is the strategy, in general terms, for the step-by-step procedure given in the next section:

1. In the text file, make sure the column destined for the Key field of the child table has unique values. (Create a new column for the Key if necessary.)

2. Save the text file (as Text), named for the child table.

3. Save the text file (as Text) again, named for the parent table.

4. In the text file named for the parent table, delete all columns that were destined for the child table, and delete the child Key field, *but do not delete the relating field for the parent table*, since it will become the Key field for the parent table.

5. Delete all rows in the text file that duplicate values of the parent Key field, so that all remaining rows have unique Key values.

6. Save the text file (as Text), still named for the parent table.

7. If the parent table is, in turn, the child of another parent, it becomes the "child table" as you repeat steps 3 through 6.

8. As each text file is completed, make a list of the fields that will be imported (including any intervening fields that you want to tell Biota to skip). You will need these lists to set up the Import Editor.

Step by Step: Preparing Text Files for Import from a Taxonomic Flatfile

Suppose your raw text file looks something like the example for Galápagos finches and mockingbirds below (but probably with many more rows and columns).

Genus	Specific Name	Family	Order
Cactospiza	heliobates	Fringillidae	Passeriformes
Cactospiza	pallida	Fringillidae	Passeriformes
Camarhynchus	parvulus	Fringillidae	Passeriformes
Camarhynchus	pauper	Fringillidae	Passeriformes
Camarhynchus	psittacula	Fringillidae	Passeriformes
Certhidea	olivacea	Fringillidae	Passeriformes
Geospiza	conirostris	Fringillidae	Passeriformes
Geospiza	difficilis	Fringillidae	Passeriformes
Geospiza	fortis	Fringillidae	Passeriformes
Geospiza	fuliginosa	Fringillidae	Passeriformes
Geospiza	magnirostris	Fringillidae	Passeriformes
Geospiza	scandens	Fringillidae	Passeriformes
Pinaroloxias	inornata	Fringillidae	Passeriformes
Platyspiza	crassirostris	Fringillidae	Passeriformes
Nesomimus	parvulus	Mimidae	Passeriformes
Nesomimus	trifasciatus	Mimidae	Passeriformes

1. Open the file in a spreadsheet application (e.g., Microsoft Excel).

2. Save the file as Species Import, so your original file (which we will assume has some other name) will remain unchanged. (This immediate "save" guards against accidentally altering your original file.)

3. Add a column for Species Code and fill it with unique values (see comments, above, on Record Codes).

Species Code	Genus	Specific Name	Family	Order
cactheli	Cactospiza	heliobates	Fringillidae	Passeriformes
cactpall	Cactospiza	pallida	Fringillidae	Passeriformes
camaparv	Camarhynchus	parvulus	Fringillidae	Passeriformes
camapaup	Camarhynchus	pauper	Fringillidae	Passeriformes
camapsit	Camarhynchus	psittacula	Fringillidae	Passeriformes
certoliv	Certhidea	olivacea	Fringillidae	Passeriformes
geosconi	Geospiza	conirostris	Fringillidae	Passeriformes
geosdiff	Geospiza	difficilis	Fringillidae	Passeriformes
geosfort	Geospiza	fortis	Fringillidae	Passeriformes
geosfuli	Geospiza	fuliginosa	Fringillidae	Passeriformes
geosmagn	Geospiza	magnirostris	Fringillidae	Passeriformes
geosscan	Geospiza	scandens	Fringillidae	Passeriformes
pinainor	Pinaroloxias	inornata	Fringillidae	Passeriformes
platcras	Platyspiza	crassirostris	Fringillidae	Passeriformes
nesoparv	Nesomimus	parvulus	Mimidae	Passeriformes
nesotrif	Nesomimus	trifasciatus	Mimidae	Passeriformes

4. Save the Species Import file, *but do not close it*. The Species Import file is now complete.

5. With pencil and paper or in another file, make a list of each column in the Species Import file, in left-to-right order, up to and including the last column (field) you will import into Biota's Species table. Be sure to include in your list any intervening columns that you do *not* intend to import, since you will need to tell Biota to skip them. You can ignore any columns to the right of the last column to be imported into the Species table.

6. Save the file again as Genus Import.

7. Delete the Species Code, Species Name, and any other fields (e.g., Author, Subgenus) destined for Biota's Species table, but *do not delete* the Genus column.

Genus	Family	Order
Cactospiza	Frinigillidae	Passeriformes
Cactospiza	Frinigillidae	Passeriformes
Camarhynchus	Fringillidae	Passeriformes
Camarhynchus	Fringillidae	Passeriformes
Camarhynchus	Fringillidae	Passeriformes
Certhidea	Fringillidae	Passeriformes
Geospiza	Fringillidae	Passeriformes
Geospiza	Fringillidae	Passeriformes
Geospiza	Fringillidae	Passeriformes
Geospiza	Fringillidae	Passeriformes
Geospiza	Fringillidae	Passeriformes
Geospiza	Fringillidae	Passeriformes
Pinaroloxias	Fringillidae	Passeriformes
Platyspiza	Fringillidae	Passeriformes
Nesomimus	Mimidae	Passeriformes
Nesomimus	Mimidae	Passeriformes

Unless you had only one species per genus, there are now duplicated generic names (shown shaded in the example above) in the Genus column, which will become the Key field for the Genus table.

8. Sort by rows on the Genus column, if necessary, to group duplicated genera by rows.

9. Manually delete the entire *row* for each duplicated Genus name, leaving just one record for each distinct Genus name.

Genus	Family	Order
Cactospiza	Fringillidae	Passeriformes
Camarhynchus	Fringillidae	Passeriformes
Certhidea	Fringillidae	Passeriformes
Geospiza	Fringillidae	Passeriformes
Pinaroloxias	Fringillidae	Passeriformes
Platyspiza	Fringillidae	Passeriformes
Nesomimus	Mimidae	Passeriformes

10. Save the Genus Import file, *but do not close it*. The Genus Import file is now complete.

11. Make a list of each column in the Genus Import file, in left-to-right order, up to and including the last column (field) you will import into Biota's Genus table. Be sure to include in your list any intervening columns that you do *not* intend to import, since you will need to tell Biota to skip them. You can ignore any columns to the right of the last column to be imported into the Genus table.

12. Save the file again as Family Import.

13. Delete Genus column and any other column destined for Biota's Genus table, but *do not delete* the Family column.

Family	Order
Frinigillidae	Passeriformes
Frinigillidae	Passeriformes
Frinigillidae	Passeriformes
Frinigillidae	Passeriformes
Frinigillidae	Passeriformes
Frinigillidae	Passeriformes
Mimidae	Passeriformes

14. Proceed just as you did for the Genus table, following steps 8. through 12, above, but reducing rows to unique Family values (duplicates shown shaded in the example above).

Family	Order
Frinigillidae	Passeriformes
Mimidae	Passeriformes

15. Continue the same protocol to create an Order import file, Class import file, and so on, up to Kingdom if necessary, keeping a list of the columns in each file.

16. The text files are now ready for importing into Biota. Go to the section "Set Up and Launch the Import or Update," pp. 490–495.

NOTE: If you follow the strategy outlined above, a text file destined for a Biota table lower in the hierarchy will generally include columns destined for Biota tables higher in the hierarchy. For example, the Species Import file (Step 3, above) includes columns for Family (destined for the Genus and Family tables), and Order (destined for the Family and Order tables). You can deal with these extraneous columns in any of three ways. The first is the "cleanest" way to deal with extraneous columns, but consider the other two options, too.

- **Reopen each text file** you made in the procedure above and delete each of the extraneous columns.

- **If the extraneous columns are all located to the right of the columns to be imported,** for each file just ignore the columns when you set up the Import Editor. Once Biota finds a value for each column you list in the Import Editor, it goes on to the next record (the next text file row).

- **If the extraneous columns are to the left of the ones you need to import,** or interspersed with them, you can use the Skip button in the Import Editor to tell Biota to skip over them (p. 492).

Step by Step: Preparing Text Files for Import from a Specimen Flatfile

Suppose your raw text file looks something like Galápagos finch example below (but probably with many more rows and columns). (With apologies to Darwin and Fitzroy, these are fictitious records, though with plausible dates and localities.)

Collected By	Date Collected	Locality Name	State/Prov	Country	Lat	Long	Genus	Specific Name
C. Darwin	18 Sep 1832	San Cristóbal I.	Galápagos	Ecuador	-0.83	-89.4	Geospiza	fortis
R. Fitzroy	18 Sep 1832	San Cristóbal I.	Galápagos	Ecuador	-0.83	-89.4	Geospiza	fortis
R. Fitzroy	18 Sep 1832	San Cristóbal I.	Galápagos	Ecuador	-0.83	-89.4	Geospiza	fuliginosa
C. Darwin	18 Sep 1832	San Cristóbal I.	Galápagos	Ecuador	-0.83	-89.4	Geospiza	scandens
C. Darwin	24 Sep 1832	Floreana I.	Galápagos	Ecuador	-1.28	-90.4	Geospiza	scandens
C. Darwin	24 Sep 1832	Floreana I.	Galápagos	Ecuador	-1.28	-90.4	Camarhynchus	psittacula
C. Darwin	24 Sep 1832	Floreana I.	Galápagos	Ecuador	-1.28	-90.4	Camarhynchus	pauper
C. Darwin	24 Sep 1832	Floreana I.	Galápagos	Ecuador	-1.28	-90.4	Camarhynchus	parvulus
R. Fitzroy	24 Sep 1832	Floreana I.	Galápagos	Ecuador	-1.28	-90.4	Certhidea	olivacea
C. Darwin	30 Sep 1832	Fernandina I.	Galápagos	Ecuador	-0.43	-91.5	Geospiza	magnirostris
C. Darwin	30 Sep 1832	Fernandina I.	Galápagos	Ecuador	-0.43	-91.5	Geospiza	fortis
C. Darwin	30 Sep 1832	Fernandina I.	Galápagos	Ecuador	-0.43	-91.5	Geospiza	difficilis
R. Fitzroy	30 Sep 1832	Fernandina I.	Galápagos	Ecuador	-0.43	-91.5	Cactospiza	heliobates
C. Darwin	9 Oct 1832	Santiago I.	Galápagos	Ecuador	-0.23	-90.8	Cactospiza	pallida
C. Darwin	9 Oct 1832	Santiago I.	Galápagos	Ecuador	-0.23	-90.8	Certhidea	olivacea
C. Darwin	9 Oct 1832	Santiago I.	Galápagos	Ecuador	-0.23	-90.8	Geospiza	difficilis
C. Darwin	9 Oct 1832	Santiago I.	Galápagos	Ecuador	-0.23	-90.8	Camarhynchus	parvulus

1. Open the text file in a spreadsheet application (e.g., Microsoft Excel).
2. Save the file (as Text) with the name Specimen Import so your original file (which we will assume has some other name) will remain unchanged. (This immediate "save" guards against accidentally altering your original file.)

3. Add a new column for Specimen Code as the new first column in the Specimen Import file.
4. Sort the rows (Specimen records), if you wish, into the order (if any) that you would like the Specimen Codes to represent.
5. Fill the Specimen Code column with unique values (see comments, pp. 97–101, on Record Codes).

Specimen Code	Collected By	Date Collected	Locality Name	State/Prov	Country	Lat	Long	Genus	Specific Name
HMSB001	C. Darwin	18 Sep 1832	San Cristóbal I.	Galápagos	Ecuador	-0.83	-89.4	Geospiza	fortis
HMSB002	R. Fitzroy	18 Sep 1832	San Cristóbal I.	Galápagos	Ecuador	-0.83	-89.4	Geospiza	fortis
HMSB003	R. Fitzroy	18 Sep 1832	San Cristóbal I.	Galápagos	Ecuador	-0.83	-89.4	Geospiza	fuliginosa
HMSB004	C. Darwin	18 Sep 1832	San Cristóbal I.	Galápagos	Ecuador	-0.83	-89.4	Geospiza	scandens
HMSB005	C. Darwin	24 Sep 1832	Floreana I.	Galápagos	Ecuador	-1.28	-90.4	Geospiza	scandens
HMSB006	C. Darwin	24 Sep 1832	Floreana I.	Galápagos	Ecuador	-1.28	-90.4	Camarhynchus	psittacula
HMSB007	C. Darwin	24 Sep 1832	Floreana I.	Galápagos	Ecuador	-1.28	-90.4	Camarhynchus	pauper
HMSB008	C. Darwin	24 Sep 1832	Floreana I.	Galápagos	Ecuador	-1.28	-90.4	Camarhynchus	parvulus
HMSB009	R. Fitzroy	24 Sep 1832	Floreana I.	Galápagos	Ecuador	-1.28	-90.4	Certhidea	olivacea
HMSB010	C. Darwin	30 Sep 1832	Fernandina I.	Galápagos	Ecuador	-0.43	-91.5	Geospiza	magnirostris
HMSB011	C. Darwin	30 Sep 1832	Fernandina I.	Galápagos	Ecuador	-0.43	-91.5	Geospiza	fortis
HMSB012	C. Darwin	30 Sep 1832	Fernandina I.	Galápagos	Ecuador	-0.43	-91.5	Geospiza	difficilis
HMSB013	R. Fitzroy	30 Sep 1832	Fernandina I.	Galápagos	Ecuador	-0.43	-91.5	Cactospiza	heliobates
HMSB014	C. Darwin	9 Oct 1832	Santiago I.	Galápagos	Ecuador	-0.23	-90.8	Cactospiza	pallida
HMSB015	C. Darwin	9 Oct 1832	Santiago I.	Galápagos	Ecuador	-0.23	-90.8	Certhidea	olivacea
HMSB016	C. Darwin	9 Oct 1832	Santiago I.	Galápagos	Ecuador	-0.23	-90.8	Geospiza	difficilis
HMSB017	C. Darwin	9 Oct 1832	Santiago I.	Galápagos	Ecuador	-0.23	-90.8	Camarhynchus	parvulus

6. Add a blank column for Species Code as the new second column in the Specimen Import file.
7. Sort the rows (Specimen records) in the text file by Genus and Specific Name (see the next table below).
8. For each conspecific series of rows (each set of conspecific specimen records), create and enter a value for the Species Code (see comments, pp. 97–101, on Record Codes.) You can use range-filling commands or mouse actions to enter the values in your spreadsheet.

Specimen Code	Species Code	Collected By	Date Collected	Locality Name	State/Prov	Country	Lat	Long	Genus	Specific Name
HMSB013	cactheli	R. Fitzroy	30 Sep 1832	Fernandina I.	Galápagos	Ecuador	-0.43	-91.5	Cactospiza	heliobates
HMSB014	cactpall	C. Darwin	9 Oct 1832	Santiago I.	Galápagos	Ecuador	-0.23	-90.8	Cactospiza	pallida
HMSB008	camaparv	C. Darwin	24 Sep 1832	Floreana I.	Galápagos	Ecuador	-1.28	-90.4	Camarhynchus	parvulus
HMSB017	camaparv	C. Darwin	9 Oct 1832	Santiago I.	Galápagos	Ecuador	-0.23	-90.8	Camarhynchus	parvulus
HMSB007	camapaup	C. Darwin	24 Sep 1832	Floreana I.	Galápagos	Ecuador	-1.28	-90.4	Camarhynchus	pauper
HMSB006	camapsit	C. Darwin	24 Sep 1832	Floreana I.	Galápagos	Ecuador	-1.28	-90.4	Camarhynchus	psittacula
HMSB009	certoliv	R. Fitzroy	24 Sep 1832	Floreana I.	Galápagos	Ecuador	-1.28	-90.4	Certhidea	olivacea
HMSB015	certoliv	C. Darwin	9 Oct 1832	Santiago I.	Galápagos	Ecuador	-0.23	-90.8	Certhidea	olivacea
HMSB012	geosdiff	C. Darwin	30 Sep 1832	Fernandina I.	Galápagos	Ecuador	-0.43	-91.5	Geospiza	difficilis
HMSB016	geosdiff	C. Darwin	9 Oct 1832	Santiago I.	Galápagos	Ecuador	-0.23	-90.8	Geospiza	difficilis
HMSB001	geosfort	C. Darwin	18 Sep 1832	San Cristóbal I.	Galápagos	Ecuador	-0.83	-89.4	Geospiza	fortis
HMSB002	geosfort	R. Fitzroy	18 Sep 1832	San Cristóbal I.	Galápagos	Ecuador	-0.83	-89.4	Geospiza	fortis
HMSB011	geosfort	C. Darwin	30 Sep 1832	Fernandina I.	Galápagos	Ecuador	-0.43	-91.5	Geospiza	fortis
HMSB003	geosfuli	R. Fitzroy	18 Sep 1832	San Cristóbal I.	Galápagos	Ecuador	-0.83	-89.4	Geospiza	fuliginosa
HMSB010	geosmagn	C. Darwin	30 Sep 1832	Fernandina I.	Galápagos	Ecuador	-0.43	-91.5	Geospiza	magnirostris
HMSB004	geosscan	C. Darwin	18 Sep 1832	San Cristóbal I.	Galápagos	Ecuador	-0.83	-89.4	Geospiza	scandens
HMSB005	geosscan	C. Darwin	24 Sep 1832	Floreana I.	Galápagos	Ecuador	-1.28	-90.4	Geospiza	scandens

9. Add a column for Collection Code as the new third column in the Specimen Import file.

10. Each Collection record specifies a particular "collecting-event." In the example, each combination of collector (Collected By—Darwin or Fitzroy), Date Collected, and Locality Name requires a unique Collection Code. Sort the rows (Specimen records) in the text file by collecting-event (see the next table below).

11. For each series of rows (each set of Specimen records) from the same collecting-event, create and enter a value for the Collection code.

Specimen Code	Species Code	Coll. Code	Collected By	Date Collected	Locality Name	State/Prov	Country	Lat	Long	Genus	Specific Name
HMSB001	geosfort	CD001	C. Darwin	18 Sep 1832	San Cristóbal I.	Galápagos	Ecuador	-0.83	-89.4	Geospiza	fortis
HMSB004	geosscan	CD001	C. Darwin	18 Sep 1832	San Cristóbal I.	Galápagos	Ecuador	-0.83	-89.4	Geospiza	scandens
HMSB008	camaparv	CD002	C. Darwin	24 Sep 1832	Floreana I.	Galápagos	Ecuador	-1.28	-90.4	Camarhynchus	parvulus
HMSB007	camapaup	CD002	C. Darwin	24 Sep 1832	Floreana I.	Galápagos	Ecuador	-1.28	-90.4	Camarhynchus	pauper
HMSB006	camapsit	CD002	C. Darwin	24 Sep 1832	Floreana I.	Galápagos	Ecuador	-1.28	-90.4	Camarhynchus	psittacula
HMSB005	geosscan	CD002	C. Darwin	24 Sep 1832	Floreana I.	Galápagos	Ecuador	-1.28	-90.4	Geospiza	scandens

Specimen Code	Species Code	Coll. Code	Collected By	Date Collected	Locality Name	State/Prov	Country	Lat	Long	Genus	Specific Name
HMSB012	geosdiff	CD003	C. Darwin	30 Sep 1832	Fernandina I.	Galápagos	Ecuador	-0.43	-91.5	Geospiza	difficilis
HMSB011	geosfort	CD003	C. Darwin	30 Sep 1832	Fernandina I.	Galápagos	Ecuador	-0.43	-91.5	Geospiza	fortis
HMSB010	geosmagn	CD003	C. Darwin	30 Sep 1832	Fernandina I.	Galápagos	Ecuador	-0.43	-91.5	Geospiza	magnirostris
HMSB014	cactpall	CD004	C. Darwin	9 Oct 1832	Santiago I.	Galápagos	Ecuador	-0.23	-90.8	Cactospiza	pallida
HMSB017	camaparv	CD004	C. Darwin	9 Oct 1832	Santiago I.	Galápagos	Ecuador	-0.23	-90.8	Camarhynchus	parvulus
HMSB015	certoliv	CD004	C. Darwin	9 Oct 1832	Santiago I.	Galápagos	Ecuador	-0.23	-90.8	Certhidea	olivacea
HMSB016	geosdiff	CD004	C. Darwin	9 Oct 1832	Santiago I.	Galápagos	Ecuador	-0.23	-90.8	Geospiza	difficilis
HMSB002	geosfort	RF001	R. Fitzroy	18 Sep 1832	San Cristóbal I.	Galápagos	Ecuador	-0.83	-89.4	Geospiza	fortis
HMSB003	geosfuli	RF001	R. Fitzroy	18 Sep 1832	San Cristóbal I.	Galápagos	Ecuador	-0.83	-89.4	Geospiza	fuliginosa
HMSB009	certoliv	RF002	R. Fitzroy	24 Sep 1832	Floreana I.	Galápagos	Ecuador	-1.28	-90.4	Certhidea	olivacea
HMSB013	cactheli	RF003	R. Fitzroy	30 Sep 1832	Fernandina I.	Galápagos	Ecuador	-0.43	-91.5	Cactospiza	heliobates

12. Save the Specimen Import file (as Text), *but do not close it*. The Specimen Import file is now complete.

13. With pencil and paper or in another file, make a list of each column in the Specimen Import file in left-to-right order, up to and including the last column (field) you will import into Biota's Specimen table. Be sure to include in your list any intervening columns that you do *not* intend to import, since you will need to tell Biota to skip them. You can ignore any columns to the right of the last column to be imported into the Specimen table.

14. Save the Specimen Import file (as Text) *twice*, first with the name Collection Import, then as Species Import. The Species Import file will remain open.

15. In the Species Import file, delete the Specimen Code column, the Collection Code Column, and all other columns destined for the Specimen, Collection, and Locality tables—*but do not delete the Species Code column*.

16. Sort by rows on the Species Code column.

Species Code	Genus	Specific Name
geosdiff	Geospiza	difficilis
geosdiff	Geospiza	difficilis
geosfort	Geospiza	fortis
geosfort	Geospiza	fortis
geosfort	Geospiza	fortis
geosfuli	Geospiza	fuliginosa

cactheli	Cactospiza	heliobates
geosmagn	Geospiza	magnirostris
certoliv	Certhidea	olivacea
certoliv	Certhidea	olivacea
cactpall	Cactospiza	pallida
camaparv	Camarhynchus	parvulus
camaparv	Camarhynchus	parvulus
camapaup	Camarhynchus	pauper
camapsit	Camarhynchus	psittacula
geosscan	Geospiza	scandens
geosscan	Geospiza	scandens

17. Unless you had only one specimen per species, there are now duplicated values in the Species Code column. Because the Species Code column will become the Key field for the Species table, the duplicate rows (shown shaded in the example above) must be deleted.

18. Manually delete the entire *row* for each duplicated Species code, leaving just one record with each distinct Species Code.

Species Code	Genus	Specific Name
geosdiff	Geospiza	difficilis
geosfort	Geospiza	fortis
geosfuli	Geospiza	fuliginosa
cactheli	Cactospiza	heliobates
geosmagn	Geospiza	magnirostris
certoliv	Certhidea	olivacea
cactpall	Cactospiza	pallida
camaparv	Camarhynchus	parvulus
camapaup	Camarhynchus	pauper
camapsit	Camarhynchus	psittacula
geosscan	Geospiza	scandens

19. Save the file (as Text), *but do not close it*. The Species Import file is now complete.

20. Follow steps 5 through 16 in the section entitled "Step by Step: Preparing Text Files for Import from a Taxonomic Flatfile" (pp. 508–512) above, to complete the text import files for Genus, Family, Order, and other higher taxon tables.

21. Close the last taxon file when it is complete.

22. Now it is time to create the text files you will use to import data into Biota's Collection and Locality tables. Open the file you saved in step 14, named Collection Import.

Specimen Code	Species Code	Coll. Code	Collected By	Date Collected	Locality Name	State/ Prov	Country	Lat	Long	Genus	Specific Name
HMSB001	geosfort	CD001	C. Darwin	18 Sep 1832	San Cristóbal I.	Galápagos	Ecuador	-0.83	-89.4	Geospiza	fortis
HMSB004	geosscan	CD001	C. Darwin	18 Sep 1832	San Cristóbal I.	Galápagos	Ecuador	-0.83	-89.4	Geospiza	scandens
HMSB008	camaparv	CD002	C. Darwin	24 Sep 1832	Floreana I.		Galápagos	Ecuador	-1.28	-90.4	Camarhynchus	parvulus
HMSB007	camapaup	CD002	C. Darwin	24 Sep 1832	Floreana I.		Galápagos	Ecuador	-1.28	-90.4	Camarhynchus	pauper
HMSB006	camapsit	CD002	C. Darwin	24 Sep 1832	Floreana I.		Galápagos	Ecuador	-1.28	-90.4	Camarhynchus	psittacula
HMSB005	geosscan	CD002	C. Darwin	24 Sep 1832	Floreana I.		Galápagos	Ecuador	-1.28	-90.4	Geospiza	scandens
HMSB012	geosdiff	CD003	C. Darwin	30 Sep 1832	Fernandina I.		Galápagos	Ecuador	-0.43	-91.5	Geospiza	difficilis
HMSB011	geosfort	CD003	C. Darwin	30 Sep 1832	Fernandina I.		Galápagos	Ecuador	-0.43	-91.5	Geospiza	fortis
HMSB010	geosmagn	CD003	C. Darwin	30 Sep 1832	Fernandina I.		Galápagos	Ecuador	-0.43	-91.5	Geospiza	magnirostris
HMSB014	cactpall	CD004	C. Darwin	9 Oct 1832	Santiago I.		Galápagos	Ecuador	-0.23	-90.8	Cactospiza	pallida
HMSB017	camaparv	CD004	C. Darwin	9 Oct 1832	Santiago I.		Galápagos	Ecuador	-0.23	-90.8	Camarhynchus	parvulus
HMSB015	certoliv	CD004	C. Darwin	9 Oct 1832	Santiago I.		Galápagos	Ecuador	-0.23	-90.8	Certhidea	olivacea
HMSB016	geosdiff	CD004	C. Darwin	9 Oct 1832	Santiago I.		Galápagos	Ecuador	-0.23	-90.8	Geospiza	difficilis
HMSB002	geosfort	RF001	R. Fitzroy	18 Sep 1832	San Cristóbal I.	Galápagos	Ecuador	-0.83	-89.4	Geospiza	fortis
HMSB003	geosfuli	RF001	R. Fitzroy	18 Sep 1832	San Cristóbal I.	Galápagos	Ecuador	-0.83	-89.4	Geospiza	fuliginosa
HMSB009	certoliv	RF002	R. Fitzroy	24 Sep 1832	Floreana I.		Galápagos	Ecuador	-1.28	-90.4	Certhidea	olivacea
HMSB013	cactheli	RF003	R. Fitzroy	30 Sep 1832	Fernandina I.		Galápagos	Ecuador	-0.43	-91.5	Cactospiza	heliobates

23. In the Collection Import file, delete the Specimen Code column, the Species Code Column, and all other columns that were destined for the Species, Genus, and higher taxon tables—*but do not delete the Collection Code column.* Columns to be deleted are shown shaded in the example above.

24. Sort by rows on the Collection Code column, if not already in order.

Coll. Code	Collected By	Date Collected	Locality Name	State/ Prov	Country	Lat	Long
CD001	C. Darwin	18 Sep 1832	San Cristóbal I.	Galápagos	Ecuador	-0.83	-89.4
CD001	C. Darwin	18 Sep 1832	San Cristóbal I.	Galápagos	Ecuador	-0.83	-89.4
CD002	C. Darwin	24 Sep 1832	Floreana I.		Galápagos	Ecuador	-1.28	-90.4
CD002	C. Darwin	24 Sep 1832	Floreana I.		Galápagos	Ecuador	-1.28	-90.4
CD002	C. Darwin	24 Sep 1832	Floreana I.		Galápagos	Ecuador	-1.28	-90.4
CD002	C. Darwin	24 Sep 1832	Floreana I.		Galápagos	Ecuador	-1.28	-90.4
CD003	C. Darwin	30 Sep 1832	Fernandina I.		Galápagos	Ecuador	-0.43	-91.5

CD003	C. Darwin	30 Sep 1832	Fernandina I.	Galápagos	Ecuador	-0.43	-91.5
CD003	C. Darwin	30 Sep 1832	Fernandina I.	Galápagos	Ecuador	-0.43	-91.5
CD004	C. Darwin	9 Oct 1832	Santiago I.	Galápagos	Ecuador	-0.23	-90.8
CD004	C. Darwin	9 Oct 1832	Santiago I.	Galápagos	Ecuador	-0.23	-90.8
CD004	C. Darwin	9 Oct 1832	Santiago I.	Galápagos	Ecuador	-0.23	-90.8
CD004	C. Darwin	9 Oct 1832	Santiago I.	Galápagos	Ecuador	-0.23	-90.8
RF001	R. Fitzroy	18 Sep 1832	San Cristóbal I.	Galápagos	Ecuador	-0.83	-89.4
RF001	R. Fitzroy	18 Sep 1832	San Cristóbal I.	Galápagos	Ecuador	-0.83	-89.4
RF002	R. Fitzroy	24 Sep 1832	Floreana I.	Galápagos	Ecuador	-1.28	-90.4
RF003	R. Fitzroy	30 Sep 1832	Fernandina I.	Galápagos	Ecuador	-0.43	-91.5

25. Unless you had only one specimen per collecting event, there are now duplicated values in the Collection Code column, which will become the Key field for the Collection table. The duplicate rows (shown shaded in the example above) must be deleted.

26. Manually delete the entire *row* for each duplicated Collection code, leaving just one record with each distinct Collection Code.

Coll. Code	Collected By	Date Collected	Locality Name	State/Prov	Country	Lat	Long
CD001	C. Darwin	18 Sep 1832	San Cristóbal I.	Galápagos	Ecuador	-0.83	-89.4
CD002	C. Darwin	24 Sep 1832	Floreana I.	Galápagos	Ecuador	-1.28	-90.4
CD003	C. Darwin	30 Sep 1832	Fernandina I.	Galápagos	Ecuador	-0.43	-91.5
CD003	C. Darwin	30 Sep 1832	Fernandina I.	Galápagos	Ecuador	-0.43	-91.5
CD004	C. Darwin	9 Oct 1832	Santiago I.	Galápagos	Ecuador	-0.23	-90.8
RF001	R. Fitzroy	18 Sep 1832	San Cristóbal I.	Galápagos	Ecuador	-0.83	-89.4
RF002	R. Fitzroy	24 Sep 1832	Floreana I.	Galápagos	Ecuador	-1.28	-90.4
RF003	R. Fitzroy	30 Sep 1832	Fernandina I.	Galápagos	Ecuador	-0.43	-91.5

27. Add a blank column for Locality Code as the new second column in the Specimen Import file.

28. Sort the rows (Collection records) in the text file by Locality Name (or whatever field or fields provide unique information on geographic localities).

29. For each series of rows from the same locality, create and enter a value for the Locality code. (You can use range-filling commands or mouse actions to do this in your spreadsheet.)

Coll. Code	Locality Code	Collected By	Date Collected	Locality Name	State/Prov	Country	Lat	Long
CD001	cristobal	C. Darwin	18 Sep 1832	San Cristóbal I.	Galápagos	Ecuador	-0.83	-89.4
RF001	cristobal	R. Fitzroy	18 Sep 1832	San Cristóbal I.	Galápagos	Ecuador	-0.83	-89.4
CD003	fernandina	C. Darwin	30 Sep 1832	Fernandina I.	Galápagos	Ecuador	-0.43	-91.5
CD003	fernandina	C. Darwin	30 Sep 1832	Fernandina I.	Galápagos	Ecuador	-0.43	-91.5
RF003	fernandina	R. Fitzroy	30 Sep 1832	Fernandina I.	Galápagos	Ecuador	-0.43	-91.5
CD002	floreana	C. Darwin	24 Sep 1832	Floreana I.	Galápagos	Ecuador	-1.28	-90.4
RF002	floreana	R. Fitzroy	24 Sep 1832	Floreana I.	Galápagos	Ecuador	-1.28	-90.4
CD004	santiago	C. Darwin	9 Oct 1832	Santiago I.	Galápagos	Ecuador	-0.23	-90.8

30. Save the Collection Import file (as Text), *but do not close it*. The Collection Import file is now complete.

31. With pencil and paper or in another file, make a list of each column in the Collection Import file, in left-to-right order, up to and including the last column (field) you will import into Biota's Collection table. Be sure to include in your list any intervening columns that you do *not* intend to import, since you will need to tell Biota to skip them. You can ignore any columns to the right of the last column to be imported into the Collection table.

32. Save the Collection Import file (as Text), with the name Locality Import.

33. In the Locality Import file, delete the Collection Code column, the and all other columns that were destined for Collection table—*but do not delete the Locality Code column*.

34. Sort by rows on the Locality Code column, if not already in order.

Locality Code	Locality Name	State/Prov	Country	Lat	Long
cristobal	San Cristóbal I.	Galápagos	Ecuador	-0.83	-89.4
cristobal	San Cristóbal I.	Galápagos	Ecuador	-0.83	-89.4
fernandina	Fernandina I.	Galápagos	Ecuador	-0.43	-91.5
fernandina	Fernandina I.	Galápagos	Ecuador	-0.43	-91.5
fernandina	Fernandina I.	Galápagos	Ecuador	-0.43	-91.5
floreana	Floreana I.	Galápagos	Ecuador	-1.28	-90.4
floreana	Floreana I.	Galápagos	Ecuador	-1.28	-90.4
santiago	Santiago I.	Galápagos	Ecuador	-0.23	-90.8

35. Unless you had only one collecting event per, there are now duplicated values in the Locality Code column, which will become the Key field for the Locality table. The duplicate rows (shown shaded in the example above) must be deleted.

36. Manually delete the entire *row* for each duplicated Locality code, leaving just one record with each distinct Locality Code.

Locality Code	Locality Name	State/Prov	Country	Lat	Long
cristobal	San Cristóbal I.	Galápagos	Ecuador	-0.83	-89.4
fernandina	Fernandina I.	Galápagos	Ecuador	-0.43	-91.5
floreana	Floreana I.	Galápagos	Ecuador	-1.28	-90.4
santiago	Santiago I.	Galápagos	Ecuador	-0.23	-90.8

37. If your text file includes latitude and longitude data, be sure that these values are in decimal degrees. See pp. 115–117, 489.

38. Save the Locality Import file (as Text), *but do not close it*. The Locality Import file is now complete.

39. With pencil and paper or in another file, make a list of each column in the Locality Import file in left-to-right order, up to and including the last column (field) you will import into Biota's Locality table. Be sure to include in your list any intervening columns that you do *not* intend to import, since you will need to tell Biota to skip them. You can ignore any columns to the right of the last column to be imported into the Locality table.

40. Close the Locality Import file.

41. The text files are now ready for importing into Biota. Go to the section "Set Up and Launch the Import or Update," pp. 490–495.

NOTE: If you follow the strategy outlined above, a text file destined for a Biota table lower in the hierarchy will generally include columns destined for Biota tables higher in the hierarchy. For example, the Species Import file (step 3, above) includes columns for Family (destined for the Genus and Family tables), and Order (destined for the Family and Order tables). You can deal with these extraneous columns in any of three ways. The first is the "cleanest" way to deal with extraneous columns, but consider the other two options, too.

- **Reopen each text file** you made in the procedure above and delete each of the extraneous columns.

- **If the extraneous columns are all located to the right of the columns to be imported,** for each file just ignore the columns when you set up the Import Editor. Once Biota finds a value for each column you list in the Import Editor, it goes on to the next record (the next text file row).

- **If the extraneous columns are to the left of the ones you need to import,** or interspersed with them, you can use the Skip button in the Import Editor to tell Biota to skip over them (p. 492).

Appendix A Biota Structure

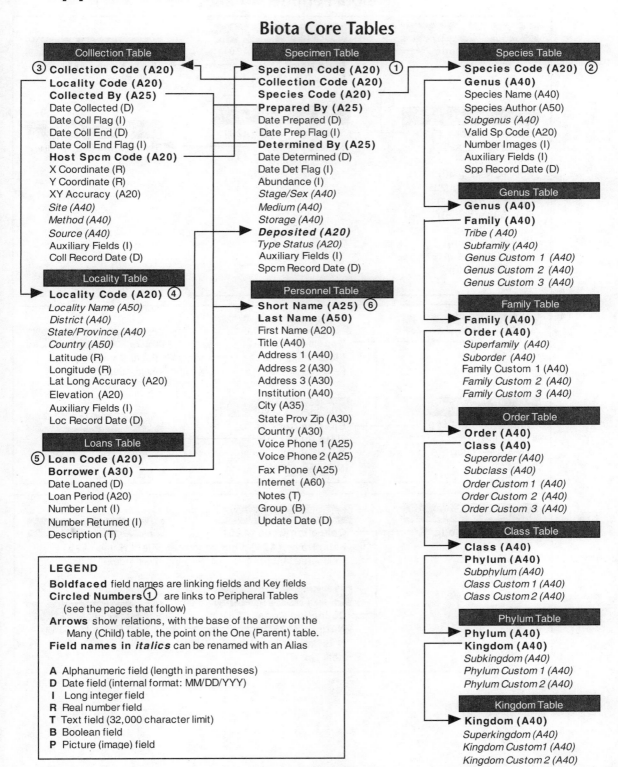

Biota Peripheral Tables

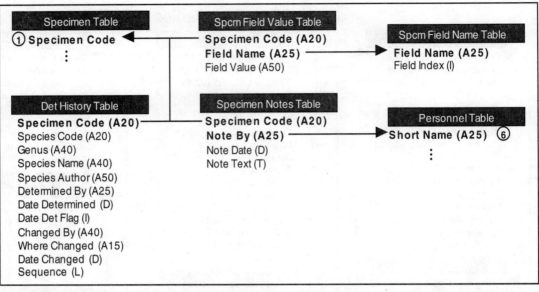
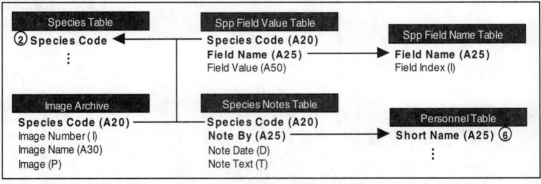
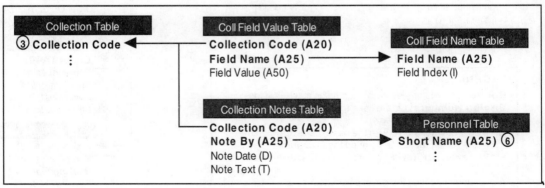

Appendix A

Biota Peripheral Tables

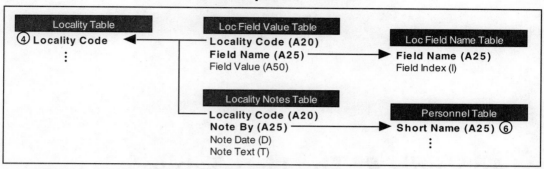

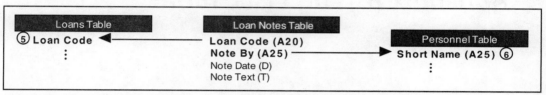

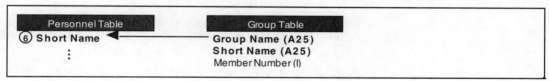

Appendix B The Lists Table

The Lists table in Biota has two special purposes. It keeps track of Core Field Aliases that you define (pp. 323–326) and it manages the Field Value Choice Lists that you create (pp. 315–322).

For most users, the only reason to import records into the Lists table is to transfer Lists records from one Biota Data File to another (to a new, blank one, for example). The procedure for exporting Lists records from one Biota Data File and importing them into another is detailed in Chapter 15, in the section "Transferring Choice Lists to a Different Biota Data File," pp. 319–322.

This appendix details the fields, default values, and required records for the Lists table. You should study this appendix if you intend to create the required records for import into the Lists table using another application (e.g., a spreadsheet application) or if you want to modify exported Lists records before importing them again.

Record Structure of the Lists Table

Unlike other Biota tables (Appendix A), each record in the Lists table has a special function.

- **Lists records must have consecutive integer values for the field [Lists] Record Number.** The first three records have special purposes and must carry the correct number in the Record Number field.

- **Record 1 of the Lists table specifies values for "Long" Field Aliases** (pp. 324–325). Biota assigns default values for fields in this record when you create a new Data File (pp. 9–10). Any Aliases you later assign replace the defaults.

- **Record 2 specifies values for "Short" Field Aliases** (pp. 324–325). Biota assigns default values for fields in this record when you create a new Data File (pp. 9–10). Any Aliases you later assign replace the defaults.

- **Record 3 specifies which Core Fields have Entry Choice Lists currently enabled.** A value of 1 means the List is enabled, a value of 0 means that the List is currently disabled (p. 316). Biota assigns a default value of 0 to fields in this record when you create a new Data File (pp. 9–10).

- **Records 4 and above contain the values you have assigned to Entry Choice Lists for Core Fields.** Biota creates these records in the Lists table automatically, as needed—a new Data File has none. For a field that has three entry choices, there will be non-blank values in Records 4, 5, and 6 of the Lists table. For a field with five choices, Lists table Records 4–8 will have values. Biota automatically deletes Lists table records (Record Number 4 and higher only) when there are no longer any Entry Choice List items in one or more of these records.

Field Structure and Field Lengths in the Lists Table

The diagrams below show the fields of the Lists table. All fields are alphanumeric with the exception of Record Number, which is an integer field.

Each alphanumeric field has the same character length as the field with the same name in Core and Peripheral tables shown in Appendix A.

Lists					
SpecimenCode	A	SpeciesCode	A	Subclass	A
StageSex	A	SpeciesName	A	OrderCustom1	A
Medium	A	SpeciesAuthor	A	OrderCustom2	A
Storage	A	Subgenus	A	OrderCustom3	A
Deposited	A	ImageName	A	Subphylum	A
CollectionCode	A	Tribe	A	ClassCustom1	A
Site	A	Subfamily	A	ClassCustom2	A
Source	A	GenusCustom1	A	Subkingdom	A
Method	A	GenusCustom2	A	PhylumCustom1	A
CollXYAccuracy	A	GenusCustom3	A	PhylumCustom2	A
LocalityCode	A	Superfamily	A	Superkingdom	A
LocalityName	A	Suborder	A	KingdomCustom1	A
District	A	FamilyCustom1	A	KingdomCustom2	A
StateProvince	A	FamilyCustom2	A	RecordNumber	I
Country	A	FamilyCustom3	A	TypeStatus	A
LocXYAccuracy	A	Superorder	A	Elevation	A

Appendix C Setting Preferences

The illustration below directs you to the options you can set using the Preferences screen. You can display this screen by choosing Preferences from the Special menu.

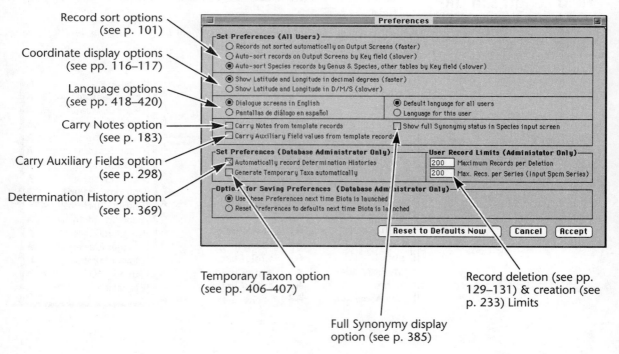

Record sort options (see p. 101)
Coordinate display options (see pp. 116–117)
Language options (see pp. 418–420)
Carry Notes option (see p. 183)
Carry Auxiliary Fields option (see p. 298)
Determination History option (see p. 369)
Temporary Taxon option (see pp. 406–407)
Full Synonymy display option (see p. 385)
Record deletion (see pp. 129–131) & creation (see p. 233) Limits

Appendix D Troubleshooting and Support

If you are having a problem with Biota, first look through the "Troubleshooting" section below for help. If you cannot find the answer there, please consult the Index and Table of Contents. If you are still unable to solve the problem, see the "Support" section of this appendix (pp. 535–537).

Troubleshooting

This section is organized by the following subheadings: Problems with Data Files, Problems with Passwords, Problems with Windows and Records, Problems with Web Pages, English and Spanish Dialogues, Other Problems, and Error Messages. Where a problem might reasonably be placed in two different to categories, you will find it listed in both.

Problems with Data Files

- **I can't get Biota to create a *new, empty* Data File. It keeps on opening the same existing file.** To create a new, empty Biota Data File, *press and hold the* OPTION *key* (Macintosh) *or* ALT *key* (Windows) while launching Biota. (If Biota is already launched, quit Biota first, then launch it again.)

  If the user password system has been activated, you will need to press and hold the OPTION key (Macintosh) or ALT key (Windows) while clicking the Connect button in the Password window.

  Be sure to hold the key down firmly until the file navigation window for your operating system appears. Click the New… button (not the Open button) to create a new Data File. Biota will take a few moments to do this, so be patient. See pp. 9–10.

- **I can't get Biota to open a *different, existing* Data File. It keeps on opening the same old file.** To choose among existing Biota Data Files, *press and hold the* OPTION *key* (Macintosh) *or* ALT *key* (Windows) while launching Biota. (If Biota is already launched, quit Biota first, then launch it again.)

If the user password system has been activated, you will need to press and hold the OPTION key (Macintosh) or ALT key (Windows) while clicking the Connect button in the Password window.

Be sure to hold the key down firmly until the file navigation window for your operating system appears, then find the Biota Data File you wish to open and double-click it (or select it and click the Open button). See pp. 8–9.

- **I deleted hundreds of records from my Data File, but the file is still the same size.** When you delete records, Biota reuses the "holes" for new records. Meanwhile, the Data File stays the same size unless you compact it using 4D Tools. See Appendix F.

- **I can't find a Save command to save the changes I have made to my Data File.** Biota saves automatically every five minutes and whenever you quit Biota. If you want to force Biota to save changes immediately, you can press the keys ⌘(COMMAND)+W (Macintosh) or CTRL+W (Windows) at any time. See pp. 10–11.

- **What is the small window, shown below, that appears in the lower left corner of my screen periodically while I am running Biota?** This window is displayed while Biota is saving changes to your Data File (see the previous question).

- **When I choose "To Load a Biota Data File" or "About the Current Data File" from Biota's File menu, I get an error message reading "The external procedure cannot be executed."**

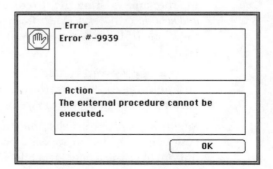

This problem is addressed in the section "Error Messages," p. 534.

- **When I try to open my Data File, I get an error message that the file is damaged and cannot be opened.** This problem is addressed in the section "Error Messages," p. 534.

- **When I try to create a new record, I get an error screen reading "No more room to save the record."** This problem is addressed in the section "Error Messages," p. 534.

Problems with Passwords

- **I can't get my user password to work.** Check each of the following.
    - ◊ Passwords are case sensitive. If you create a password or are assigned a password that includes uppercase and/or lowercase letters, you must always reproduce the case of each letter precisely.
    - ◊ Be sure the CAPS LOCK key is not on.
    - ◊ Make sure you are entering the correct User Name in the Password window. (User Names are not case sensitive.)
    - ◊ Make sure you are opening the correct copy of Biota. User passwords are saved with the Biota application file, not the Data File (pp. 415–424).

    If none of these suggestions solve the problem, ask the database Administrator to assign you a new password (pp. 415–420). If you *are* the Administrator, install a fresh copy of Biota and reestablish the user Password system (pp. 415–420) or contact Biota Support for help (see pp. 535–537).

- **I have forgotten my user password.** Ask the database Administrator to assign you a new password (pp. 415–420). If you *are* the Administrator, install a fresh copy of Biota and reestablish the user Password system (pp. 415–420) or contact Biota Support for help (see pp. 535–537).

- **When I try to open a Data File, Biota asks for a Data File password.**
    - ◊ Make sure you are launching the right copy of Biota and the right Data File. When the Data File password link is activated for the Data File, a copy of the Data File password is stored in the Biota application file. The copy in the Biota application file must precisely

match the copy of the password in the Data File. See pp. 430–431.

◊ Ask the database Administrator to open the Data File or supply the password; the Data File Password will not be requested again until a new copy or a new version of Biota is installed. See pp. 425–431.

◊ If you *are* the Administrator, and you have forgotten or lost the Data File Password contact, Biota Support for help (see pp. 535–537).

Problems with Biota Windows and Records

- **How can I find out which windows are currently open?** Use Windowshade (Macintosh, System 7.5 and later) or the Minimize button (Windows) to minimize open windows that you are not currently using but do not want to close. There is no menu list of open windows.

- **I get a cursor that looks like a page behind a another page, and I can't get a window to respond:**

  Another window requires attention before you can proceed. (You clicked in an inactive window.) To return to the active window, press any key on the keyboard. (The keystroke is not recorded.)

- **In input screens, the Carry and Delete buttons are dimmed, and I can't enter or change any data.** The User Name you signed on with when you launched Biota is not authorized to create, edit, or delete records. Ask your database Administrator to change your access privileges if you need to be able to carry out these actions. See pp. 415–420.

- **What is the small window, shown below, that appears in the lower left corner of my screen periodically while I am running Biota?** This window is displayed while Biota is automatically saving changes to your Data File (see pp. 10–11).

- **When I try to open a record in an input screen, I get a message that "The record is already being edited."** See the discussion of *record locking* in the "Error Messages" section, p. 533.

  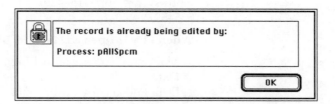

1878
Sylvester de Brown
Mycosyphon Institution
Flore, New York 36:142

WLThomas@gateway.net

Problems with Web Pages

- **When I open a Biota-produced Web page in a browser and click on a hyperlinked taxon name, nothing happens.** See below.

- **When I open a Biota-produced Web page in a browser, the wrong taxa are displayed.** When Biota creates Web page disk files, the file names are assigned automatically, and the new files are placed in the same folder as the Biota application. If a Web file of the same name already exists in the Biota folder, the new file is not created. *Be sure to move or delete all Web files from the Biota folder before you use the Create Web Pages command.* See p. 475.

- **When I open a Biota-produced Species Web page in a browser, the images are not displayed. Only the image placeholders appear.** The image files have probably not been converted to the GIF format. See p. 481.

English and Spanish Screens and Dialogues

- **Some of the screens and dialogues are in Spanish. I want them all in English.**
  - ◊ *Either:* The Language setting in the Preferences screen (Special menu) is set to Spanish. Change the setting to English. See pp. 418–420.
  - ◊ *Or:* If the user Password system has been enabled, your User Name may have been assigned a special startup procedure that sets the Language toggle to Spanish. Ask the database Administrator to change the startup procedure to specify English. See pp. 418–420.

- **Only some of the screens and dialogues are in Spanish. I want all screens and menus in Spanish.** When the Language toggle has been set to Spanish, most instructions, dialogues, and error messages are displayed in Spanish—with the exception of screens and dialogues accessible only to the Administrator, which are currently presented only in English. Menus and commonly used input and output screens are also in English only, since these are viewed repeatedly and are thus much more easily learned than instructions, dialogues, and error messages.

  If it would be helpful for future versions of Biota to include Spanish versions of particular screens or messages that are currently only in English, please contact Biota Support (see pp. 535–537).

Other Problems

- **When I try to display images, I get a "Quick Time" logo instead of an image.** The image has been compressed using Apple's QuickTime.
  - ◊ *Macintosh:* Install QuickTime (System 6.x) or activate the QuickTime extension in your System folder (System 7.0 and later). See p. 349.
  - ◊ *Windows:* The image cannot be displayed unless it is first decompressed and resaved without compression, using Biota for Macintosh on a Macintosh machine (see p. 349). The Data File must be converted (Appendix G) to move it between platforms.

- **When I load a Record Set Pointer file from the File menu, the wrong records (or no records) are displayed.**
  - *Either:* The Record Set Pointer file was saved for a different Biota table (see pp. 16–19).
  - *Or:* Some of the records referenced by the Record Set Pointer file have been deleted since the time the Record Set Pointer file was created.
- **In the Biota menu bar, menu names are cut off at the right side (Macintosh only).** This anomaly, which occurs only in certain Macintosh models and with certain monitor sizes, is caused by the Microsoft OLE extension, which apparently interferes with narrow menu font substitution in smaller screens. There is no harm in ignoring the truncated menu names, but if you want to correct the problem, disable the Microsoft OLE extension (for example, by using the Extensions Manager control panel) and restart your computer.
- **My PowerMacintosh sometimes crashes when I launch Biota.** Try turning off Virtual Memory in the Memory Control Panel (Apple menu).
- **My Macintosh quits Biota spontaneously and the operating system displays a message that an error of Type 1 or Type 15 has occurred.** You need to increase the RAM allocated by Biota. See p. xxiv.

Error Messages

Three kinds of error messages may appear: Biota error messages, operating system error messages, and 4th Dimension error messages.

- **Biota error messages** are treated fully in the appropriate chapters of this book. In other words, if you get an error message while you are importing records, look in Chapter 26 ("Importing Data") in the section "Displays and Error Messages During Record Importing or Updating." Biota error messages are also listed alphabetically in the index of this book, under "Error Messages."
- **Error messages displayed by your operating system** are not treated here. Please consult the documentation that came with your computer or operating system upgrade.
- **4th Dimension Error messages** regarding problems that you can correct yourself are listed in this section.

NOTES:

a. If you get a 4th Dimension error message *not* listed below, you may have discovered a bug in Biota's code. See examples and instructions in the "Bug Reports" section, pp. 535–537.

b. Although *table* is the standard term in relational database design, 4th Dimension error messages use the word *file* for internal relational database tables (Specimen, Locality, Personnel, etc.).

- **4D: Locked Record message.** Record locking, a feature of all database applications that allow multitasking, protects records from being edited in two places at once.

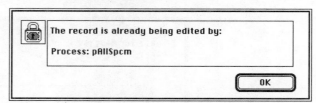

Each time you choose a command from a Biota menu on the menu bar a new *process* is launched. Certain buttons also launch separate processes (e.g., Print buttons and Child Records buttons).

If you see a message like the one above, *it means that you are trying to display, in an input screen, a record that is also the current record in another Biota process that still has an open window.*

◊ **If you are using BiotaApp (or Biota4D under 4th Dimension)** the other open window is on your own screen.

◊ **If you are running Biota under 4D Server,** the other open window may be on your screen or on the screen of another concurrent user. In the latter case, the first line of the Locked Record message will read "The Record is already being edited by *User Name*," where User Name is the Password screen sign-on name of the other user.

The name of the process that appears in the locked record error window is a shortened version of the menu command or the name of the button that launched the process. (In the example above, the process was launched by the All Specimens command in the Find menu.)

When you click the OK button in the Locked Record error window, the record you were attempting to display will nonetheless be displayed, *but in read-only mode.* If you need to edit the record, find the offending open window and close it. Then dismiss the record in the current process, reopen it (to set it in read-write mode), and edit it.

- **4D: "Out of memory" error message.** If you see the message below, quit Biota (you may have to wait for a process to be completed) and increase the memory allocated to Biota. See pp. xxiv.

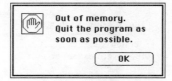

- **4D: When I choose "To Load a Biota Data File" or "About the Current Data File" from Biota's File menu, I get an error message reading "The external procedure cannot be executed."**

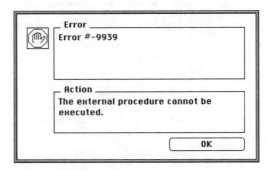

These two commands require the an *external procedure* to access the operating system and return the name of the Data File. You must install the folder Mac4DX (Macintosh) or Win4DX (Windows), which contains the external procedure, in the same folder with Biota for these commands to function correctly. You can find the Mac4DX or Win4DX file on one of the Biota distribution disks. You must relaunch Biota for the change to take effect. Meanwhile, no other tools or features of Biota are affected. See p. 11.

- **4D: When I try to open my Data File, I get an error message that the file is damaged and cannot be opened.** A disk error has occurred, possibly as a result of a power failure or other improper shutdown while data were being saved.

See Appendix F for instructions on using 4D Tools to attempt recovery on a *copy* of the damaged Data File. With the Data File backed up to another disk, you may also wish to use a generic disk repair utility to diagnose the disk for more general problems.

If all else fails, you will have to use the most current backup of your Data File instead. (Always back up regularly!) But first make a backup of the backup, for safety's sake—on a different disk, if possible.

- **4D: When I try to create a new record, I get an error screen reading "No more room to save the record."**

Either your hard disk is full, or (if you have segmented the Data File) the maximum size has been reached in all existing segments of the Data File. See "Adding a New Segment to an Existing Data File," in Appendix F, pp. 548–549.

Biota Support

Biota support starts with this book, which has deliberately been written in much more detail than many manuals, to provide you with full, rapid access to the information you need.

"How To" Questions

This book is the first place to go to find out if Biota can do something you want done. Please use the Table of Contents and the Index to see if you can find the answer you need.

Often, there is more than one way to do a task with Biota, but usually there is a fast way and a slow way. If some task seems to take longer than you think it should, try looking up the topic in the Table of Contents and or the Index for alternative approaches.

Visit the Biota Web Site for Help and Information

The Biota Web site, at http://viceroy.eeb.uconn.edu/biota, is the first place to turn for post-release updates to Biota documentation, suggestions for efficient use, warnings, workarounds for any bugs that have been discovered since Biota was released, and announcements of Biota upgrades and future plans.

Support by Electronic Mail

Registered Biota owners may request help with otherwise insoluble problems through electronic mail. Please do not telephone. If you do not have access to e-mail, please use the postal system.

E-mail address: colwell@uconnvm.uconn.edu

Postal address: Robert K. Colwell
Department of Ecology and Evolutionary Biology
University of Connecticut, U-42
Storrs, CT 06269-3042
U. S. A.

Bug Reports

Every effort has been made to beta-test Biota thoroughly, and all known programming bugs have been corrected. In any program this complex (more than 6000 objects, 900 procedures, and 180 layouts), however, users invariably prove that completely bug-free behavior is only a theoretical possibility.

If Biota does something you did not expect, first try consulting this book, including the section on "Troubleshooting" (pp. 527–535) earlier in this appendix, to see if the behavior is normal.

If something happens that is clearly not intended, you may have found a bug. If a 4th Dimension error window that is not discussed on pp. 527–535 appears (see below for two examples), you have almost certainly found a bug.

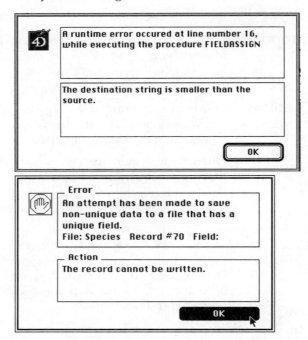

If you believe you have discovered a bug, please record the following:

1. **A description of the problem.** To be most useful the description of the problem should include a full account of the steps that led up to it. If at all possible, try making it happen again until you are sure what steps reliably produce the problem. If the problem cannot be reproduced, it cannot be fixed.

2. **Any data that were involved.** If the problem seems to concern a particular data record, please write down exactly what was in the record.

3. **Version of Biota.** From the File menu in Biota, select About Biota. Record the Version and Date of the copy of Biota that you are using, and indicate whether it is BiotaApp (the stand-alone version) or Biota4D (running under 4th Dimension or 4D Server).

4. **Your computer model and processor.**

5. **The operating system and version number you are using.** (Macintosh users: Select About This Macintosh from the Apple menu, while in the Finder.)

Send this information by e-mail (preferably) or by post to:

E-mail address: colwell@uconnvm.uconn.edu

Postal address: Robert K. Colwell
Department of Ecology and Evolutionary Biology
University of Connecticut, U-42
Storrs, CT 06269-3042
U. S. A.

Suggestions for Improving Biota

Biota's current tools and features evolved under the constant guidance of Biota users. Suggestions for improvement are warmly welcomed. Please use the contact information in the previous section.

Appendix E Using Biota4D with 4D Server or 4th Dimension

NOTES: Before reading this Appendix, please review "Biota Versions and Flavors," pp. xxiii–xxiv.

Macintosh: Running Biota4D (for Macintosh) under 4D Server or 4th Dimension on a Macintosh or PowerMacintosh is fully supported in Biota's initial release. Under 4D Server, you can run 4D Client on Macintosh or PowerMacintosh machines, or a mixture or both. The inclusion of Windows clients with a Macintosh server will be supported when Biota for Windows is released.

Windows: Until Biota for Windows is released, you cannot run Biota under either 4th Dimension for Windows or 4D Server for Windows. This Appendix, however, covers the Windows platform, in anticipation of the release of Biota for Windows.

About Biota4D

Unlike BiotaApp, which is an executable application that incorporates a licensed copy of 4D Engine, Biota4D is a program file that requires a separate database engine to run—either 4D Server or 4th Dimension. You must purchase your own copy of one of these commercial applications to run Biota4D (see "Sources and Prices for 4D Server and 4th Dimension," p. 540).

BiotaApp and Biota4D are compiled from precisely the same code and therefore share all the tools and features detailed in this book. Biota4D, however, allows you to run Biota in client-server mode on a local network or over the Internet.

Running Biota4D under 4D Server

If you need to provide direct, simultaneous access to a Biota Data File for two or more users, you can run Biota4D under 4D Server. The 4D Server software, Biota4D, and your Data File all reside on a *server* computer, which can be either a Macintosh (68xxx, PowerMacintosh, or a PowerMac Clone) or a Windows machine (once Biota4D for Windows is released).

The server computer must be networked with one or more *client* computers. Any network that can run the TCP/IP protocol, including local networks or even the Internet (see p. 484), can be used.

NOTE: In a pure Macintosh environment, you can use the AppleTalk protocol (ADSP) instead of TCP/IP, if you prefer, over a LocalTalk, Ethernet, or Token Ring network, or by using AppleTalk Remote Access with a modem. LocalTalk networks and modem access, however, are too slow for any intensive use.

The client machines connected to the server may include (if you wish) a mixture of Macintosh and Windows clients (but see the notes at the beginning of the appendix). When you purchase 4D Server, for either Macintosh or Windows platforms, you also receive two versions of the required client software: 4D Client for Macintosh and 4D Client for Windows.

The design of 4D Server/Client is a true client-server architecture (i. e., not simply file-sharing), with direct, "homogeneous" communications between server and client (no interapplication driver to slow things down) and local object caching on the client machine. The server performs all computationally intensive tasks and uses sophisticated data-caching techniques to minimize disk access delays.

The manuals you receive with your purchase of 4D Server provide full installation and operations instructions.

4D Server: Important Caveats and Recommendations

- **Do not expect to use the same machine to run both 4D Server and 4D Client.** Although the 4D Server/Client software permits you to set up 4D Server so that you can use 4D client on the server machine (while running 4D Client on other machines on the network as well, if you wish), communication between Client and Server on the same machine is many times slower than between Client on one machine and Server on another—however counterintuitive that may seem.

  On the other hand, the server machine need not be dedicated solely to running 4D Server. You can use other applications on the server machine while 4D Server operates in the background. Of course, sharing the server CPU in this way use will slow down access by clients, and it increases the risk that the server will crash, since something may go wrong with the other applications you are using.

- **Biota always runs faster in single-user mode**—either as BiotaApp or as Biota4D under 4th Dimension (see the next section)—than it runs under 4D Server, accessed only from Client machines. (This comparison assumes that the single-user machine and the server machine are either the same machine or equally fast machines.) *For this reason, you may wish to own either BiotaApp or 4th Dimension for intensive, single-user access to a Biota Data File* when it is not being used under 4D Server. (If you own 4th Dimension, you can use it with your copy of BiotaApp in single-user mode, as explained below.)

- **Running under 4D Client, Biota speeds up as you use it.** When you first launch 4D Client on a client machine, each "object" (for example, a layout) that the client requests is saved in the client machine in ".RES" or "REX" files (see the 4D Server documentation for technical details). The next time the object is needed by that client, whether in the current session or any future one, it is loaded from the local file, rather than over the network from the server.

  Further, a second request for the same records (or subsets of records), within a session, are filled faster than the first request, thanks to data caching on the server machine.

  NOTE: Whenever you install an updated version of Biota4D, you should delete all existing .RES and .REX files on each client machine, since new ones are created for the new version.

Running Biota4D under 4th Dimension (Single User)

If you own 4th Dimension, you can run Biota4D in single-user mode. The tools, features, and performance of Biota4D under 4th Dimension are identical with BiotaApp. There is no functional advantage to running Biota4D under 4th Dimension, but if you own 4th Dimension, there is no functional reason not to do so.

NOTE: Biota4D for Macintosh is compiled in "fat code" to allow both Macintosh and PowerMacintosh clients to connect to 4D Server (see the previous section). Therefore, the disk file for Biota4D is substantially larger than for BiotaApp—and will be larger still once the Windows code is also included in the cross-platform version for Macintosh servers.

Sources and Prices for 4D Server and 4th Dimension

You can buy ACI-US products from a number of commercial vendors, including PC/Mac Zone (1-800-403-9663), PC/Mac Connection (1-800-800-3333), some campus bookstores, and other retail outlets.

If you are affiliated with an educational institution, be sure to check around for educational discount prices, which are available for some products from some vendors. If you own any other database software, be sure to check for competitive upgrades as well. (Discount prices change too rapidly to justify any specific details here.)

At the time of this writing, average prediscount prices from mail order houses are: 4th Dimension, $650; 4D Server, $900 for two users, $970 for five additional users, $1775 for ten additional users.

Appendix F Data File Backup, Recovery, Compacting, and Segmenting

Backup Strategies for Data Files

It is absolutely essential to back up Biota Data Files on a regular basis.

If a power failure or hardware failure occurs, your Data File may be damaged. Records may be accidentally deleted, in spite of warnings and safeguards built into Biota, or an entire Data File might be accidentally deleted from your hard disk.

The simplest way to back up a Data File is to copy it (or each of its segments, if it is a segmented file) regularly to separate disks—preferably to removable disks or cartridges that can be stored in a different building(s) from the original Data File. A voltage spike, lightening strike, fire, or flood might easily destroy all disks connected to your computer or even all disks in the same building.

Of course, you can also use commercial backup software to facilitate record-keeping for multiple backups.

If you use 4D Server (see Appendix E), an integrated backup system (4D Backup) comes with the server software.

Checking and Recovering Damaged Data Files

In some cases, a damaged Data File can be repaired using 4D Tools, a separate 4th Dimension utility application that you will find on one of the Biota distribution disks.

If you see a 4D error message that a Data File is damaged and cannot be opened, try using 4D Tools to repair the Data File.

NOTE: If the user Password system has been activated (Chapter 23), only the Administrator can use 4D Tools to repair Biota Data Files.

1. *Make a backup of the damaged Data File,* preferably on a different disk.
2. **Find and launch 4D Tools.** An Open File window for your operating system appears, reading "Select the structure file to open."

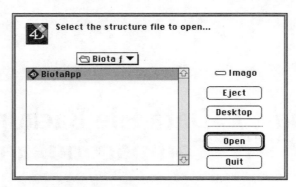

3. **Find and select Biota** (*not* the damaged Data File, which will not even be displayed in this window).
4. **Press and hold the OPTION key (Macintosh) or the ALT key (Windows) while you click the Open button in the Open File window.**
   - **If the user password system has been activated,** the Password window will appear.

     Enter the Administrator's user name and password and click the Connect button *while continuing to press the OPTION or ALT key.* An "Open data file" window will appear (below).
   - **If the user password system has not been activated,** an "Open data file" window will appear.

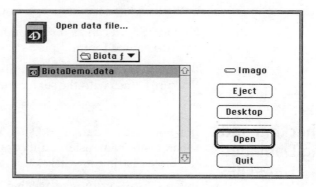

5. **Find and open the damaged Data File.**
6. **From the Utility menu in 4D Tools, choose Check & Recover.** An option screen appears.

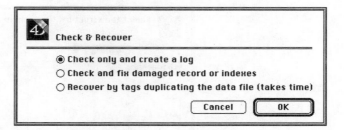

7. **Choose an option.**

   - **Check only and create a log.** If you choose this option (the default), 4D Tools will check the Data File for damage. A text file called "Journal," in the same folder with the Data File, describes any damage found. Neither Biota nor the Data File is modified.

   - **Check and fix damaged records or indexes.** If you choose this option, 4D Tools will attempt to repair any damage it finds. If it cannot do so successfully, a message appears recommending the next option.

   - **Recover by tags, duplicating the Data File.** With this option selected, 4D Tools attempts to build an entirely new copy of the Data File (named *DataFilename*.temp) based on "tags" stored with each record when it was created.

WARNINGS:

a. Use the recover-by-tags option only if the previous option fails.

b. Recovery by tags may take several hours for a large Data File.

c. If you choose to recover by tags, you must have a single mounted volume with at least enough space available to duplicate the Data File.

d. A Data File that has been recovered by tags may contain records that were previously deleted.

Compacting Data Files by Using 4D Tools

When you delete records, Biota reuses the "holes" for new records. Meanwhile, the Data File stays the same size on the disk.

If you want to compact a Data File after deleting many records, use this procedure.

1. **Make a backup of the Data File to be compacted,** preferably on a different disk. 4D Tools leaves the original Data File untouched, but an extra backup is a good idea, just to be on the safe side.

2. **Make sure space is available on your hard disk for a duplicate copy of the Data File to be compacted.** The process of compacting creates a new copy of the Data File.

3. **Find and launch 4D Tools.** An Open File window for your operating system appears, reading "Select the structure file to open."

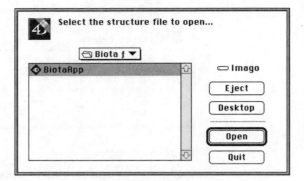

4. **Find and select Biota** (*not* the Data File to be compacted, which will not even be displayed in this window).
5. **Press and hold the OPTION key (Macintosh) or the ALT key (Windows) while you click the Open button in the Open File window.**
    - **If the user password system has been activated,** the Password window will appear.

        Enter the Administrator's user name and password and click the Connect button *while continuing to press the OPTION or ALT key.* An "Open data file" window will appear (below).
    - **If the user password system has not been activated,** an "Open data file" window will appear.

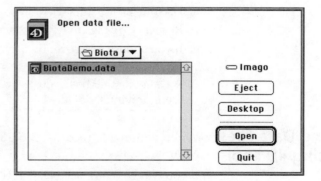

6. **Find and open the Data File to be compacted.**
7. **From the Utility menu in 4D Tools, choose Compact.** A Save file window appears, with the instruction "Create a data file."

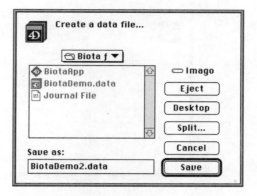

8. **Enter a name for the new, compacted Data File.**

9. **Click the Save button.** 4D Tools creates a compacted version of the Data File, displaying progress indicators as it compacts the records for each Biota table.

10. **Open the compacted Data File and test it carefully** before accepting it as the current working copy.

Segmenting Data Files

Biota Data Files can be segmented to allow expansion beyond the capacity of a single disk drive or simply to partition a Data File into smaller units. 4th Dimension allows you to segment each Data File into a maximum of 64 segments of up to two gigabytes each. Thus, the maximum size of a single Biota Data File is 128 gigabytes.

To the user, a segmented Data File behaves just the same as a unitary Data File. Addressing of data in the segments is handled transparently by 4D Engine, 4th Dimension, or 4D Server.

To use a Data File with segments on two or more hard disks, you must have the disk drives accessible simultaneously to the same computer.

Segmenting a New Data File

If you expect a new Data File to become very large in the course of time, you may want to segment it from the outset.

1. **Create a new, empty Biota Data File,** following the instructions on pp. 9–10.

2. **When the "Open a Data File" window appears (below), click the New button.** (Macintosh screens are shown here. The Windows screens differs in appearance but not in function.)

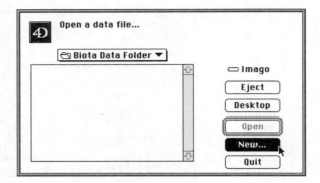

The "Create a data file" window appears.

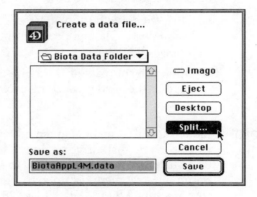

3. **Click the Split button** (above) instead of the Save button. (A default name will appear in the "Save as" box, but you can ignore it for now.) The Data Segment Manager appears.

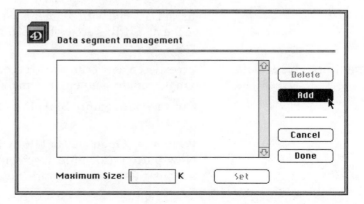

4. **Click the Add button (above) to create a data segment.** A Save File window appears.

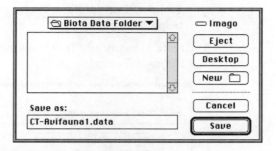

5. **Enter the segment name.**
6. **Click the Save button.** The Data Segment Manager reappears, showing the new data segment.

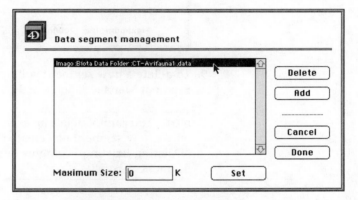

7. **To set a maximum size for the segment** (e. g., the size of your hard disk, minus a margin of safety):

   a. **Select the segment name** in the Data Segment Manager (shown above).

   b. **Enter a numerical value (in KB) in the Maximum Size entry area.** The maximum is 2,000,000 KB (2 gigabytes).

   c. **Click the Set button** to register the maximum value.

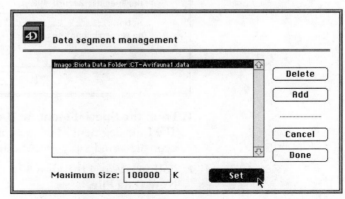

The size limit for the segment now appears in the segment list (see below).

8. **Create as many additional segments as you wish** by following steps 4 through 7, above.

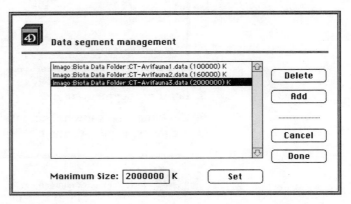

9. **To delete a new segment** you have just created, select it in the Data Segment Manager and click the Delete button.

NOTE: You cannot delete an existing segment with the delete button, only a new segment just created. To delete an existing segment, see "Deleting Existing Segments or Splitting an Existing Data File into Segments," p. 549.

Adding a New Segment to an Existing Data File

If a Data File (or the latest Data File segment) becomes as large as you want it to be, you can add a new segment. If you get the error message below, you *must* add either add a new segment or increase the size of an existing segment.

1. **From the Special menu in Biota, choose Data Segment Manager.** The Data Segment Manager appears. If the user password system has been activated, only the Administrator can use this command.

2 **Follow steps 4 through 8 in the previous section, "Segmenting a New Data File."**

NOTE: You can use the Data Segment Manager to *reduce* the maximum size limit for an existing segment, but if you enter a value less than the *current* size of the segment, the value is increased automatically to the current size to protect the data in the segment.

Deleting Existing Segments or Splitting an Existing Data File into Segments

WARNING: *Never* delete data file segments directly by deleting the files themselves in the operating system environment, even if they have not been used. Always use the method below, instead.

If you reduce the size of a Data File and want to delete a segment, decide you no longer need a Data File segmented, or wish to split an existing Data File into segments, you can use 4D Tools to reconfigure the Data File segmentation.

1. **Follow the instructions in the section "Compacting Data Files by Using 4D Tools"** (pp. 543–544) **up to step 7.**
2. **In the Save file window with the instruction "Create a data file," click the Split button.**

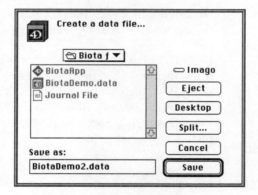

The Data Segment Manager appears.

3. **Follow steps 3 through 9 in the section "Segmenting a New Data File" to reconfigure the segmentation (from scratch) as you wish.**

Appendix G Biota Data File Conversion: Macintosh and Windows

Note: Until Biota For Windows is released, the procedures described in this appendix have no practical use.

Biota Data Files can easily be converted between Macintosh and Windows formats. (The process takes only a few seconds, regardless of file size.) With the exception of QuickTime compressed images (which at this time can be opened only in the Macintosh operation system; p. 349), Biota Data Files, once the appropriate conversion is applied, are completely cross-compatible.

Data and Resources

Like all Macintosh files, Biota Data Files on the Macintosh platform have two internal parts—the *Data Fork* and the *Resource Fork*—which together appear to the user as a single disk file in the Macintosh environment.

In the Windows operating system, in contrast, a Biota Data "File" actually becomes two separate files with the same name but different extensions:

- A large file with the extension ".4DB," listed as a file of type "Data" in the Windows 95 "Details" file listing format. This Windows file corresponds precisely to the Macintosh Data Fork.

- A small file with the extension ".RSR," listed as a file of type "Resource" in the Windows 95 "Details" file listing format. This Windows files corresponds precisely to the Macintosh Resource Fork.

To open a Biota Data File in Windows, the .4DB and the corresponding .RSR file must both be in the same folder (directory).

Data File Conversion: Step by Step

1. **Back up the Data File to be converted.**
   - *Macintosh:* Make a copy of the Data File to be converted and move the copy to a different folder or directory, for safety's sake.
   - *Windows:* Make a copy of both the .4DB Data File to be converted and also its associated .RSR file and move the copies to a different folder or directory, for safety's sake.

2. **Find and launch the utility program 4D Transporter** (which came on of the Biota distribution disks). An option screen appears.

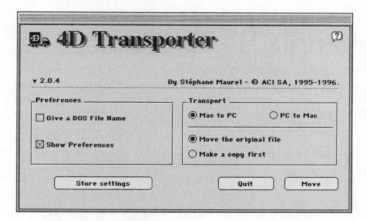

3. **Choose "Mac to PC" or "PC to Mac"** in the Transport panel.

4. **Choose a copy option in the Transport panel.** If you made a backup in step one, choose "Move the original file." Otherwise, or as an extra precaution, choose "Make a copy first." (In the Transporter window, "move" means "convert.")

5. **If you chose "Mac to PC," the "Give a DOS File Name" checkbox is enabled. If you want a filename that conforms to DOS restrictions, check this box** and Transporter will invent something based on the consonants in your Macintosh file name. If you leave this box unchecked, your Macintosh filename will be preserved in the Windows version (with extensions added).

6. **Click the Move button** in the 4D Transporter window. The file navigation window for your operating system appears.

7. **In the file navigation window, find and select the Biota Data File to be converted.** Transporter creates the new file (PC to Mac) or files (Mac to PC). The Transporter window then reappears, to allow further conversions.

8. **Click the Quit button** when you are done.

Appendix H Barcodes

Barcodes speed data entry (pp. 144, 229–230), record retrieval (pp. 236–237), and specimen loan management (pp. 328–333, 341–345). Biota can be set up to look for and recognize barcodes (pp. 105–108) and to strip and replace barcode alphanumeric prefixes for automated specimen series entry and retrieval (p. 108).

Barcodes

If the barcode labels you want to use do not need to be particularly minute, you can print your own barcodes (for example, in Code 39) on acid free, plain or adhesive-backed stock using a commercially available barcode font and an ordinary text editor and laser printer.

For very small labels (e.g. 0.56 x 0.35 inches), Intermec's proprietary Code 49 (a miniaturized, two-dimensional code), preprinted on 4 mil polyester stock, has become a de facto standard in entomology for pin labels (no adhesive backing). Project ALAS (see the Preface) uses the same labels, but with a permanent adhesive backing, for slide-mounted material (see pp. 251–254).

The cost of preprinted entomological barcodes from Intermec (as of this writing) is approximately $500 for setup, plus about $25 per thousand labels. Subsequent orders are less expensive since the setup is saved by Intermec. Inquire with Intermec at 1-800-221-8314 regarding preprinted barcodes.

Barcode Scanners

To use barcodes, you need a barcode "reader"—a *visible laser scanner*. The scanner simply acts as an alternative input device to the keyboard.

A ballpark price for a stand-mounted scanner from Intermec that will read Code 49 (as well as many other codes, including Code 39) is $1500–$2000, depending on the model and the computer you want to use it with. For Intermec *hardware*, you must work through regional offices. Consult your local yellow pages, or call 1-800-829-8959 for a local reference.

Appendix I Biota Menus: Quick Reference

This appendix briefly describes the function of each item in the Biota menus and directs you to the appropriate pages of this book for full details. The menus and their items (commands) are listed in the order they appear on the screen.

NOTE: The notation used for each item is **Menu Name → Menu Item**.

Access Privileges for Menu Items

If the user password system has been activated, each user is assigned to an access privilege level (pp. 415–420). Depending on the privileges you have been assigned, certain menu commands may be unavailable to you. If so, Biota posts an explanatory message when you select the item. A complete tabulation of commands and the privilege level required to use them appears on pp. 416–417.

The Menu Bar

Because Biota supports simultaneous procedures (multitasking), the menu bar does not change. All menus are present and active at all times.

| File | Edit | Input | Series | Tree | Find | Display | Labels | Im/Export | Loans | Special |

The File Menu

File	Edit	Input	Series	Tree
To Load a Biota Data File...				
Save Record Set Pointer File...				
Load Record Set Pointer File...				
Print...				⌘P
About Biota...				
About the Current Data File...				
Quit Biota				⌘Q

- **File → To Load a Biota Data File.** Displays an information screen explaining how to load a different Data File or create a new Data File while launching Biota. See p. 89.

- **File → Save Record Set Pointer File.** Creates a disk file with pointers to the records in the current Record Set for any Core Table or for the Determination History table. See pp. 16–18.

- **File → Print.** Prints (pp. 266–280) or exports to disk file (p. 141) a report for the current Selection of records displayed in an output screen. (This command is disabled if no records are displayed.)

- **File → Load Record Set Pointer File.** Loads the Selection of records indicated by the record pointers in a saved Record Set Pointer File. See pp. 18–19.

- **File → About Biota.** Displays the copyright notice, version, and release date for your copy of Biota. See p. 3.

- **File → About the Current Data File.** Displays the name and location (path) of the Data File that is currently open, plus a scrolling list of the number of records in each Biota table in the Data File. See p. 11. (If you get an error message when you choose this command, see p. 534.)

- **File → Quit Biota.** Choose this command to shut down Biota. See pp. 10–11.

The Edit Menu The Edit Menu contains the standard text editing commands for your operating system. See your operating system manual.

The Input Menu

- **Input → Specimen.** Presents the Specimen input screen. See pp. 143–152.

- **Input → Collection.** Presents the Collection input screen. See pp. 152–159.

- **Input → Locality.** Presents the Locality input screen. See pp. 160–164.

- **Input → Species.** Presents the Species input screen. See pp. 164–168.

- **Input → Genus, Family, Order, Class, Phylum or Division, or Kingdom.** Presents the input screen for the corresponding table. See pp. 168–172.

- **Input → Personnel.** Presents the Personnel input screen. See pp. 172–179.

- **Input → Project Name.** Presents the input screen for information on the project or sponsoring institution represented by the Data File. Project information is used on reports, loan forms, and herbarium labels produced from the Data File. See pp. 177–179.

The Series Menu

- **Series → Input Specimen Series.** Presents an input screen for the efficient creation of sets of Specimen records that share Collection data. See pp. 228–234.

- **Series → Input & Identify Spcm Series.** Presents an input screen for the efficient creation of sets of Specimen records that share Collection and determination (identification) data. See pp. 228–234.

- **Series → Find Specimen Series.** Presents a query tool for finding Specimen records efficiently by Specimen Code. The Specimen Codes may be either sequential or nonsequential. (The same command appears in the Find menu.) See pp. 234–242.

- **Series → Find & Identify Specimen Series.** Presents a query and record update tool for finding Specimen records efficiently by Specimen Code and for adding determination (identification) data and/or certain other fields in the Specimen table. The Specimen Codes may be either sequential or nonsequential. See pp. 234–242.

The Tree Menu

- **Tree → Lower Taxa for Higher Taxa.** Presents a query tool for finding all records in the Specimen table or a taxon table that are linked to a single record or a Record Set in a higher taxon table (e.g., all Specimens of a Species or a Genus, all Genera for a set of Orders). (The same command appears in the Find menu.) See pp. 201–205 and 52–56.

- **Tree → Higher Taxa for Lower Taxa.** Presents a query tool for finding all records in the Species table or a higher taxon table that are linked to a Record Set for the Specimen table or any lower taxon table (e.g., all Species for the Specimen Record Set; all Orders for the Genus Record Set). (The same command appears in the Find menu.) See pp. 205–208.

- **Tree → Kingdom Down.** Displays a listing of all Kingdom records. Double-click a Kingdom to display all its Phyla (Divisions). Double-click a Phylum to display all its Classes, a Class to display all its Orders, etc.—down to all Specimens for a Species. See pp. 187–192 and 48–51.

- **Tree → Phylum or Division Down.** Displays a listing of all Phylum (Division) records. Double-click a Phylum to display all its Classes, a Class to display all its Orders, etc.—down to all Specimens for a Species. See pp. 187–192 and 48–51.

- **Tree → Class Down.** Displays a listing of all Class records. Double-click a Class record to display all its Orders, an Order to display all its Families, etc.—down to all Specimens for a Species. See pp. 187–192 and 48–51.

- **Tree → Order Down, Family Down, Genus Down, or Species Down.** Displays a listing of all records for the specified table. Double-click the record for a taxon to display records for its subtaxa at the next lower taxonomic level—down to all Specimens for a Species. (These commands work in the same way as the Kingdom Down, Phylum or Division Down, and Class Down commands, above.) See pp. 187–192 and 48–51.

The Find Menu

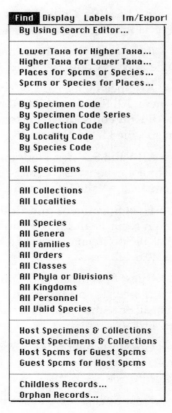

- ♦**Find → By Using Search Editor.** Presents the Search Editor, a general purpose query tool for finding records in a particular table based on the content of fields in those records or the content of fields in linked records of other tables. See pp. 192–200.

- ♦**Find → Lower Taxa for Higher Taxa.** Presents a query tool for finding all records in the Specimen table or a taxon table that are linked to a single record or a Record Set in a higher taxon table (e.g., all Specimens of a Species or a Genus, all Genera for a set of Orders). (The same command appears in the Tree menu.) See pp. 201–205 and 52–56.

- **Find → Higher Taxa for Lower Taxa.** Presents a query tool for finding all records in the Species table or a higher taxon table that are linked to a Record Set for the Specimen table or any lower taxon table (e.g., all Species for the Specimen Record Set; all Orders for the Genus Record Set). (The same command appears in the Tree menu.) See pp. 205–208.

- **Find → Places for Spcms or Species.** Presents a query tool for finding all records in the Collection table or Locality table that are linked to a Record Set for the Specimen table or the Species table (e.g., all

Localities for the Specimen Record Set; all Collections for the Species Record Set). See pp. 208–210.

- **Find → Spcms or Species for Places.** Presents a query tool for finding all records in the Specimen table or Species table that are linked to a Record Set for the Collection table or the Locality table (e.g., all Species for the Locality Record Set; all Specimens for the Collection Record Set). See pp. 210–213.

- **Find → By Specimen Code.** Finds and displays a Specimen record based on single Specimen Code. See pp. 213–215.

- **Find → By Specimen Code Series.** Presents a query tool for finding groups of Specimen records efficiently by Specimen Code. The Specimen Codes may be either sequential or nonsequential. (The same command appears in the Series menu.) See pp. 234–242.

- **Find → By Collection Code.** Finds and displays a Collection record based on single Collection Code. See pp. 213–215.

- **Find → By Locality Code.** Finds and displays a Specimen record based on single Locality Code. See pp. 213–215.

- **Find → By Species Code.** Finds and displays a Species record based on single Species Code. See pp. 213–215.

- **Find → All Specimens, All Collections, All Localities, All Species, All Genera, All Families, All Orders, All Classes, All Phyla or Divisions, All Kingdoms, or All Personnel.** Finds and displays all records for the specified table. See pp. 185–186.

- **Find → All Valid Species.** Finds and displays all Species records that are not junior synonyms (i.e., all Species records for which the Species Code equals the Valid Species Code). See pp. 380–381.

- **Find → Host Specimens and Collections.** Finds and displays host Specimen records and linked host Collection records based on a search of all Specimen records or a search of the Specimen Record Set. See pp. 401–402.

- **Find → Guest Specimens and Collections.** Finds and displays guest Specimen records and linked guest Collection records based on a search of all Specimen records or a search of the Specimen Record Set. See pp. 402–404.

- **Find → Guest Spcms for Host Spcms.** Finds and displays all guest Specimen records that are linked to host Specimen records in the Specimen Record Set. See p. 404.

- **Find → Host Spcms for Guest Spcms.** Finds and displays all host Specimen records that are linked to guest Specimen records in the Specimen Record Set. See p. 404.

- **Find → Childless Records.** For a specified parent table, finds all records that are not linked to any child record in a Core table. (For example, you could find all Species records that are not linked to any Specimen records.) See pp. 215–216.

- **Find → Orphan Records.** For a specified child table, finds all records that are not linked any parent record in a Core table. (For example, you could find all Specimen records that are not linked to any Species record.) See pp. 216–218.

The Display Menu

- **Display → [Table Name].** Displays the current Record Set for the specified table. See pp. 15–16.

The Labels Menu

- **Labels → Collection Labels.** For each record in the current Collection Record Set, prints or exports (to a text file) a label that includes most fields from the Collection record and its linked Locality record. See pp. 246–248.

- **Labels → Pin Labels: Locality.** For each record in the current Specimen Record Set, prints a standard entomological locality label for a pinned specimen, including appropriate fields from linked Collection and Locality records and (optionally) the Specimen Code. See pp. 248–251.

- **Labels → Pin Labels: Determination.** For each record in the current Specimen Record Set, prints a standard entomological determination label for a pinned specimen, including appropriate fields from the Specimen record and from linked Species and Genus records. See pp. 248–251.

- **Labels → Vial Labels: Locality.** For each record in the current Specimen Record Set, prints a standard locality label for a fluid preserved specimen, including appropriate fields from linked Collection and Locality records and (optionally) the Specimen Code. See pp. 248–251.

- **Labels → Vial Labels: Determination.** For each record in the current Specimen Record Set, prints a standard determination label for a fluid-preserved specimen, including appropriate fields from the Specimen record and from linked Species, Genus, and Family records. See pp. 248–251.

- **Labels → Slide Labels: Locality.** For each record in the current Specimen Record Set, prints a standard locality label for a slide mounted specimen, including appropriate fields from linked Collection and Locality records and (optionally) the Specimen Code. See pp. 251–254.

- **Labels → Slide Labels: Determination.** For each record in the current Specimen Record Set, prints a standard determination label for a slide mounted specimen, including appropriate fields from the Specimen record and from linked Species, Genus, and Family records. See pp. 251–254.

- **Labels → Herbarium Labels.** For each record in the current Specimen Record Set, prints one or more standard herbarium labels, including appropriate fields from the Specimen record and from linked Collection, Locality, Species, Genus, and Family records. See pp. 255–258.

- **Labels → Species Labels.** For each record in the current Species Record Set and/or for Species records found using the Species Labels screen search tools, prints a label including appropriate fields from the Species record and from the linked Genus record, plus (optionally) the determiner's name. See pp. 259–261.

- **Labels → Design & Print Custom Labels.** For each record in the current Record Set for any Core table, prints a label that you design using the Label Editor, using fields from the selected table and linked records in parent tables. See pp. 261–264.

The Import/Export (Im/Export) Menu

- **Im/Export → Import by Tables and Fields.** Presents the Import Editor, which imports delimited, column-by-row text files into the fields you specify for any table in Biota's structure. See Chapter 26.

- **Im/Export → Export by Tables and Fields.** Presents the Export Editor, which exports delimited, column-by-row text files based on the fields you specify for any table in Biota's structure. See pp. 434–443.

- **Im/Export → Export Notes.** For the current Species, Specimen, Collection, Locality, or Loans Record Set, exports (to a text file) the full record for each attached Note, certain fields from each parent record in the Record Set, and appropriate fields from other linked Core tables. See pp. 446–448.

- **Im/Export → Export Auxiliary Fields.** For the current Species, Specimen, Collection, or Locality Record Set, exports (to a tab-delimited text file) Auxiliary Field values in matrix format (standard or transposed) or NEXUS format. See pp. 308–311, 448–449.

- **Im/Export → Export Images.** For each record in the current Species Record Set, exports either the first linked image (you can specify the order of images, pp. 359–363) or all linked images, creating a separate disk file for each image exported. See pp. 449–451.

- **Im/Export → Export Taxonomic Flatfile.** Exports a tab-delimited text file table with the Record Set for a lower taxonomic level as rows and information on the higher classification of those taxa as columns, based on the range of taxonomic levels you specify. See pp. 451–453.

- **Im/Export → Export Specimen Flatfile.** Exports a tab-delimited text file table with the current Specimen Record Set as rows and fields from any linked Core table as columns. See pp. 454–457.

- **Im/Export → Export Custom Flatfile.** For any Core table, exports a text file based on the current Record Set with selected fields from the focal table and related tables, using a custom format that you design using the Quick Report Editor. See pp. 457–462.

- **Im/Export → Export Specimens Examined List.** For the current Species Record Set, exports (to a text file) draft text for a Specimens Examined section of a taxonomic monograph, based on linked Specimen records. See pp. 463–469.

- **Im/Export → Colls x Species Table.** Based on the Collection, Specimen, or Species Record Set, exports a tab-delimited text file table, with Collections as rows and Species as columns, specifying the incidence (presence/absence) or abundance or Specimens of each Species in each Collection. See pp. 469–472.

- **Im/Export → Create Web Pages.** Exports hierarchically linked Web pages, ready for posting on a Web server, for any range of taxonomic levels, based on Record Sets that you define. You can include any fields from the Specimen, Collection, or Locality tables, as well as images and host information. See Chapter 25.

The Loans Menu

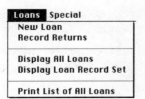

- **Loans → New Loan.** Creates a new Loan record, marks Specimen records and produces documents and text files for a new Specimen loan, based on the current Specimen Record Set and/or on ad hoc entry of Specimen Codes. See pp. 328–333.

- **Loans → Record Returns.** Updates an existing Loan record and marks Specimen records to record the return of loaned specimens, based on the current Specimen Record Set, the Specimens in the Loan, and/or on ad hoc entry of Specimen Codes. See pp. 339–345.

- **Loans → Display All Loans.** Finds and displays all records for the Loans table. See pp. 337–339.

- **Loans → Display Loan Record Set.** Displays the current Record Set for the Loans table. See pp. 337–339.

- **Loans → Print List of All Loans.** Prints a list of all records for the Loans table, with pertinent information from other tables. See pp. 333–337.

The Special Menu

Special
- Empty Record Sets
- Find and Replace...
- Change Password
- Edit User Password System...
- Edit Data File Password Link
- Preferences...
- Field Value Defaults...
- Entry Choice Lists...
- Core Field Aliases...
- Record Code Prefixes...
- Make Guest Collection Records
- Specimen Count by Taxon
- Clear All Synonymies
- All Determination Histories
- Display Determination Hists
- Det Hists for the Spcm Rec Set
- Show Shortcut Screen

- **Special → Empty Record Sets.** Empties the Record Set for each Core table and for the Determination History table. See pp. 12–16.

- **Special → Find and Replace.** Presents the Find and Replace tool, which allows you to replace a specified text string with a new string in a specified field in a specified Biota table, for all records or for selected records. See pp. 220–225.

- **Special → Change Password.** Changes a validated current user password to a specified new password (pp. 413–415). The Administrator can enable or disable the user password system using this command (pp. 409–412).

- **Special → Edit User Password System.** Presents the Password Editor, which the Administrator can use to assign or change user names, user passwords, access privilege levels, and default language settings for users. See pp. 415–424.

- **Special → Edit Data File Password Link.** Enables or disables the Data File Password Link or changes a validated current Data File Password Link to a new one. See pp. 425–431.

- **Special → Preferences.** Presents the Preferences screen. See Appendix C for a graphical guide to page references for each option.

- **Special → Field Value Defaults.** Allows you to set default values for data entry in many fields. See pp. 313–315.

- **Special → Entry Choice Lists.** Allows you to enable or disable an Entry Choice List for rapid entry of repetitive data in many fields. See pp. 315–322.

- **Special → Core Field Aliases.** Allows you to rename many Core fields. An Alias appears on all input and output screens and in certain reports instead of the corresponding Internal Field Name. See pp. 323–326.

- **Special → Record Code Prefixes.** Allows you to specify default prefixes for Specimen, Species, Collection, or Locality Record Codes and to set parameters for using barcodes. See pp. 102–108.

- **Special → Make Guest Collection Records.** Automatically creates a new Collection record for the "guests" of each Specimen in the current Specimen (host) Record Set. See pp. 398–401.

- **Special → Specimen Count by Taxon.** Prints a report, organized by the taxonomic hierarchy, showing the number of Specimens for each Species and for each higher taxon in the Specimen Record Set. See p. 269.

- **Special → Clear All Synonymies.** Sets the Valid Species Code equal to the Species Code for all Species records. See p. 391.

- **Special → All Determination Histories.** Finds and displays all records for the Determination History table. See pp. 375–377.

- **Special → Display Determination Histories.** Displays the current Record Set for the Determination History table. See pp. 375–377.

- **Special → Det Hists for the Spcm Rec Set.** Displays all Determination History records for the current Specimen Record Set. See pp. 375–377.

- **Special → Show Shortcut Screen.** Displays the Shortcut screen, which offers buttons for several common Biota menu commands. The Shortcut screen cannot be customized.

Index

@ wildcard character, 39–40, 96
4D Server
  caveats and recommendations, 539–540
  running Biota4D under, 538–540
  sources and prices for, 540
4D Tools
  compacting Data Files using, 543–545
  deleting segments or segmenting an existing Data File using, 549
4th Dimension (Single User)
  running Biota4D under, 540
  sources and prices for, 540

Abundance field (Specimen table), 152, step 6
Accept button in input screens, 121
access privilege levels, 79, 415–417, 420–424
  for menu items, 554
  Read Export, 415–417
  Read Only, 415–417
  Read Write Export, 415–417
  Super User, 415–417
ad hoc queries, *See* finding records based on content
Add Class button, 127
Add Collection button, 127
Add Family button, 127
Add Genus button, 127
Add Order button, 127
Add Phylum button, 127
Add Record buttons, 127
Add Species button, 127
Add Specimen button, 127
adding new information to records, *See* updating
Administration access privilege level, 415–417

Administrator, database, 415–417
agreement, loan, *See* loans, invoices
Aliases, Core field, 65, 323–326
  advantages and disadvantages of using Aliases, 323
  checking, 325–326
  clearing all Aliases and resetting to defaults, 326
  displaying, 66, 325–326
  help, 66
  setting, 324–325
  using 147, 152, 155, 158, 161, 167
Aliases, field name, *See* Aliases, Core field
All button, 441, 493
all records for a table, finding and displaying, 185–186
alphanumeric fields
  exporting, 435
  importing, 488
Append button, 175–176, 440, 493
approximate determinations, *See* temporary taxa
ASCII files
  exporting data to, *See* exporting
  importing data from, *See* importing
Assign button, 102–103
assigning Record Codes automatically during data entry, 102–103
audit trail, Specimen determination, *See* determination histories
authorization, *See* access privilege levels
automatic assignment
  of field values, *See* Field Value Defaults
  of Record Codes, *See* Record Codes, automatic assignment
automatic field value entries, *See* Field Value Defaults

Auxiliary Field Blank Value Error, 505
Auxiliary Field Names
  creating, editing, and ordering, 283–295
  editing, 288–291
  reordering alphabetically, 291–293
  reordering in a specific order, 293–295
Auxiliary Field Setup Error, 505
Auxiliary Field Value Key Match Error: updating existing records, 506
Auxiliary Field Values, exporting, *See* exporting Auxiliary Field Values
Auxiliary Fields, 66, 281–326
  adding new, 284–288
  carrying, 298
  Core fields and Auxiliary Fields, 281–282
  creating and displaying, 66
  creating new Auxiliary Fields, 284–288
  displaying in the Standard format: records as rows, Auxiliary Fields as columns, 300–302
  displaying in the Transposed format: Auxiliary Fields as rows, records as columns, 302–304
  displaying, for a Selection of records, 298–304
  editing the list of Auxiliary Field Names, 288–291
  entering and displaying data in Auxiliary Fields, 295–304
  entering new data (Auxiliary Field Values), 295–298
  exporting, 308–311, 448–449
  exporting in the NEXUS format, 310–311
Field Name Editor, opening, 283–284
Field Name Editor, using, 283–295

Index

how Auxiliary Fields work, 282–283
names of Auxiliary Fields, creating, editing, and ordering, 283–295
new, 284–288
printing Auxiliary Fields in matrix (row-by-column) format, 305
printing Auxiliary Fields in Standard Triplet format (Record Code, Auxiliary Field Name, Auxiliary Field Value), 305–306
printing Auxiliary Fields in Transposed Triplet format (Auxiliary Field Name, Record Code, Auxiliary Field Value), 307–308
printing Auxiliary Fields, 304–308
reordering Auxiliary Field Names alphabetically, 291–293
reordering Auxiliary Field Names in a specific order, 293–295

backup strategies for Data Files, 541
barcodes, 552–553 (Appendix H)
  scanners (readers), 552–553
  using, for Record Codes, 70, 552–553
before you begin, xxi–xxv
biodiversity data and relational databases, 1–4, 64–65
Biota
  and the Internet, 473–484
  bug reports, 535–537
  citing, xxv
  client-server mode, 64, 439–540
  customizing, 281–326
  data dictionary, 521–523 (Appendix A)
  Data Files, *See* Data Files
  data model, 521–523 (Appendix A)
  data structure, 521–523 (Appendix A)
  flavors, *See* versions and flavors of Biota
  for Macintosh, *See* versions and flavors of Biota; operating systems
  for Windows, *See* versions and flavors of Biota; operating systems
  idiosyncrasies, 60–61
  installing, xxv, 21–22
  introducing, 1–57
  opening the Demo Data File, *See* Demo Data File, installing and opening
  overview of tools and features, 63–87
  structure (tables, fields, and relational links), 521–523 (Appendix A)
  suggestions for improving, 537
  support, 535–537
  tutorial, 21–57
  versions, *See* versions and flavors of Biota
  Website, xxi, 535
Biota4D
  description, 538
  running under 4D Server, 539
  running under 4th Dimension, 540
Blank Key Error, 499
Boolean Field Error, 501–502

Boolean fields
  exporting, 437
  importing, 490
borrowing Specimens, *See* loans
bottom-up
  strategy, for exporting Web pages, 474–475
  taxonomic searches, *See* finding records for higher taxa based on lower taxa
bug reports, 535–537
button
  Accept, 121
  Add Class, 127
  Add Collection, 127
  Add Family, 127
  Add Genus, 127
  Add Kingdom, 127
  Add Order, 127
  Add Phylum, 127
  Add Record, 127
  Add Species, 127
  Add Specimen, 127
  All, 441, 493
  Append, 175–176, 440, 493
  Assign, 102–103
  Cancel, 121
  Carry, 69–70, 127–128, 183, 298
  Carry, *See* Carry button
  Child Records, 30–31, 125–126
  Classes, 125–126
  Collections, 125–126
  Families, 125–126
  Field Name/Alias toggle, 222–223, 439–440, 491–492
  Full Record, 32–33, 124–125
  Genera, 125–126
  Insert, 175–176, 441, 493
  Look Up, *See* Look Up buttons
  Look Up by Species, *See* Look Up by Species button
  Look Up Collection, *See* Look Up Collection button
  Look Up Locality, *See* Look Up Locality button
  Look Up Species, *See* Look Up Species button
  navigation (first record, previous record, next record, last record), 27, 121
  option (radio), keyboard equivalents, 91
  Orders, 125–126
  Phyla, 125–126
  Print, in input screen, 266
  Print, in output screen, 266–268
  radio, keyboard equivalents, 91
  Remove, 175–176, 441, 493
  Skip, 492–493
  Species, 125–126
  Specimens, 125–126
  Today, keyboard equivalents for, 91–92

Cancel button in input screens, 121
captured images, *See* images

Carry button, 69–70, 127–128
  carrying Auxiliary Fields, 298
  carrying Notes, 183
  using an existing record as a template for a new record, 127–128
  using, to treat an existing record as a template for a new record, 69–70
changing
  field names, *See* Aliases, Core field
  your user password, 413–415
character matrices in the NEXUS format, exporting, *See* exporting character matrices in the NEXUS format
checking and recovering damaged Data Files, 541–543
child records, *See* parent and child records
Child Records buttons, 30–31, 125–126
childless records, finding, 215–216
Choice Lists, Entry, *See* Entry Choice Lists
citing Biota, xxv
Class Down, *See* finding records using the taxonomic hierarchy
Class input, 168–172
Classes button, 125–126
Codes, Record, *See* Record Codes
cohorts, 109–111
Collection Code
  assigning automatically during data entry, 102–103
  finding records by, *See* finding records by Record Code
collection date ranges, 113–114
Collection input, 152–159
collection labels, *See* labels, collection
Collection lists, based on Specimen or Species records, *See* finding records for Collections or Localities (places) based on Specimens or Species records
Collection Notes, 179–183
Collections button, 125–126
Collections × Species matrices, exporting, *See* exporting Collections × Species matrices
color images, *See* images, color
colors, screen, 89–90
column heading row option, for importing and exporting text files, 442, 486–487
columns
  resizing in Quick Reports, 272–273
  setting up, in the Quick Report layout, 270–271
commands, menu, 554–563 (Appendix I)
compacting Biota Data Files using 4D Tools, 543–545
compound junior synonyms, *See* synonymy
compression, image, 349
concepts, key, 5–19
consecutive Specimen records, *See* Specimen record series

566 INDEX

contracts, Specimen loan, *See* loans, invoices
coordinates, geographical: latitude and longitude, 115–117
copying
  records, 127–128
  Image records, 355–357
Core field Aliases, *See* Aliases, Core field
Core fields and Auxiliary Fields, 281–282
Core tables and Core fields, 65
correcting errors in records, *See* updating
counting Specimens by taxon, 269
creating
  an empty Biota Data File, *See* Data Files, creating empty
  records, limits, *See* records, creating, record creation limits
  new records "on the fly" (tutorial lesson), 44–48, *See also* records, creating new, "on the fly"
  new records one table at a time (tutorial lesson), 34–44, *See also* records, creating new
cross-hierarchy searches, *See* finding records: sequential, cross-hierarchy searches
current selection, *See* selections of records
cursor, "hidden window," 61
custom
  fields, *See* Auxiliary Fields
  flatfiles, exporting, *See* exporting custom flatfiles
  labels, 261–264
customizing
  Biota, 281–326
  default field entries, *See* Field Value Defaults

data
  biodiversity, 1
  entering, *See* input
  exporting, *See* exporting
  fields, *See* fields, data
  importing, *See* importing
  input, *See* input
  lot-based, 64
  records, *See* records
  special types, 109–117
  specimen-based, 64
  types, special, 109–117
data and resources (Data File conversion between Macintosh and Windows), 550
Data File password link, *See* passwords, Data File password link
Data File, Biota, 8–11
  backup, recovery, compacting, and segmenting (Appendix F), 541–543
  checking and recovering damaged Data Files, 541–543
  compacting using 4D Tools, 543–545
  conversion between Macintosh and Windows, 550–551 (Appendix G)

creating empty, 9–10
creating new, 9–10
damaged, checking and recovering, 541–543
finding out which is open, 11
opening, 8–9, 60–61
password link, *See* passwords, Data File password link
saving changes in, 10–11, 60
security, *See* passwords, Data File password link
segmenting, *See* segmenting Data Files
spanning more than one hard disk, *See* segmenting Data Files
to find out which Data File is currently open, 11
Data Segment Manager, adding a new segment to an existing Data File using, 548–549
databases, relational, and biodiversity data, 1–4
date fields, 112–113
  exporting, 435–436
  importing, 488–489
date flag fields, 112–113
  exporting, 435–436
  importing, 488–489
Date Format Error, 500–501
dates, partial, 112–113
  exporting, 435–436
  importing, 488–489
dates, 111–114
  collection date ranges, on labels and in exported text files, 114
  collection date ranges, 113–114
  date displays on output screens, 112
  date formats in Biota, 111–112
  entering, 112
  on labels, 111–112
  partial, 112–113, 435–436
  partial, importing and exporting, 113, 435–436, 488–489
  sorting by day, month, or year using formulas, 136–138
default entries for fields, *See* Field Value Defaults
default
  buttons, 92
  entry order for fields, 90
  field value entries, *See* Field Value Defaults
  names for intermediate taxon fields, 110
  Record Code prefixes and lengths, changing, 102–103, 105–108
  values for fields, *See* Field Value Defaults
deleting
  a record in an input screen, 122–123
  a record or a group of records in the output screen, 29–30, 129–131
  controls and limits, 80, Appendix C
  Image records, 358–359
  Notes, 181–183

delimiters
  end-of-field, 441, 493–494
  end-of-record, 434, 441, 493–494
Demo Data File
  contents of, 23
  installing and opening, 21–22
determination histories, 80–81, 367–377
  changing a determination in a Genus record, 374–375
  changing a determination in a Species record, 374
  changing a determination in a Specimen record series, 373–374
  changing a determination in an individual Specimen record, 373–374
  changing a determination using the Synonymy tool, 375
  deleting Determination History records, 375–377
  disabling the Determination History system, 369
  displaying Determination History records for editing, 375–377
  displaying the Determination History for a Specimen record, 369–372
  displaying, editing, or deleting Determination History records, 375–377
  editing Determination History records, 375–377
  enabling the Determination History system, 369
  exporting and importing Determination History records, 377
  format of Determination History records, 367–368
  how and when changes in determination are recorded, 372–375
determination, changing
  by using the Synonymy tool, 375
  in a Genus record, 374–375
  in a Species record, 374
  in a Specimen record series, 373–374
  in an individual Specimen record, 373–374
determination labels, *See* labels, determination
determinations, updating, in Specimen series (ordered or unordered), *See* updating Specimen determinations
Display Menu, 559
displaying
  and editing existing records (tutorial lesson), 23–34, *See also* records, displaying; records, editing
  Auxiliary Fields for a Selection of records, 298–304
  data in Auxiliary Fields, 295–304
  Determination History for a Specimen record, 369–372
  Image records, 355–357
  images (thumbnail) in the Species output screen, 72–73, 363–365

INDEX

latitude and longitude, setting the display format, 116–117, 526
   Loan records, 337–339
   Record Sets, 71
   records in a standard output screen, 119–120
   synonymy status of a Species record, 381–384
DNA-source links, *See* host–guest relations
DNA-Specimen links, *See* host–guest relations
documentation, xxi
double-bordered default buttons, 92
drawings, *See* images
Duplicate Auxiliary Field Value Key Error: importing new records, 505
Duplicate Key Error
   while creating or updating records, 7
   while importing new records, 498–499
duplicate specific names, checking for, 81
duplicating records, 127–128

Edit Menu, 555
editing Notes, 181–183
Editor
   Export, *See* exporting using the Export Editor
   Field Name, 283–295
   Group, 174–177
   Import, *See* importing using the Import Editor
   Label, 261–264
   Password, 415–424
   Quick Report Graph, 279–280
   Quick Report, 269–279
   Search, 192–200
   Sort, 131–139
efficiency
   of data input and updating, 3–4
   of relational searches, 4
emptying Record Sets, *See* Record Sets, emptying all
English and Spanish screens and dialogues, 531
enlargement of printed output, 243–245
entering data, *See* input
entering records, *See* input
entering images, 350–355
entomological pin labels, 248–251
entry areas, 90
Entry Choice Lists, 315–322
   activating or deactivating, 316
   adding items, 317–318
   adding, deleting, or modifying items, 317–318
   copying Choice Lists to a different Biota Data File, 319–322
   deleting items, 317–318
   editing, 317–318
   modifying items, 317–318
   saving changes in existing Lists, 318
   saving new Lists or recording changes in existing Lists, 318
   transferring Choice Lists to a different Biota Data File, 319–322
   undoing all changes made in Choice Lists during the current Biota session, 318–319
   using, 40–41, 69, 92, 102, 319
   using, to enter Record Code prefixes, 102
entry order, default, *See* default entry order for fields
equivalents, keyboard buttons, onscreen, *See* keyboard equivalents
errors
   Auxiliary Field Blank Value Error, 505
   Auxiliary Field Setup Error, 505
   Auxiliary Field Value Key Match Error: updating existing records, 506
   Blank Key Error, 499
   Boolean Field Error, 501–502
   Date Format Error, 500–501
   Duplicate Auxiliary Field Value Key Error: importing new records, 505
   Duplicate Key Error: creating or updating records, 7
   Duplicate Key Error: importing new records, 498–499
   Field Length Error, 500
   file damaged error message, 534
   Invalid Date Error, 501
   locked record message ("The record is already being edited"), 533
   messages, 532–535
   missing External Procedure (4D Error, –9939), 534
   Missing Key Error: importing new records, 497–498
   Missing Key Error: updating existing records, 498
   Mixed Characters Error, 500
   "No more room to save the record" error message (4D Error, 98), 534, 548–549
   No Unique Key Update Error, 502–503
   "Out of memory" 4D error message, 533
   Too Many Fields Error, 497
   Unknown Key Error, 503
   Update Key Match Error: updating existing records, 499
   guest Collection records, while creating automatically, 400–401
Excel files, importing data from, *See* importing
existing record as a template for a new one, *See* Carry button
exporting, 84–87, 433–472
   alphanumeric fields, 435
   Auxiliary Field Values, 85–86, 308–311, 448–449
   Auxiliary Fields Values in the NEXUS format, 310–311
   by tables and fields, using the Export Editor, 84–87, 434–443
   character matrices in the NEXUS format, 86, 310–311
   collection date ranges in exported text files, 114
   Collection (Locality) label data, 246–248
   Collections × Species incidence or abundance matrices, 86, 469–472
   Collections-by-Species table, example, as exported by Biota, 472
   column heading row option, 442
   Custom Flatfile tool, when to use, 457–458
   custom flatfiles, 86, 457–462
   custom flatfiles, examples, as exported by Biota, 462
   date fields, 435–436
   date flag fields, 435–436
   dates, partial, 113, 435–436
   Determination History records, 377
   Export Editor, 84–87, 434–443
   field types and field lengths, 435–437
   geographical coordinates in exported text files, 117
   image file names, how Biota assigns to exported image files, 450–451
   images, 86, 355–357, 449–451
   images, using the Export Images tool, 450
   images, in groups, to disk files, 365
   images, options for, 449
   Key Fields and the Export Editor, 435
   latitude and longitude fields, 436–437
   loan records, 333–337
   Notes, 85, 443–448
   Notes, all records for a Notes table, 444
   Notes records based on the content of parent records, 445–446
   Notes records based on their own content, 445
   Notes records linked to records in the current parent-table Record Set, 443–444
   Notes records using the Export Notes tool, 446–448
   Notes records with the Export Editor, 443
   quadrat or sample data, *See* exporting Collections × Species matrices
   specimen flatfiles, 85, 454–457
   Specimens Examined lists for publications, 86, 463–469
   Specimens Examined lists, examples, as exported by Biota, 468–469
   Specimens Examined lists, what Biota exports, 463–465
   taxonomic flatfiles, 85, 451–453
   text data for custom locality/collection labels, *See* labels, custom, exporting text for
   text files, based on records in an output screen, 139–141
   text files, based on records in an output screen, 141
   text flatfile of Specimens loaned, *See* loans, exporting a text flatfile of Specimens loaned

using the Export Editor, 84–87, 434–443
Web pages: choosing a strategy for creating Web pages, 474–475
Web pages: step by step, 475–483
Web pages, 86–87, 473–483,
express route, 59–87

Families button, 125–126
Family Down, *See* finding records using the taxonomic hierarchy
Family input, 168–172
features
  operational, 63–64
  overview, 63–87
Field Length Error, 500
field lengths, 521–523 (Appendix A)
field name Aliases, *See* Aliases, Core field
Field Name Editor (for Auxiliary Fields), 283–295
Field Name/Alias button, 222–223, 439–440, 491–492
field names, ordering in certain editors, 61
field types, 521–523 (Appendix A)
Field Value Defaults, 69, 313–315
  setting, 313–315
  using, 49, 152, 158, 161
fields
  Auxiliary, *See* Auxiliary Fields
  changing the name of, *See* Aliases, Core field
  Core, 65
  custom, *See* Auxiliary Fields
  database, 6, 65
  default names for intermediate taxon fields, 110
  entry order, default, *See* default entry order for fields
  field lengths, 521–523 (Appendix A)
  field types, 521–523 (Appendix A)
  Key, *See* Key fields
  linking, *See* linking fields; parent and child records
  new, *See* Auxiliary Fields
  renaming, *See* Aliases, Core field
  user-defined, *See* Auxiliary Fields
figures, *See* images
File Menu, 554–555
file names for image files, how Biota assigns to exported images, 450–451
Files, Biota Data, *See* Data File, Biota
files, Record Set pointer, *See* Record Set pointer files
Find and Replace tool, 77, 220–225
Find Menu, 557–559
finding records, 185–218
  all records for a table, 185–186
  based on content, 72, 192–200, *See also* Search Editor
  by Record Code, 75–76, 213–215
  childless and orphan records, 215–218
  childless records, 80, 215–216
  Collections for a higher taxon, *See* finding records, sequential, cross-hierarchy searches

Collections or Localities (places) based on Specimens or Species records, 74–75, 208–210
general tools for finding records, 71–73
guest Specimens and guest Collections, 402–404
guest Specimens for host Specimens, 404
higher taxa based on lower taxa, 74, 205–208
higher taxa for Localities or Collections, *See* finding records, sequential, cross-hierarchy searches
host and guest Specimens and Collections, 215, 401–404
host Specimens and host Collections, 401–402
host Specimens for guest Specimens, 404
in one table based on a set of records in another table, 52–56, 73–75, 200–213, 200–213
individual, by Species Code, Collection Code, or Locality Code, 213–215
Localities for a higher taxon, *See* finding records, sequential, cross-hierarchy searches
lower taxa based on higher taxa, 73–74, 187, 201–205
orphan records, 80, 216–218
sequential, cross-hierarchy searches, 75, 213
series, Specimen record, *See* Specimen record series, finding
Specimen records by Record Code, 75, 213–215
Specimens or Species based on Collection or Locality records (places), 75, 210–213
Tree menu, finding and updating records using, 48–51, 187–192, 546–547
using the Search Editor, 192–200
using the Search Editor, based on fields from related tables, 197–200
using the taxonomic hierarchy, 71, 48–51, 187–192, 546–547
valid species, 558 (Find menu, Find All Valid Species), 380–381
flatfiles, *See* text files
custom, exporting, *See* exporting custom flatfiles
fluid collection labels, 248–251
foreign keys, *See* linking fields
formats, date, 111–112
forms, loan, *See* loans, invoices
formulas, using, in a Quick Report column instead of a field name, 272
Full Record buttons, 32–33, 124–125
displaying records with the Full Record button, 124–125
editing records displayed with a Full Record button, 125

Genera button, 125–126
generic names, legitimate duplicate, 169–170
Genus Down, *See* finding records using the taxonomic hierarchy
Genus input, 168–172
Genus name, duplicate, 169–170
geographical coordinates, 115–117
  entry and display of coordinate data in the Locality input screen, 116
  longitude and latitude on printed labels and in exported text files, 117
  options for recording geographical coordinates in Biota, 115–116
  setting the display format for latitude and longitude in the Locality output screen, 116–117, 526
Geographical Information Systems (GIS), 115–117
Geographical Positioning Systems (GPS), 115–117
getting started, xxi–xxv
GIF images, 348–349,
  on exported Web pages, 481
GIS, 115–117
global changes in record fields, *See* Find and Replace tool
global replacement of one value by another, *See* Find and Replace tool
GPS, 115–117
graphics, *See* images
graphs: using the Quick Report Graph Editor, 279–280
Group Editor, 174–177
Group Personnel records: reordering members of a Group, 177
Group Personnel records, 174–177

hardware and software requirements, xxiii–xxv
hardware requirements, xxiv
herbarium labels, 255–258
herbivore-plant links, *See* host–guest relations
hidden window cursor, 61
hierarchy, taxonomic, 109–111
history, determination, *See* determination histories
host records, *See* host–guest relations
host–guest relations, 67, 393–404
  creating a Host Specimen link, 396–397
  creating guest Collection records automatically, 398–401,
  error messages while creating guest Collection records automatically, 400–401
  example, 394–395
  finding guest Specimens and guest Collections, 402–404
  finding guest Specimens for host Specimens, 404
  finding host and guest records, 401–404

finding host Specimens and host Collections, 401–402
finding host Specimens for guest Specimens, 404
guest Collection record, recording information in, 395–396
guest Collection records, creating automatically, 398–401
host information in guest Specimen labels and printed reports, 396
how Biota handles links between Specimens, 393–396
in exported Web pages, 479–480
labels, guest Specimen, host information, 396
linking host and guest records, 396–397
reports, printed, host information in, 396
how to use this book, xxi–xxiii
HTML, 473–474
  translation of special characters for exported Web pages, 478

identification
  approximate, *See* temporary taxa of Specimens, *See* determination
idiosyncrasies of Biota, 60–61
images, 67, 72–73, 347–365
  changing the order of Image records for a Species, 359–363
  characteristics in Biota, 347–349
  color, 348
  comparing in the Image input screen, 365
  compression, 349
  copying an Image record, 355–357
  creating a new Image record by pasting or importing an image, 350–355
  deleting, 358–359
  displaying, 355–357
  displaying thumbnail images in the Species output screen, 28, 72–73, 363–365
  displaying, printing, exporting, or copying an Image record, 355–357
  exporting, 86, 355–357, 449–451, *See* exporting images, *for detailed headings*
  exporting an Image record, 355–357
  exporting groups of images to disk files, 365
  file formats, 348–349
  in exported Web pages, 480–481
  printing an Image record, 355–357
  reordering Image records for a Species, 359–363
  size and shape, 348
  sources, 347–348
Import/Export (Im/Export) Menu, 560–561
importing, 84, 485–520
  alphanumeric fields, 488
  ASCII files, using the Import Editor, 84, 485–520
  Auxiliary Field Blank Value Error, 505

Auxiliary Field Setup Error, 505
Auxiliary Field Value Key Match Error: updating existing records, 506
Auxiliary Fields: importing new records and updating existing records, 503–506
Auxiliary Fields: importing or updating records in Field Name Tables, 504
Auxiliary Fields: importing or updating records in Field Value Tables, 504
Blank Key Error, 499
Boolean Field Error, 501–502
column heading row option, 486–487
data files, using the Import Editor, 84, 485–520
date fields, 488–489
date flag fields, 488–489
Date Format Error, 500–501
dates, partial, 113, 488–489
Determination History records, 377, 502–503
displays and error messages during record importing or updating, 495–502
Duplicate Auxiliary Field Value Key Error: importing new records, 505
Duplicate Key Error: importing new records, 498–499
error messages, during record importing or updating, 495–502
Excel files, using the Import Editor, 84, 485–520
Field Length Error, 500
field types and field lengths, 435–437
Invalid Date Error, 501
Key Fields, 487–488
latitude and longitude fields, 489
Lists table records, 506–507
Missing Key Error: importing new records, 497–498
Missing Key Error: updating existing records, 498
Mixed Characters Error, 500
No Unique Key Update Error, 502–503
Notes records, 502–503
text files, preparing for import, 486–490, 507–520
text files, preparing for import from a specimen flatfile, 507, 512–520
text files, preparing for import from a taxonomic flatfile, 507–512
progress indicator, 495
redoing an aborted import (updating existing records), 496
setting up and launching the import or update, 490–495
space characters, automatic stripping of initial and terminal, 487
spreadsheet files, using the Import Editor, 84, 485–520
text files, preparing, 486–490, 507–520
text files, using the Import Editor, 84, 485–520

Too Many Fields Error, 497
undoing an aborted import (new records), 495–496
Unknown Key Error, 503
Update Key Match Error: updating existing records, 499
updating existing records using the Import Editor, *See* updating, using the Import Editor to update existing records
updating existing records: redoing an aborted import, 496
updating records by importing information from text files, *See* updating
individual records, finding by Record Code, *See* finding records, individual
information types, 64–65
input, 67–70, 143–183
  Auxiliary Field values, 295–298
  Collection records, 152–159
  dates, 112
  efficiency of, 3–4
  entering data, guidelines, 89–96
  Genus, Family, Order, Class, Phylum, and Kingdom records, 168–172
  Group Personnel records, 174–177
  higher taxon input screens, differences among, 168–179
  images, 350–355
  latitude and longitude, 116
  Locality records, 160–164
  Menu, 555–556
  Notes, Carrying, 183
  Notes, new, 179–181
  Notes, viewing, editing, or deleting, 181–183
  Notes, 179–183
  Personnel and Project Name records, 172–179
  Personnel records, Group, 174–177
  Personnel records, Individual, 173–174
  Project Name record, 177–179
  screens, 119–141
  Species records, 164–168
  Specimen records, 143–152
Input Menu, 555–556
Insert button, 175–176, 441, 493
installing Biota, xxv
intermediate taxonomic levels, 109–111
Internet
  4D client-server access to Biota Data Files, 484
  Biota and the, 473–484, *See also* exporting Web pages
  dynamic access to Biota Data Files, 483–484
  options, with Biota, 483–484
introducing Biota, 1–57
Invalid Date Error, 501
invoices, loan, *See* loans, invoices

junior synonyms, *See* synonymy

key concepts, 5–19
Key Fields, 6–7
    exporting, 435
    importing, 487–488
keyboard equivalents
    for on-screen buttons, 91–92
    for option window (radio) buttons, 91
    for Today buttons, 91–92
Kingdom Down, *See* finding records using the taxonomic hierarchy
Kingdom input, 168–172

Label Editor, 261–264
labels, 243–264
    collection (locality), information included on, 246–247
    collection (locality), printing or exporting, 246–248
    collection date ranges on, 114
    custom, designing and printing, 261–264
    custom, exporting text for, 82
    data options, 245–246
    determination, for pinned specimens, 248–251
    determination, for slide-mounted specimens, 251–254
    determination, for specimens in vials, 248–251
    determination, temporary taxa on, 408
    geographical coordinates on, 117
    guest Specimen, host information, 396
    herbarium folder, *See* labels, species
    locality, for pinned specimens, 248–251
    locality, for slide-mounted specimens, 251–254
    locality, for specimens in vials, 248–251
    Menu, 559–560
    option windows, 245–246
    printing procedures, 243–245
    sort options, 245
    species, 83, 259–261
    specimen, herbarium, information included on, 255–256
    specimen, herbarium, 255–258
    specimen, pin and vial labels, information included on, 251–254
    specimen, slide labels, information included on, 251–253
    specimen, 82
    temporary taxa on determination labels, 408
    unit tray, *See* labels, species
Labels Menu, 559–560
Lambert coordinates, 115–117
latitude and longitude fields, 115–117
    exporting, 436–437
    importing, 489
latitude, 115–117
launching a password-protected copy of Biota, 413
legitimate duplicate generic names, 169–170

lending Specimens, *See* loans
linked records
    "on-the-fly" creation of, 44–48, 93–95
    table-by-table creation of, 95
linking fields, 2–3
    entering data in, 92–95
    Look Up button for entering Record codes, *See* Look Up buttons
    wildcard data entry for, *See* wildcard data entry for linking fields
Lists table, 524–525 (Appendix B)
    field structure and field lengths, 525
    record structure, 524–525
Lists, Entry Choice, *See* Entry Choice Lists
living organisms, 64
loading Record Set pointer files, 18–19
Loan Codes, 328–329
Loan Notes, 179–183
loans, Specimen, 83, 327–345
    agreements, *See* loans, invoices
    displaying an existing loan, 337–339
    exporting a text flatfile of Specimens loaned, 83–84
    exporting loan records, 83, 333–337
    forms, 83, 333–337
    how Biota keeps track of Specimen loans, 327–328
    invoices, 83, 333–337
    Loan Codes, 328–329
    Menu, 561
    new loan, recording, 83, 328–333
    previewing loan records, 333–337
    previewing, printing, and exporting loan forms (invoices) and records, 333–337
    printing a list of Specimens loaned, 83, 333–337
    printing loan records, 333–337
    recording returns, 83
    returned Specimens, recording, 83, 339–345
    returned Specimens, recording, using the loan records screens, 341–345
    returned Specimens, recording, using the Specimen Record Set, 339–341
Locality Codes, 99–100
    assigning automatically during data entry, 102–103, 160
    finding records by, *See* finding records by Record Code
Locality input, 160–164
    entry and display of latitude and longitude data, 116
locality labels, *See* labels, locality; labels, collection
Locality lists, based on Specimen or Species records, *See* finding records, Collections or Localities (places) based on Specimens or Species records
Locality Notes, 179–183
longitude, 115–117

Look Up
    buttons, for entering Record Codes in linking fields, 69–70
    by Species button, for entering Host Specimen Code in guest Collection records, 397
    Collection button, for entering Collection Code in Specimen records, 36–37, 147–149
    Locality button, for entering Locality Code in Collection records, 154–156
    Species button, for entering Species Code in Specimen records or the Species label setup screen, 44–46, 145–146, 260–261
lot-based data, 64

MacClade format, *See* exporting character matrices in the NEXUS format
Macintosh
    Biota Data File conversion to Windows format, 550–551
    Biota versions and flavors, *See* versions and flavors of Biota
    data and resources, 550
    keyboard equivalents, *See* keyboard equivalents
manuals, if you hate them, 59–87
maps, 115–117
menu bar, 554
menu commands, (Appendix I), 554–563
    access privileges for, 554
menus, quick reference (Appendix I), 554–563
Missing Key Error
    importing new records, 497–498
    updating existing records, 498
Mixed Characters Error, 500
morphospecies, *See* temporary taxa
moving dialog windows (Macintosh), 61
moving down the table hierarchies from an input screen, 125–126
moving up the table hierarchies from an input screen, 124–125
multitasking, 63, 71

navigation buttons, 27, 121
new fields, *See* Auxiliary Fields
NEXUS format, *See* exporting character matrices in the NEXUS format
No Unique Key Update Error, 502–503
Notes
    deleting, 181–183
    editing, 181–183
    exporting, *See* exporting Notes
    input, 179–183
    new, 179–181
    viewing, 26–27, 181–183
number fields, importing, 489
numbers, sorting records by alphanumeric fields that contain numbers, 139
numeric fields, importing, 489

on your own—what to try next (tutorial lesson), 57
"on-the-fly" creation of linked records, 44–48, 93–95
opening
 a password-protected copy of Biota, 413
 a password-protected Data File with a new copy or new version of Biota, 430–431
operating systems, xxiv, 63–64, *See also* versions and flavors of Biota
 Biota Data File conversion between Macintosh and Windows, 550–551 (Appendix G)
 data and resources, 550
operational features, 63–64
option (radio) buttons, keyboard equivalents, 91
Order Down, *See* finding records using the taxonomic hierarchy
Order input, 168–172
ordered Specimen series, *See* Specimen record series
ordering of Biota table and field names in certain editors, 61
Orders button, 125–126
organ-donor links, *See* host–guest relations
organisms, living, 64
orphan junior synonyms, *See* synonymy
orphan records, finding, 216–218
output screens, 71, 119–141
 date displays on, 112
 displaying records in, 119–120
 returning to, from an input screen, 121
overview: Biota tools and features, 63–87

page setup for printing, 243–245
parasite records, *See* host–guest relations
parent and child records, 7–8
parent records, automatic prompting for entry, 67
partial dates, 112–113
Password Editor, using, *See* passwords, Password Editor
passwords
 access privilege levels, 415–417, 420–423, *See* access privilege levels *for detailed headings*
 activating and deactivating the user password system, 410–411
 activating the Data File password link, 426–428
 adding a new user record, 420
 access groups, assigning users to and removing users from, 420–423
 changing the Data File password, 428–429
 changing your user password, 413–415
 Data File password link, activating, 426–428
 Data File password link, deactivating, 429–430

Data File password link, 79, 425–431
Data File password, changing, 428–429
deactivating the Data File password link, 429–430
deactivating the user password system, 411–412
editing a user's password and profile, 418–420
format and length of passwords, 415
high security, if you need, 432
launching a password-protected copy of Biota, 413
length and format of passwords, 415
moving user names, user passwords and access group assignments to a new copy of Biota, 424–425
opening a password-protected copy of Biota, 413
opening a password-protected Data File with a new copy or new version of Biota, 430–431
opening and closing the Password Editor, 417–418
Password Editor, using, 415–424
pitfalls to beware and features to ignore in the Password Editor, 424
transferring passwords and user records to a new copy of Biota, 424–425
user names, 415–417
user passwords, 78, 409–425
users, user names, passwords, and access privilege levels, 415–417
using the Data File link with backup files, 431
PAUP format, *See* exporting character matrices in the NEXUS format
Peripheral tables, 65
Personnel
 input, 172–179
 records, Group, 174–177
 records, Group: reordering members of a Group, 177
 records, Individual, 173–174
Phyla button, 125–126
Phylum input, 168–172
Phylum or Division Down, *See* finding records using the taxonomic hierarchy
picklists, *See* Entry Choice Lists
PICT images, 348–349
 on exported Web pages, 481
pictures, *See* images
pin labels, 248–251
plant specimen labels, 255–258
platforms, *See* versions and flavors of Biota; operating systems
preferences, setting, 526 (Appendix C)
previewing loan records, 333–337
Print button, *See* button, Print
print preview, 243–245, 333–337
printing
 an individual record, 266
 Auxiliary Fields, 304–308

custom reports based on the records in an output screen, 266–268, *See also* Quick Reports
custom reports, *See* Quick Reports
Image records, 355–357
loan records, 333–337
procedures, 243–245
Quick Reports, *See* Quick Reports
Record Set Pointer File reports, 266
Record Set reports, 266
records, individually, 266
reports and records, 81, 265–280
reports based on a Record Set Pointer File, 266
reports based on a selection of records, 266–268
reports based on the active Record Set for a table, 266
Specimen count by taxon reports, 269
privilege levels, user, *See* access privilege levels
problems, *See also* errors, support
 Biota windows and records, 530
 Carry and Delete buttons dimmed, 530
 Data File commands from the File menu (4D Error, –9939, missing External Procedure), 534
 Data File size after deletions, 528
 Data Files, 527–529
 file damaged error message, 534
 finding Data Files, 527–528
 hidden window cursor, 530
 "how to" questions, 535
 juggling windows, 530
 locked record message ("The record is already being edited…"), 533
 menu name truncation, 532
 "No more room to save the record" error message (4D Error, 98), 534, 548–549
 "Out of memory" error message, 533
 passwords, 529–530
 QuickTime, 531
 Record Set Pointer files, 532
 saving changes, 528
 system crashes, 532
 Web pages, 531
Project Name record, 177–179
quadrat or sample data, exporting, *See* exporting Collections × Species matrices
queries, *See* finding records
queries, ad hoc, *See* finding records based on content

Quick Reports, 141, 269–280
 columns, resizing, 272–273
 columns, setting up, 270–271
 designing and printing reports with the Quick Report Editor, 269–278
 dismissing the Quick Report Editor without printing, 277–278
 displaying all values in a sorted column, 273–274

example of a Quick Report, 278–279
formulas, using, in a column instead of a field name, 272
framing, fonts, and text styles, 276
graphs: using the Quick Report Graph Editor, 279–280
hiding a row or column in the printed report, 276
loading or saving a Quick Report layout, 277
page headers and footers, adding to the report, 276–277
printing, 277–278
resizing a column, 272–273
saving or loading a Quick Report layout, 277
sort criteria, adding or removing from the layout, 273
statistics, summary, computing and displaying for all records in the selection, 274
statistics, summary, computing and displaying for sorted groups of records in the selection, 275–276
quick start, 59–61

radio buttons, keyboard equivalents, 91
ranges, collection date, 113–114
RDMS, 1–4
  and biodiversity data, 1–4
Read Export access privilege level, 415–417
Read Only access privilege level, 415–417
Read Write Export access privilege level, 415–417
ReadMe file, xxi
Record Codes, 97–108
  abbreviated barcode prefixes, substituting for long barcode prefixes in the Data File, 107–108
  Assign button, using to assign Record Codes, 102–103
  assigning automatically during data entry, 70, 102–103
  barcode entry of, See barcodes
  Collection Codes, See Collection Codes
  default prefixes and lengths, changing, 102–103, 105–108
  finding records by, See finding records by Record Code
  format of automatically assigned Record Codes, 102
  guidelines and suggestions for designing Record Code systems, 98–101
  Loan Codes, See Loan Codes
  Locality Codes, See Locality Codes
  prefixes, automatic Record Code prefix recognition and substitution, 106–108
  prefixes, setting default prefixes for recognizing Specimen and Species Record Codes, 105–108

  prefixes, using Entry Choice Lists to enter, 102
  sorting records by, 101
  Species Codes, See Species Codes
  Specimen Codes, See Specimen Codes
  unified Record Code systems: Specimen, Collection, and Locality Codes, 99–101
  why does Biota require and display Record Codes?, 87–98
record entry, parent, automatic prompting for, See parent records, automatic prompting for entry
Record Set options screen, 13–15, 33
Record Set pointer files, 16–19
  and the current Record Set, 19
  loading, 18–19
  saving, first method, 16–17
  saving, second method, 17–18
Record Sets, 12–16, 67
  criterion, 474–475
  combining, 13–15
  difference, 13–15
  displaying, 15–16, 71
  emptying all, 16
  how Biota uses them, 15
  intersection, 13–15
  manipulating, 13–15
  options screen, 13–15
  union, 13–15
recording loan returns, 339–345
records
  adding new information, See updating records
  child, See parent and child records
  copying, 127–128
  counting, See statistics, summary, computing and displaying in Quick Reports
  creating, see input
  creating, record creation limits, 80, Appendix C
  deleting a group of records from the output screen, 129–131
  deleting a record displayed in the input screen, 122–123
  deleting, controls and limits and, 80, Appendix C
  displaying, 23–34
  displaying in a standard output screen, 119–120
  duplicating, 127–128
  editing and saving changes to a record, 121–122
  editing records displayed with a Full Record button, 125
  editing, 23–34
  entering, See input
  existing record, using as a template for a new record, 127–128
  finding, See finding records
  in output and input screens, working with, 119–141

  linked, See linked records
  new selections, 128–129
  parent, See parent and child records
  printing a single record displayed in an input screen, 122
  queries, See finding records
  retrieval and manipulation, See finding records
  saving changes to a record, 121–122
  searching for, See finding records
  selection techniques, 28, 128
  selections of, See selections of records
  sorting by Record Codes, 101
  sorting dates by day, month, or year using formulas, 136–138
  sorting numbers in an alphanumeric field, 139
  sorting, using fields from related tables, 134–136
  sorting, 24–25, 131–139
  Sub-Selections, 28–29, 128–129
  template, using an existing record as a template for a new record, 127–128
  updating, See updating records
  viewing, editing, printing, or deleting from an input screen, 120–123
recovering damaged Data Files, 541–543
recursive Specimen relations, See host–guest relations
reduction (scaling) of printed output, 243–245
related tables, using fields from, for sorting records, 134–136
relational databases, 1–4
relational integrity, maintaining, 80, See also finding records, orphan records, childless records, updating
relational links, 521–523 (Appendix A)
relational model, 1–4
relations (tables), database, See tables, database
Remove button, 175–176, 441, 493
renaming
  fields, See Aliases, Core field, See Aliases, Core field
  intermediate taxon fields, 111
reordering Image records for a Species, 359–363
replacing text globally, See Find and Replace tool
reports, 139–141
  custom, See Quick Reports
  printed, host information in, 396
  list of Specimens loaned, See loans, printing a list of Specimens loaned
  printing procedures, 243–245
  printing, based on records in an output screen, 139–141
  printing, See printing reports and records
restoring damaged Data Files, 541–543
retrieval of records, See finding records
returned Specimens, recording, 339–345

saving
  changes in a Biota Data File, 10–11
  Quick Report layouts, 277
  Record Set pointer files, 16–18
  Search Editor queries, 194
scanned images, *See* images
screens
  colors and textures, 89–90
  input, *See* input screens
  output, *See* output screens
  Spanish, *See* Spanish dialog screens
search and replace, *See* Find and Replace tool
Search Editor, 192–200
searches, efficiency of relational, 4
searching for records, 185–218
security, 78, 409–432, *See also* passwords, access privilege levels
  high security, if you need, 432
segmenting Data Files, 545–549
  adding a new segment to an existing Data File using the Data Segment Manager, 548–549
  deleting existing segments using 4D Tools, 549
  over multiple disks, 545–549
  segmenting a new Data File, 545–548
  splitting an existing Data File into segments using 4D Tools, 549
selection techniques for records (consecutive and nonconsecutive), 28, 128
Selections of records, 11–12
senior synonyms, *See* synonymy
Sequence Numbers, 102
sequential Specimen Codes, 99
Series Menu, 556
series, Specimen record, *See* Specimen record series
server-client mode, using Biota in, 64, 538–540
sets, record, *See* Record Sets
setting
  default entries (Field Value Defaults), *See* Field Value Defaults
  Field Value Defaults, *See* Field Value Defaults
  Preferences, 526
Shortcut Screen, 563
shortcuts, keyboard, *See* keyboard equivalents
simultaneous display of records from any number of tables, 63, 71
simultaneous procedures, 63, 71
Skip button, 492–493
slide labels, 251–254
software requirements, xxiii–xxv
sort criteria, adding or removing from the Quick Report layout, 273
Sort Editor, 131–139
sorting
  by Record Code, 101

records in the Quick Reports, 273
  records, 131–139
Spanish dialog screens, 79
spanning multiple hard disks with a Biota Data File, *See* segmenting Data Files
Special Menu, 552–563
special tools and features, 327–432
Species button, 125–126
Species Codes, 99–101
  assigning automatically during data entry, 102–103, 164
  automatic Species Code prefix recognition and substitution, 106–108
  finding records by, *See* finding records by Record Code
  prefixes, setting default prefixes for recognizing Species Codes, 105–108
Species Down, *See* finding records, using the taxonomic hierarchy
Species input, 164–168
Species labels, 83, 259–261
Species lists, based on Collection or Locality records, 210–213
species lists, site-based, with no specimens, 65
Species names, duplicate name checking, 81
Species Notes, 179–183
specific epithets, duplicate name checking, 81
specific names, duplicate name checking, 81
Specimen Codes, 98, 99–101, 144
  finding records by, *See* finding records by Record Code
  sequential, 99–101
  alphanumeric, using with the Series tools, 105–106
  assigning automatically during data entry, 102–103
  prefix recognition and substitution, automatic, 106–108
  prefixes, setting default prefixes for recognizing Specimen Codes, 105–108
Specimen counts, by taxon, 269
Specimen determination histories, *See* determination histories
Specimen flatfiles, exporting, *See* exporting Specimen flatfiles
Specimen input, 143–152
Specimen labels, *See* labels, Specimen
Specimen lists, based on Collection or Locality records, *See* finding records for Specimen or Species based on Collection or Locality records (places)
Specimen loans, *See* loans
Specimen Notes, 179–183
Specimen record series
  alphanumeric Specimen Codes with the Series tools, 105–106
  creating, finding, and updating, 70, 227–242

entering records that do not include Species data, 227–234
entering records that include Species data, 227–234
finding sets of Specimen records with consecutive Specimen Codes, 236–238
finding sets of Specimen records with Specimen Codes in any order, 75–76, 236–238
input 227–234,
sequential Specimen Codes, 99
updating determinations and other Specimen information (ordered or unordered series), *See* updating Specimen determinations
using the Find Specimen Series and Find and Identify Specimen Series tools, 234–242
using the Input Specimen Series and Input and Identify Specimen Series tools, 227–234
Specimen Record Set, using to record loan returns, 339–341
Specimen relations, recursive, *See* host–guest relations
specimen-based data, 64
Specimens button, 125–126
Specimens Examined lists, exporting, *See* exporting Specimens Examined lists
spreadsheet files, importing data from, *See* importing
starting up a password-protected copy of Biota, 413
Sub-Selection of records, creating, 128–129
subclasses, 109–111
subfamilies, 109–111
subgenera, 109–111
suborders, 109–111
subselections, 28–29, 128–129
subsets of records, 28–29, 128–129
subspecies, 109–111
subtaxa and supertaxa, 109–111
suggestions for improving Biota, 537
superfamilies, 109–111
superorders, 109–111
support, Biota, 535–537
synonyms, Species, *See* synonymy
synonymy, 77–78, 379–391
  clearing all Synonymies, 391
  compound junior synonyms, 380–381
  declaring a Species a junior synonym and transferring its Specimens, 388–390
  Determination History recorded for Specimens of a Species declared a junior synonym, 375, 390
  disabling full synonymy display, 385
  enabling full synonymy display, 385
  finding and displaying all valid species, 558 (Find menu, Find All Valid Species)

how Biota keeps track of Species synonymies, 380–381
junior synonyms, 380–381, 388–390
orphan junior synonyms, 380–381
senior synonyms, 380–381
status display, 385–387
status display, after using the Synonymy screen, 387
status display, if full synonymy display is enabled, 386–387
status display, if full synonymy display is not enabled, 385–386
synonymy status of a Species record, displaying, 381–384
valid species, finding and displaying all, 558 (Find menu, Find All Valid Species)

table name, ordering in certain editors, 61
table-by-table creation of linked records, 95
tables
 Core, 65
 criterion, 201, 205, 208–209, 210–211, 474–475
 database, 5, 65
 Peripheral, 65
 target, 201, 205, 208–209, 210–211, 474–475
taxonomic flatfiles, exporting, *See* exporting taxonomic flatfiles
taxonomic hierarchy, finding records using, *See* finding records using the taxonomic hierarchy
taxonomic levels, intermediate (subtaxa and supertaxa), 109–111,
 default names for intermediate taxon fields, 110
 defining additional intermediate levels, 111
 how Biota handles intermediate taxa, 110
 renaming intermediate taxon fields, 111
 updating intermediate taxa, 111
template for a new record, using an existing record as, *See* Carry button
temporary determinations, *See* temporary taxa
temporary taxa, 405–408
 automatic creation of temporary taxon records, 406–408
 Biota's convention for temporary taxon records, 406

disabling automatic creation of temporary taxon records, 406–408
eliminating unused temporary taxon records, 407–408
enabling automatic creation of temporary taxon records, 406–408
how Biota creates temporary taxon records, 407
on determination labels, 408
tentative determinations, *See* temporary taxa
terminological conventions, xxii–xxiii
text fields
 exporting, 437
 importing, 489
text files, 1–2
 exporting data to, *See* exporting
 importing data from, *See* importing
textures, screen, 89–90
TIFF images, 348–349
Today buttons, keyboard equivalents for, 91–92
Too Many Fields Error, 497
tools and features, special, 327–432
tools, overview, 63–87
top-down
 strategy, for exporting Web pages, 474
 taxonomic searches, *See* finding records for lower taxa based on higher taxa
Tree menu, 48–51, 186–192, 187–192, 546–547, 556–557
tribes, 109–111
troubleshooting and support, 527–537 (Appendix D)
troubleshooting, 527–535, *See also* problems *for detailed headings*
Tutorial, Biota, 21–57
typographic and terminological conventions, xxii–xxiii

Unknown Key Error, 503
unordered Specimen series, *See* Specimen record series
Update Key Match Error: updating existing records, 499
updating
 child records by changing a parent record, automatically, 79–80, 218–220
 determination of Specimens for a Species declared a junior synonym, 388–390
 efficiency of, 3–4
 intermediate taxon fields, 111
 junior synonyms, Specimens of, 388–390

linking fields in child records, automatic updating, 79–80, 218–220
records by importing information from text files
records, 76–78, 218–225
relational links to child records, 79–80, 218–220
Specimen determinations and other Specimen information; ordered Specimen series, 76–77
Specimen determinations and other Specimen information; unordered Specimen series, 76
using the Find and Replace Tool, based on record content, 77, 220–225
existing records, using the Import Editor to update, 225, 485–506
user access privilege levels, *See* access privilege levels
user password system, *See* passwords, user
user-defined custom fields, *See* Auxiliary Fields
UTM coordinates 115–117

valid species, 380–381
 finding and displaying, 558 (Find menu, Find All Valid Species)
validation, data input, 6–7, 111–114, 115–116, 218–220, 487–490, 495–507
verification, user, *See* security
versions and flavors of Biota, xxiii–xxiv, 63–64
vial labels, 248–251

Web, World Wide, *See* World Wide Web
wildcard data entry for linking fields, 35–36, 39–40, 68, 96
Windows, Microsoft
 Biota Data File conversion to Macintosh format, 550–551
 Biota versions and flavors for, *See* versions and flavors of Biota
 data and resources, 550
 keyboard equivalents, *See* keyboard equivalents
windows, dialog, moving (Macintosh), 61
windows, multiple, *See* multitasking
World Wide Web
 Biota Website, xxi
 exporting Web pages, 86–87, 473–483, *See* exporting Web pages *for detailed headings*